Electron Microscopy of Proteins

Macromolecular Structure and Function

Electron Microscopy of Proteins

Macromolecular Structure and Function

Volume 4

edited by

JAMES R. HARRIS

Principal Scientific Officer
and
Head, Research and Quality Control Department,
North East Thames
Regional Transfusion Centre,
Crescent Drive, Brentwood,
Essex CM15 8DP, UK

1983

Academic Press

A Subsidiary of Harcourt Brace Jovanovich, Publishers

LONDON   NEW YORK
Paris   San Diego   San Franciso
São Paulo   Sydney   Tokyo   Toronto

ACADEMIC PRESS INC. (LONDON) LTD.
24/28 Oval Road
London NW1

United States Edition published by
ACADEMIC PRESS INC.
111 Fifth Avenue
New York, New York 10003

Copyright © 1983 by
ACADEMIC PRESS INC. (LONDON) LTD

All Rights Reserved
No part of this book may be reproduced in any form by photostat, microfilm, or any other means,
without written permission from the publishers

British Library Cataloguing in Publication Data

Electron microscopy of proteins.
Vol. 4
1. Proteins 2. Electron microscopy
I. Harris, J.R.
547.19′245   QP551

ISBN 0-12-327604-7

Printed in Great Britain
at the Alden Press
Oxford

Contributors to Volume 4

C. A. L. S. Colaco *Imperial Cancer Research Fund, Lincoln's Inn Fields, London WC2A 3PX, England*

R. Craig *Department of Anatomy, University of Massachusetts Medical School, 55 Lake Avenue North, Worcester, Massachusetts 01605, USA*

M. J. Dickens *MRC Cell Biophysics Unit, Kings College, 26–29 Drury Lane, London WC2B 5RL, England*

W. H. Evans *National Institute for Medical Research, Mill Hill, London NW7, England*

P. Knight *Muscle Biology Division, ARC Meat Research Institute, Langford, Bristol BS18 7DY, England*

E. J. O'Brien *MRC Cell Biophysics Unit, Kings College, 26–29 Drury Lane, London WC2B 5RL, England*

D. M. Shotton *Department of Zoology, University of Oxford, South Parks Road, Oxford OX1 3PS, England*

Foreword

The four chapters of this volume of the series "The Electron Microscopy of Proteins" provide a most interesting account of the contribution this technique has made—and is continuing to make—to the study of some of the protein molecules which carry out their functions as part of organized structural assemblies in cells. The best known of these are the proteins first studied in muscle but now found in many types of cells—actin, myosin and the other proteins associated with them. These are described in the two chapters which cover "Actin and Thin Filaments" and "Myosin Molecules, Thick filaments and the Actin-Myosin Complex" respectively. The third chapter is concerned with "The Proteins of the Erythrocyte Membrane". As in muscle, the function of these proteins is primarily a mechanical one, in this case to provide a strong but very flexible inner layer to the cell membrane, closely coupled to the lipid bilayer, which will enable the red cells to change their shape very readily and allow easy passage throughout the circulatory system and at the same time perform their transporting functions optimally. The fourth and final chapter deals with another set of proteins whose function is carried out as part of a cell membrane, and is entitled "Plasma Membrane Intercellular Junctions. Morphology and Protein Composition".

These chapters provide a good picture of how the processes of discovery have taken place in areas of increasing experimental difficulty, and with techniques of increasing sophistication and discrimination. In the case of muscle (and other motile systems) this history follows closely the course of technical developments in electron microscopy, for the simple reason that many of these developments originated from the requirements and the possibilities of the muscle work, or were very closely linked to them. The first step was the elucidation of the basic structure of striated muscle which relied heavily on the development of adequate methods of fixation, staining and sectioning for high-resolution electron microscopy. The next step was to visualize the elements of the structure—in particular the thick and thin filaments—in sufficient detail to begin to see how they might be assembled from their respective principal protein components, namely myosin and actin. This involved the partial separation of the protein assemblies present in the intact muscle cells so that they could be examined by the negative-staining technique, and this combination of techniques has provided an obvious but very useful general procedure which has been widely applied to many biological systems.

At the same time, the major protein components of these assemblies—actin, myosin and tropomyosin—were studied in purified form, by negative staining and also by the shadow-casting technique, and the general features of their assembly into structures having the general functional requirements for a sliding filament system were worked out.

It then became apparent that even more detailed studies of the structure of these protein assemblies might provide important clues about their detailed function. In particular, more information was needed about the configurations of the crossbridges and the possible changes in them when they attached to actin and developed the sliding force; about the exact position of tropomyosin in the thin filaments and changes in it associated with the operation of the on-off regulation mechanism; and about structural changes in myosin in systems where myosin-based regulation was involved. These studies have continued up to the present and have made extensive use of image analysis techniques, in particular three-dimensional image reconstruction from electron micrographs. The helical or near-helical symmetry with which actin and myosin are arranged in their respective filaments has rendered them excellent subjects for the application and development of such techniques. And in all of the work, muscle has been particularly favourable material for the development of various electron-microscopic techniques since it allowed an independent check to be made on the plausibility of the structural conclusions—namely, whether or not they were compatible with the rich and detailed X-ray diffraction patterns given by intact muscles. In themselves, these X-ray patterns often do not have a direct and complete structural interpretation, but they can often be used to test the validity of structural models based on other evidence.

In the future, it is likely that the application of rapid-freezing techniques, to arrest fine-structural events in defined states so that they can be studied in the electron microscope, will become increasingly important. The corresponding use of high time resolution X-ray diffraction techniques using synchrotron radiation, which is now possible on muscle (because of the ready availability of large oriented specimens), make it likely that this material will continue to provide particularly favourable test material and be one of the first to benefit from these developments. The information so gained should also be of great benefit in understanding the behaviour of other motile systems in cells involving actin-myosin interactions.

The study of the structural basis of the behaviour of the erythrocyte membrane is a less tractable problem since the detailed nature of the structure is much more difficult to see in the intact cells. Nevertheless, it is easy to obtain large amounts of pure material, and this has made it possible to follow the same general pathway as in the muscle work, namely to identify many of the individual protein components, to characterize them in the electron microscope, to see how they will assemble together and to arrive at a plausible model for the original structure. Such a model has important implications about the control of the mechanical behaviour of membranes in many other types of cells.

The study of the structure and composition of intercellular junctions is an even more difficult problem to approach and is at a somewhat earlier stage of development. However, there is no doubt that as improved isolation procedures are developed for the various types of junction and their components, the application of electron microscopy to examine the protein molecules involved, and to investigate the way in which they will reassemble, should lead to corresponding advances in our understanding of these systems too.

H. E. Huxley
Medical Research Council,
Laboratory of Molecular Biology, Cambridge

Preface

Volume 4 of the series so far completes the survey of the subject, essentially as initially planned. Although the original aim was to produce three specialist volumes covering the electron microscopy of soluble, membrane and fibrous proteins, for reasons of a technical and editorial nature this plan was not adopted. Thus, all the books have been published containing a fairly diverse range of subject matter, but within the overall framework indicted by the title "Electron Microscopy of Proteins". Volume 4 differs slightly from the previous volumes in that it contains only four chapters, two on muscle proteins and two on membranous proteins. Both the lengthy chapters on muscle proteins achieve a very high standard of review, and include much recent data of the authors and their colleagues. The first chapter is on Actin and Thin Filaments, by E. J. O'Brien and M. J. Dickens, of the MRC Cell Biophysics Unit, Kings College, University of London. The second chapter deals with Myosin Molecules, Thick Filaments and the Actin-Myosin Complex, by R. Craig of the Department of Anatomy, University of Massachusetts, Worcester, and P. Knight of the ARC Meat Research Institute, Langford, Bristol. In an even longer chapter, D. M. Shotton of the Department of Zoology, University of Oxford, discusses in considerable depth the Proteins of the Erythrocyte Membrane. The concise final chapter on Plasma Membrane Intracellular Junctions is by C. A. L. S. Colaco of the Imperial Cancer Research Fund, London and W. H. Evans of the MRC National Institute for Medical Research, London.

The chapters included in Volume 4 maintain the extremely high standard set by those in the previous volumes in the series, both with respect to the quality of the electron micrographs and the textual discussion of the subject matter. Despite some considerable delay in the publication of Volume 4, it has been a great pleasure to liaise with the authors who have contributed to this work.

Several of the chapters included in the proposal as originally presented to Academic Press, such as those on blood plasma proteins, mitochondrial and chloroplast proteins, enzymes, endoplasmic reticulum and plasma membrane proteins, have not appeared in the series so far. It is, nevertheless, hoped that these and other topics will be incorporated into future volumes. The immediate plan is, however, to produce Volume 5 as a specialist book on Viral Proteins. I am pleased to announce that I am proposing to share the editorial responsibilities in the future with Professor R. W. Horne of the School of Biological Sciences, University of East Anglia. Following discussion with Professor Horne and the editorial staff of Academic Press, it has been decided that the series should take on the subtitle "Macromolecular Structure and Function", since under the title "Electron Microscopy of Proteins" alone, the books have tended to be wrongly classified by librarians and booksellers under microscopy and electron microscopy techniques, rather than under the broad application of electron microscopy in present-day biochemically-based biological research.

The continued support of Dr J. F. Harrison, Director of the North East Thames Regional Transfusion Centre, has been a great encouragement, as has the interest shown in the series by my scientific colleagues. I would like to thank the staff of Academic Press for their co-operation throughout the production of the book and for their enthusiastic support for the continuation of the series.

August 1983

J. R. Harris
North East Thames
Regional Transfusion
Centre

Contents of Volume 4

CONTRIBUTORS TO VOLUME 4	v
FOREWORD	vii
PREFACE	ix
CONTENTS OF VOLUME 1	xii
CONTENTS OF VOLUME 2	xiii
CONTENTS OF VOLUME 3	xiv
1. Actin and Thin Filaments—E. J. O'BRIEN and M. J. DICKENS	1
2. Myosin Molecules, Thick Filaments and the Actin-Myosin Complex—R. CRAIG and P. KNIGHT	97
3. The Proteins of the Erythrocyte Membrane—D. M. SHOTTON	205
4. Plasma Membrane Intercellular Junctions. Morphology and Protein Composition—C. A. L. S. COLACO and W. H. EVANS	331
SUBJECT INDEX	365

Contents of Volume 1

1. Arthropodan and Molluscan Haemocyanins—E. F. J. van Bruggen, W. G. Schutter, J. F. L. van Breemen, M. M. C. Bijlholt and T. Wichertjes

2. The Nuclear Envelope and the Nuclear Pore Complex: Some Electron Microscopic and Biochemical Considerations—J. R. Harris and P. Marshall

3. Intermediate Filaments—P. M. Steinert

4. Protein Synthesis in Prokaryotes and Eukaryotes: The Structural Bases—J. A. Lake

5. Electron Microscopy of Glycoproteins—H. S. Slayter

6. Coated Vesicles—C. D. Ockleford

7. Structure–Function Relationships in Cilia and Flagella—F. D. Warner

Subject Index

Contents of Volume 2

1. Multienzyme Complexes—R. M. Oliver and L. J. Reed

2. Nonenzymic Proteins—J. R. Harris

3. Bacterial Appendages—D. G. Smith

4. Plasma Lipoproteins—I. Pasquali-Ronchetti, M. Baccarani-Contri and C. Fornieri

5. The Electron Microscopy of Fibrous Proteins of Connective Tissue—A. Serafini-Fracassini

6. High Resolution Electron Microscopy of Unstained, Hydrated Protein Crystals—W. Chiu

7. Specialized Membranes—S. Knutton

Subject Index

Contents of Volume 3

1. The Structure of Algal Cell Walls—K. ROBERTS, G. J. HILLS and P. J. SHAW

2. Bacterial Cell Walls and Membranes—U. B. SLEYTR and A. M. GLAUERT

3. Chromatin and Chromosomal Proteins—J. R. PAULSON

4. The Extracellular Haemoglobins and Chlorocruorins of Annelids—S. N. VINOGRADOV, O. H. KAPP and M. OHTSUKI

5. Electron Microscopy of Amyloid—A. S. COHEN, T. SHIRAHAMA and M. SKINNER

6. Tubulin and Associated Proteins—L. A. AMOS

SUBJECT INDEX

1. Actin and Thin Filaments

E. J. O'BRIEN AND M. J. DICKENS

Medical Research Council Cell Biophysics Unit, King's College, 26–29 Drury Lane, London, England

I. Introduction	2
II. Muscle Actin	3
A. F-actin	3
B. F-actin Paracrystals	5
C. X-ray Diffraction of F-actin	10
D. Actin Microcrystals	13
E. Structure of Actin: DNase I	18
III. Tropomyosin	19
A. Crystals, Fibres and Tactoids	21
B. The Coiled-coil	25
C. Crystal Structure	28
D. Association with Actin	30
IV. Troponin	34
A. Isolation and Components	34
B. Interaction with Tropomyosin	35
C. The 40 nm Stripe	39
V. Role of Tropomyosin and Troponin in Regulation	45
A. Tropomyosin Movement	45
B. Influence of Troponin and Ca^{2+}	46
C. Actin-tropomyosin Interaction	52
D. Attachment Site of Myosin	56
VI. Z-disc	57
A. Z-disc in Insects and Vertebrates	58
B. α-actinin	63
VII. Actin and Associated Proteins in Other Cells	63
A. Non-muscle Actin	64
B. Non-muscle Tropomyosin	65
C. Troponin-like Proteins	67
D. Thin-filament Assemblies	67
E. Actin-binding Proteins	73
VIII. Summary and Discussion	75
A. Actin	75

B. Tropomyosin 81
C. Troponin 82
D. Role of Tropomyosin and Troponin in Regulation 83
E. Z-disc 86
F. Actin and Associated Proteins in Other Cells 87
IX. Acknowledgement 88
X. References 89

I. INTRODUCTION

Although not as yet the subject of a best seller, the double-helical structure of filamentous actin is just as familiar, at least among biologists, as its more famous nuclear counterpart. The appearance of actin as two intertwined strings of beads is of compelling elegance rivalling that of DNA, and the very widespread occurrence of this protein throughout nearly all cell types implies a biological role of comparable importance. The two actin strands do not separate, however, so that in this important respect there is no analogy; it is in regulation rather than replication that the strands assume a functional significance, as we will discuss in some detail later.

In muscle, actin is associated with tropomyosin, an elongated molecule that forms end-to-end chains running in the grooves between the two actin strands. In vertebrate skeletal muscle a third protein, troponin, is attached periodically along the actin-tropomyosin complex. This is the component which is sensitive to calcium ions and which leads to the control of actomyosin activity. The thin filament assembly is illustrated in Fig. 1, and was first described in this form by Ebashi *et al.* (1969). In this diagram the actin subunits are spherical and the other components are similarly apolar, but in reality of course the structure has polarity. In vertebrate skeletal muscle the filaments are oppositely directed from the Z-line, a feature which is crucial for the

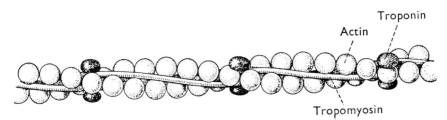

Fig. 1. Model of the thin filament proposed by Ebashi *et al.* (1969). Tropomyosin molecules lie end-to-end in the long-pitch helical grooves of actin, and troponin molecules are attached at intervals of 38·5 nm along the filament. For each troponin molecule to make equivalent contacts with the tropomyosin:actin complex, a small axial displacement should be incorporated between the pair of molecules in each repeat (reprinted from Ebashi *et al.*, 1969, courtesy of I. Ohtsuki and Cambridge University Press).

operation of the contractile apparatus and which has been vividly demonstrated by the addition of heavy meromyosin which "decorates" the filaments with arrowheads (Huxley, 1963).

II. MUSCLE ACTIN

A. F-actin

Actin may exist either as filaments (F-actin) or as the individual globular subunits (G-actin) according to ionic conditions. F-actin stimulates the splitting of ATP by myosin, whereas G-actin has a very small effect (Offer et al., 1972). The polymerization of G- to F-actin takes place rapidly in the presence of 0·1 M KCl, with a dramatic viscosity increase. During the reaction the molecule of ATP that is bound to each G-actin is hydrolysed to ADP and remains tightly bound to the filament. The function of this hydrolysis is not known, however, and G-actin-ADP will in fact polymerize slowly, as will G-actin associated with ATP analogues like AMP-PP(S) (adenosine 5'-(3-thio) triphosphate). During polymerization analogues that are hydrolysable are incorporated, whereas those that are not are displaced from the nucleotide binding site (Mannherz et al., 1975).

The globular and filamentous aspects of the actin polymer were first observed in the electron microscope by Jakus and Hall (1947), who prepared filaments of variable length and width by varying pH. Astbury et al. (1947) used KCl to induce polymerization, and in micrographs of gold shadowed specimens described the appearance of fibres as the result of the "joining together of small corpuscular bodies" (a phrase worthy of Isaac Newton). In the absence of salt a dense background of these corpuscular bodies covered the grid. Rozsa et al. (1949) used chromium or palladium for the shadowing and showed that the filaments packed side-by-side into bundles, with a conspicuous cross-striation at axial intervals of about 30 nm.

The development of the technique of negative staining enabled a dramatic improvement in the quality of the image to be obtained and Hanson and Lowy (1963) showed clearly the twin strands of F-actin, each strand consisting of a single series of regularly spaced subunits of approximately spherical shape (Fig. 2). They estimated that the number of subunits per turn of the helix described by each strand was 13 or nearly 13 and that the spacing of the subunits in each strand was 5·65 nm. The total width of the two strands at the widest point of the filament was measured as 8·0 nm. Huxley (1963) also observed the double-stranded helix of globular subunits, in F-actin and also in natural thin filaments from rabbit striated muscle, but obtained a lower value for the F-actin width, 6·0 to 7·0 nm. Because of the particular interest attached

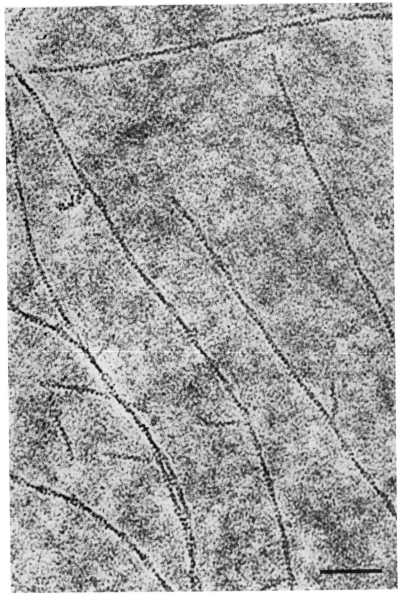

Fig. 2. A reversed-contrast print of an electron micrograph of F-actin, negatively stained with uranyl acetate and suspended in a stain-filled hole in the supporting film. The helical form of the filament and the globular shape of the subunits are clearly revealed. Scale marker indicates 50 nm (reprinted from Hanson and Lowy, 1963, courtesy of Academic Press Inc.).

to the actin helical repeat and its relationship to that of myosin in the thick filament, Hanson (1967) made careful measurements of the distance between successive points where the actin strands cross over each other. Filaments from two sources were analysed: natural filaments from the cross-striated adductor muscle of the scallop *Pecten maximus* and synthetic filaments made from rabbit skeletal muscle. The cross-over repeat differed in the two preparations and also varied with the stain used. With uranyl acetate, the scallop filaments had an average repeat of about 38·5 nm and those formed from rabbit actin about 37 nm, but there was a considerable spread of the results. Another important parameter of the filament proved simpler to establish and Depue and Rice (1965) showed by the mica-replication technique that the two strands wound round each other in a right-handed sense (Fig. 3). A much larger width for the filament than that indicated by negative staining or shadowing was given by Heuser and Kirschner (1980) from observations of actin filaments in platinum replicas of freeze-dried cytoskeletons of cells. These filaments did not show the periodic variation of width from the intertwining of the two strands (Fig. 3), but the authors noted an axial repeat of 5·5 nm in surface roughness that they considered characteristic of F-actin. The filament width was measured as 9·5 nm.

B. F-actin Paracrystals

The discovery by Hanson (1968, 1973) that F-actin could be induced to form paracrystalline arrays by the addition of divalent cations provided a preparation for the electron microscope that was well suited to diffraction analysis and thus stimulated a more detailed investigation of the actin polymer (Fig. 4). Moore *et al.* (1970) obtained optical diffraction patterns from electron micrographs of actin paracrystals that were a significant improvement over similar patterns obtained from single filaments (Klug and Berger, 1964). By applying the technique of optical filtering in which only the periodically repeating features of the structure are reconstructed and the background "noise" is eliminated (Klug and Berger, 1964), the subunit structure of F-actin was clearly resolved (O'Brien *et al.*, 1971; Hanson *et al.*, 1972; Fig. 5). The strongest layer-lines in the diffraction pattern could be indexed as the zeroth, first and sixth orders of a repeat of about 35 nm, indicating that helical symmetry in the filament comprised 13 subunits in 6 turns of the genetic helix, that is, the helix joining all subunits.

A diagram illustrating the different helices present in F-actin is given in Fig. 6. Less obvious than the twin strands, which may be referred to as the "long-pitch" helical strands, are the two genetic helices. The left-handed one has a pitch of 5·9 nm and the right-handed 5·1 nm. If the structure is imagined projected outwards radially from the helix axis onto a cylinder and the

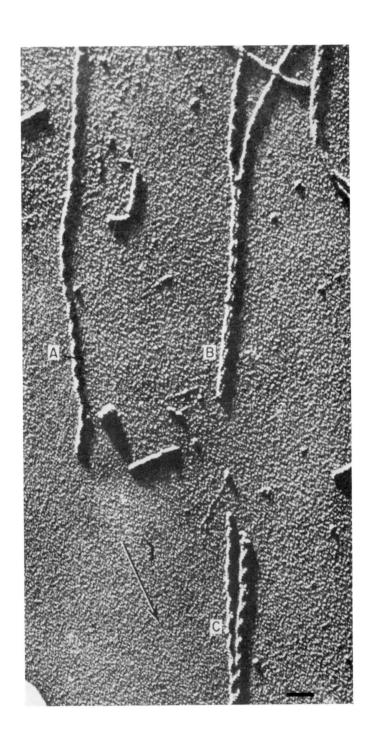

cylinder is cut lengthwise and opened out flat, the radial projection is obtained as shown in Fig. 6b. Sets of parallel lines corresponding to the long-pitch and genetic helices are drawn in. The advantage of this projection is that it can be related more simply to the diffraction pattern than can the helix itself. Thus the reflection on the first layer-line of axial spacing 35·5 nm, marked u in Fig. 6c, derives from the long-pitch helices, and the line joining this reflection to the centre of the diffraction pattern is normal to the near-longitudinal rows in the helical projection. Similarly, the reflections on the sixth and seventh layer-lines, marked v and w, derive from the genetic helices and define vectors perpendicular to the lines of axial spacing 5·9 and 5·1 nm in the helical projection. Other reflections may be considered as arising from other sets of lines. As the filament has a "back" and "front" whereas the helical projection is drawn as one-sided, the diffraction pattern contains for each reflection another of identical intensity and radial position on the other side of the meridian. An equivalent set of reflections is produced by considering vectors of opposite sense, which leads to the lower half of the diffraction pattern shown. The optical diffraction pattern shown in Fig. 5a is further complicated by the effect of interference between the filaments that form the paracrystal, so that the pattern is sampled on vertical row-lines, but the reflections corresponding to those marked u and v in Fig. 6c are prominent. Reflection w is very weak, however, indicating that the lines corresponding to the 5·1 nm helix are much less marked. These features indicate that the actin subunit is elongated along the direction of the 5·9 nm helix (O'Brien et al., 1971).

The electron micrograph of the helical filament presents many different views of the actin subunit, and the information describing and relating these views is embodied in the diffraction pattern. If the amplitudes and phases of the reflections in the diffraction pattern are recorded, a Fourier-Bessel transformation of these data produces a three-dimensional reconstruction of the micrograph image (DeRosier and Klug, 1968). This procedure carries the reconstruction a further step from that obtained by optical filtering, as shown in Fig. 5b, and was first carried out for actin by Moore et al. (1970). To obtain amplitudes and phases in the diffraction pattern, the Fourier transform of an actin filament image converted into digital form by densitometry was calculated. The filament was an average obtained by deconvolution of an actin paracrystal. Figure 7 shows the side-on and end-on views of the reconstructed filament. The subunit measures about 5·5 nm axially, 3·5 nm radially and 5·0 nm tangentially with a volume of approximately 45 nm^3, about 90% of the

---

Fig. 3. Electron micrograph of a replica of F-actin, showing that the long-pitch helical strands are right-handed. The shadow was at an angle of 5:1 and in the direction indicated by the arrow. The particles labelled A, B and C are made up of one, two and three filaments respectively. Scale marker indicates 50 nm (reprinted from Depue and Rice, 1965, courtesy of R. H. Rice and Academic Press Inc.).

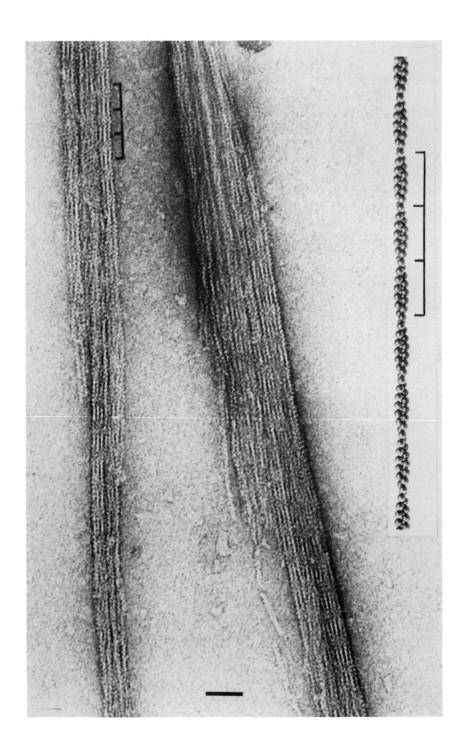

expected volume for a globular protein of molecular weight 41 700 as established from the actin sequence determined by Elzinga and Collins (1972) and Collins and Elzinga (1975). The connections between subunits along the genetic helices are less prominent than those along the long-pitch helices. The size of the subunit and the densities of the connections between subunits depend critically on the level at which the zero level of density is set. This level is particularly subject to uncertainty when the equatorial data are omitted from the reconstruction, as in this case. Moore *et al.* considered these data should be omitted because they were affected by departures of the stain envelope surrounding the filament from cylindrical symmetry. Instead of deconvoluting an actin paracrystal to obtain an average single-filament image, Spudich *et al.* (1972) selected individual filaments by masking out their neighbours. Since each filament has two sides which give diffraction spectra that are resolved from each other (Fig. 6) these spectra can be used to calculate two reconstructions, which usually differ as the stain distribution is in contact with the supporting grid on one side. Spudich *et al.* presented reconstructions from separate sides and from several filaments and took an average. The dimensions of the actin subunit were very similar to those found by Moore *et al.* Although there was some variation in the individual reconstructions in the extent of bonding along the genetic and long-pitch helices, the subunit shape was similar in each.

The paracrystals discovered by Hanson comprised actin filaments packed side-by-side, but other forms have been found. At acid pH near the isoelectric point, Kawamura and Maruyama (1970) found three types of aggregate. One of these was very similar to the Hanson type, but the others were formed of filaments arranged criss-cross in a diamond-shaped net. The paracrystals were studied in more detail by Yamamoto *et al.* (1975) who suggested that the nets were formed from sets of parallel actin filaments laid over each other at an angle of about 30°. The unit cell in the two net forms was the same but in one there was twice the amount of material, that is, two actin filaments rather than one. The unit-cell edge, 34 nm, corresponded to the distance between cross-over points of the long-pitch helical strands of actin. Strzelecka-Golanszewska *et al.* (1978) showed that net-like paracrystals of actin were formed in the presence of the ions of manganese or the alkaline earth metals at millimolar concentrations, but at higher levels of these ions the paracrystals were of the Hanson type. Polyamines have been shown to induce the polymerization of actin, and their effect on the state of aggregation of G- and F-actin may be related to their occurrence in living cells, particularly growing

Fig. 4. Paracrystals of purified actin, negatively stained with 1% uranyl acetate, after fixation with 1% OsO$_4$, pH 7·0. The cross-over repeat of the long-pitch helical strands is indicated at the top, and to the right is a model of the filament. Scale represents 50 nm (reprinted from Hanson, 1973, courtesy of The Royal Society).

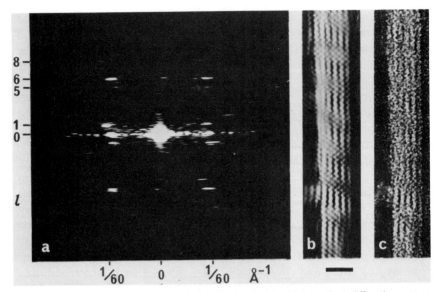

Fig. 5. A negatively-stained paracrystal of purified actin (c) with its optical-diffraction pattern (a) and optically-filtered image (b). The indexing of the diffraction pattern is consistent with a helical structure having 13 subunits in 6 turns of the genetic helix (13/6 symmetry, Fig. 6). Scale marker in (b) indicates 30 nm (reprinted from Hanson et al., 1972, courtesy of Cold Spring Harbor Laboratory).

ones (Oriol-Audit, 1978). Dickens and Oriol-Audit (1980) showed that millimolar concentrations of three naturally occurring polyamines induced paracrystal formation. Net-like paracrystals were formed in the presence of putrescine (Fig. 8) and Hanson-type in the presence of spermine or spermidine. Polylysine, a synthetic polycation, is also effective and Fowler and Aebi (1982) obtained a considerable variety of paracrystal appearance (though basically of the Hanson rather than the net-type), with both muscle and non-muscle actin. Optical diffraction patterns indicated that the actin symmetry in all paracrystals was almost the same, with between 13/6 and 28/13 subunits per turn. The various paracrystal appearances could be simulated by superimposing arrays of actin filaments, with various axial and lateral translations between filaments and filament arrays incorporated. In these model arrays all filaments had the same polarity and ran parallel to the paracrystal axis.

C. X-ray Diffraction of F-actin

Studies of F-actin structure by X-ray diffraction have developed in step with those by electron microscopy. They have both complemented the microscopy and related it to the situation in the intact muscle. Astbury *et al.* (1948)

prepared films and Cohen and Hanson (1956) fibres of F-actin at room humidity, and X-ray diffraction photographs of these specimens showed meridional and near-meridional reflections extending to high angles. On the equator, Cohen and Hanson observed a strong reflection at 5·5 nm, arising from the interfilament separation, which is at a similar position to that of the equatorial reflection observed in optical diffraction patterns of electron micrographs of actin paracrystals (Fig. 5a).

In muscle the intensities of the reflections arising from the actin-containing filaments are modified by the presence of the regulatory proteins, but their positions are determined by the actin helical symmetry. Selby and Bear (1956) indexed the actin-related reflections as arising from the type of net shown in Fig. 6b or from a very similar net with an axial repeat of 40·6 nm, which would correspond to a helix with 15 subunits in 7 turns rather than 13 in 6. Since a helix and its radial projection give very similar diffraction patterns at low and moderate diffraction angles, as explained for Fig. 6, Selby and Bear left the ambiguity for Hanson and Lowy to resolve by microscopy seven years later. Although the helical nature was thus established, the parameters of this helix for thin filaments in muscle have remained difficult to interpret unequivocally. In X-ray diffraction patterns from resting frog sartorius muscle, Huxley and Brown (1967) measured the axial separation of the near-meridional reflections with spacings about 5·9 and 5·1 nm, and concluded that the actin helix had a cross-over repeat of between 36 and 37 nm with between 13/6 and 28/13 subunits per turn of the left-handed genetic helix. Millman *et al.* (1967) considered from X-ray diffraction patterns of resting toad sartorius muscle and of relaxed anterior retractor muscle of *Mytilus edulis* that the presence of a reflection at axial spacing 40 nm could be ascribed to actin with a cross-over repeat of this value. This would imply a helical symmetry closer to 15/7 subunits per turn. Lowy and Vibert (1967) and Vibert *et al.* (1972a, b) recorded patterns from several muscle types and gave results similar to those of Huxley and Brown (1967). In patterns from insect flight muscle in rigor the layer-lines index accurately on an actin cross-over repeat of 38·5 nm (Miller and Tregear, 1972) leading to an assignment of 28/13 symmetry. Generally, in muscle at rest, however, most evidence favours a repeat somewhat smaller, about 37 nm, with a slightly tighter wound helix, non-integral, with between 28/13 and 13/6 subunits/turn.

The measurements made on X-ray diffraction patterns determined the average helical parameters of the actin filament in the muscle being studied. That there might be a variation of the helical parameters was suggested by Millman *et al.* (1967) in order to explain the much larger axial width of the reflections from actin compared with those from the myosin filaments. This idea was developed by Egelman *et al.* (1982) and Egelman and De Rosier (1983), who put forward a model in which the actin subunits had a constant

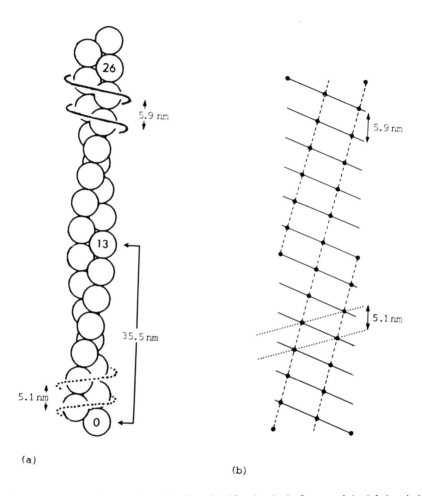

Fig. 6. (a, above). Diagram of actin helix, with 13 subunits in 6 turns of the left-handed genetic helix. The paths of the left- and right-handed genetic helices are indicated by the solid and dotted curves, respectively. (b, above). The radial projection of the 13/6 helix. The dashed, solid and dotted lines are the projections of the long-pitch and left- and right-handed genetic helices, respectively. (c, opposite). The central region of the diffraction pattern from the 13/6 helix. The numbers and scale lines at the right mark the layer-lines, which are at orders of the reciprocal of the distance between successive cross-over points of the long-pitch helices (35·5 nm in (a)). The letters u, v and w mark the reflections arising from the long-pitch and left- and right-handed genetic helices respectively in (a), and may also be considered as arising from the projections of these helices in (b).

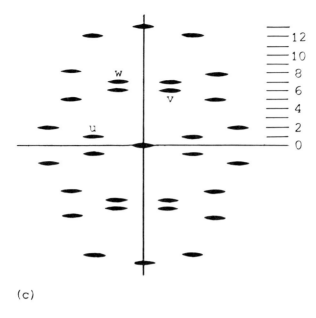

(c)

axial separation but a variable and randomized azimuthal separation, this disorder being cumulative along the filament. An angular rotation of ±10° was considered necessary to explain the optical diffraction patterns obtained from electron micrographs of single actin filaments, though a deviation of ±6° was sufficient to account for the variation of the measurements of the actin cross-over repeat made by Hanson (1967). Egelman et al. (1982) considered that the filaments within actin paracrystals were stabilized in a particular conformation by interfilament interactions, and would therefore not have exact helical symmetry.

D. Actin Microcrystals

Although X-ray diffraction patterns from oriented gels or fibres of F-actin extend to high angles of diffraction, particularly in the axial direction (Fig. 19), the prospect of using them to obtain atomic resolution of the actin subunit is remote. There are no general methods available for determining the phases of reflections in fibre-type diffraction patterns, and most analyses of these patterns involve model-building of some kind. It has long been recognized that if G-actin were modified so that it no longer tended to polymerize to the filamentous form, it might be induced to form three-dimensional crystals, on which well-established protein crystallographic methods, such as isomor-

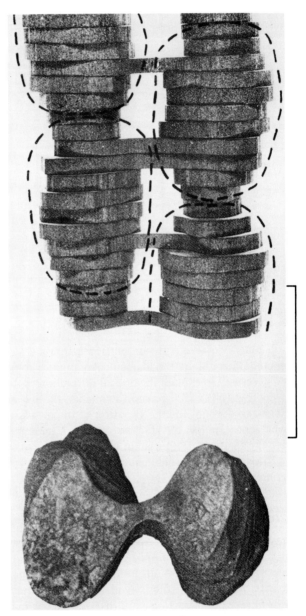

Fig. 7. Side and end-on views of a model of a portion of F-actin, built from data obtained by three-dimensional reconstruction from electron micrographs of paracrystals. In the side view, the probable outlines of the actin subunits are shown by the dashed lines. Scale marker indicates 5 nm (reprinted from Moore et al., 1970, courtesy of H. E. Huxley and Academic Press Inc.).

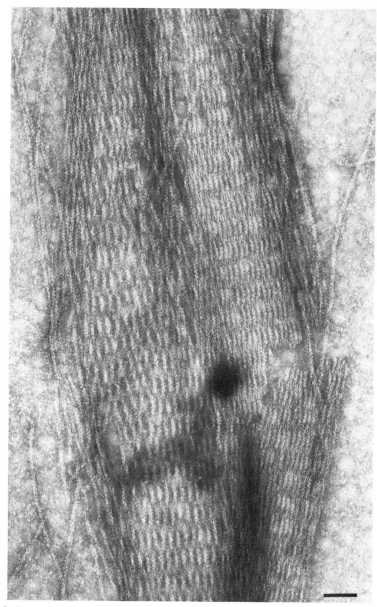

Fig. 8. Net-type paracrystal of actin, induced by putrescine and negatively stained. Each strand of the net appears to comprise one filament. The layers of filament are superimposed at different angles in different regions of the paracrystal. Scale marker indicates 50 nm. Unpublished micrograph of M. J. Dickens.

phous replacement, could be employed. The problems of this approach are firstly that the modification to G-actin should be severe enough to remove one of its most important attributes, without rendering its structure so abnormal that it is no longer worthwhile undertaking the investigation, and secondly that the crystals produced should be large enough for X-ray analysis.

The use of trivalent lanthanide ions to inhibit filament formation seems to have been successful on the first of these counts though not on the second. dos Remedios and Dickens (1978) obtained electron micrographs of negatively stained microcrystals formed from actin in the presence of gadolinium ions, and showed that they were distinct from F-actin aggregates. When the incubation with gadolinium was followed by the addition of 0·1 M KCl, tubular assemblies about 200 nm wide were formed. G-actin after treatment with and then removal of gadolinium was capable of forming F-actin, indicating that the protein was not denatured (dos Remedios and Barden, 1977). Dickens (1978) recorded micrographs of shadowed as well as negatively stained actin tubes (Fig. 9). Optical diffraction patterns of the negatively stained specimens indicated that the unit cell of the lattice was near-rectangular with dimensions about 5·5 by 6·6 nm, and optical filtering, to generate a one-sided image of the tube, showed that each unit cell contained two pear-shaped objects, each about 5·5 by 3·3 nm, with dimensions consistent with those found for the subunit in F-actin. Confirmation of the tubular nature of the aggregates was provided by dos Remedios *et al.* (1980) from electron micrographs of sectioned material. The tubes were revealed as about 120–130 nm in diameter and comprising a single layer of actin subunits. The diameter varied when trivalent lanthanide ions other than gadolinium were used, however, increasing with ionic radius. The increase could be correlated with the number and orientation of the nearly tranverse strands of subunits seen in negatively-stained specimens (Fig. 9), so that the size of the subunit itself was not considered to vary significantly with the particular lanthanide used. Some lanthanide ions did not induce tube formation. These bound at a higher stoichiometry to actin, 7:1 rather than 5 or 6:1, and the lack of polymerization was attributed to the production of a net positive charge on the monomer.

Aebi *et al.* (1980) used actin from a different source, amoeba rather than rabbit, and varied the ionic strength to produce not only the tubular form but sheet-like structures, some of which were derived from opened-up tubes. The different assemblies were constructed from the same fundamental two-dimensional lattice, called the "basic sheet", which shadowing revealed to have one side smooth and regular and the other coarse and irregular. This feature indicated that the actin monomer was asymmetric in shape in the direction perpendicular to the plane of the sheet and that all the molecules were packed with the same polarity relative to this direction. Computer filtering of

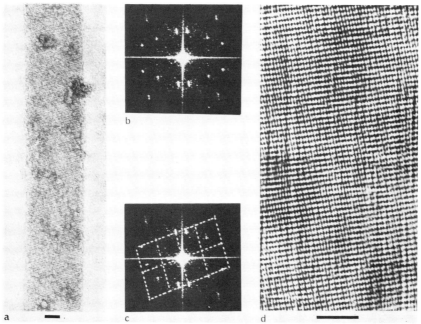

Fig. 9. (a) Negatively-stained actin tube formed in the presence of gadolinium ions. The structure has uniform width, and the strands of subunits forming the walls of the tube can be seen. (b) Optical diffraction pattern of (a). (c) The same pattern with, superimposed, the lattice joining reflections arising from one side of the tube. The dimensions of the lattice correspond to spacings of about 5·5 and 6·6 nm. (d) The optically filtered image of a part of (a), in which the subunits are seen as pear-shaped, with two subunits in one unit cell of the lattice. Scale bars in (a) and (d) indicate 50 nm (reprinted, with modification, from Dickens, 1978, courtesy of Blackwell Scientific Publications).

micrographs of negatively-stained basic sheets confirmed the dimensions for the monomer given by Dickens (1978) and, showed in addition, that the elongated shape was divided into a major and minor peak (Fig. 10). Virtually the same monomer dimensions, 5·5 × 3·3 nm, were given by Barden *et al.* (1981) from a computer-averaged micrograph of an opened-up tube, but the two-zone division was not observed. The specimen was prepared from skeletal muscle actin in the presence of ions of the trivalent lanthanide, praseodymium. In a further study of amoeba actin, Aebi *et al.* (1981) showed how two basic sheets could overlap in different ways to give the distinctive appearances seen in specimens more than one layer thick. In the tubular variant the sheets overlapped with the rough surfaces together, but in the other forms, described as "rectangular" and "square" because of the general appearance of the superimposition pattern, the smooth surfaces were in contact. Computer filtering of the different forms showed, in each case, an elongated actin

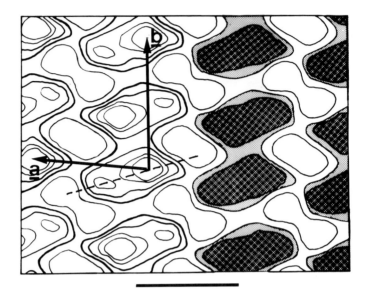

Fig. 10. The computer-averaged structure obtained from negatively-stained sheets of amoeba actin formed in the presence of gadolinium ions. The presence of a two-fold axis of symmetry perpendicular to the plane of the sheet was assumed in the averaging procedure. The actin monomer is elongated in the direction indicated by the broken line, and is divided into a larger and a smaller domain. a and b are the lattice vectors, of amplitude 5·65 and 6·55 nm respectively. In the tubular form of gadolinium-actin (Fig. 9) the b-axis runs parallel to the axis of the tube. On the right-hand side of the Fig., the actin dimers and monomers are indicated by the light and dark shaded regions, respectively. Scale marker indicates 5 nm (reprinted from Aebi et al., 1980, courtesy of U. Aebi and Macmillan Journals Ltd).

molecule, though there was some variation in the internal detail revealed, particularly in the extent and distribution of the major and minor peaks. In this analysis, as in that of Aebi et al. (1980) the pairs of molecules within each unit cell were assumed to be related by a two-fold axis (Fig. 10). With this symmetry, adjacent rows of monomers were antiparallel, a situation different from that occurring in the actin filament where the two long-pitch helical strands have the same polarity. Aebi et al. (1981) noted this distinction, but considered nevertheless that the long dimension of the monomer in the sheets should be related to the 5·5 nm axial repeat of the subunits of F-actin. In Fig. 10 the antiparallel rows of monomers are the ones running across the page. The head-to-tail packing along each row would therefore resemble the arrangement of subunits along each of the long-pitch helices of the filament.

E. Structure of Actin: DNase I

Greater success in obtaining macroscopic crystals suitable for X-ray diffraction studies has been achieved where another protein rather than a counter ion was used to complement the actin monomer and prevent filament formation. The two proteins used, profilin and DNase I, each bind to G-actin in a 1:1 complex which may be crystallized fairly easily, but the initial promise that such crystallization would soon lead to an atomic view of the actin monomer has not yet been realized, as heavy-atom derivatives giving good data at high resolution have not been obtained.

Actin associated with a low molecular weight protein, subsequently called profilin (Carlsson *et al.*, 1977) was isolated from calf spleen and crystallized by Carlsson *et al.* (1976). X-ray diffraction patterns obtained from the crystals showed data to at least 0·35 nm resolution, but a full analysis of the crystal structure has still to be announced, though we believe this may be imminent. The crystallization of a complex of skeletal muscle actin with pancreatic DNase I was reported by Mannherz *et al.* (1977), together with an X-ray diffraction pattern at 0·5 nm resolution. Similar data were given by Sugino *et al.* (1979) for the complexes of chicken gizzard and of *physarum* actin with DNase I. From an analysis using six heavy-atom derivatives of the skeletal muscle actin, Suck *et al.* (1981) presented an electron-density map at 0·6 nm resolution in which the two proteins could be distinguished, although there was some uncertainty in the position of the boundary separating them. The two regions delineated had volumes consistent with the molecular weights, and experiments in which the structural effects of modifying the proteins, for example, changing the bound Ca^{2+} in DNase I for Ba^{2+}, further clarified the assignment. The overall dimensions of the complex were $8·7 \times 6·7 \times 4·3$ nm (Fig. 11), and of the actin component $6·7 \times 4·0 \times 3·7$ nm, encompassing a large and small domain. Both actin domains contained density regions resembling a central β-pleated sheet surrounded by two or three α-helices near the surface of the molecule. The highest density occurred in the cleft between the two domains, suggesting this as the site for the bound ATP. The repeat distance along one axis of the crystal was 5·64 nm, similar to the axial separation of actin monomers along the strands of F-actin (Fig. 6), but a model of the filament using this crystal contact had a diameter of 12 nm, which was considered too large. The diameter could be reduced if the actin monomers from the crystal were arranged with their long axis parallel rather than perpendicular to the filament axis.

III. TROPOMYOSIN

For a protein of the thin filament, tropomyosin seems a misnomer. It was so called by its discoverer Bailey (1946, 1948), who noted its highly asymmetric shape and suggested accordingly that it was a "prototype of myosin". Bailey's

Fig. 11. Model of the actin:DNase I complex at 0·6 nm resolution obtained by X-ray diffraction analysis of crystals. The lower part of the model represents actin, which is elongated and divided into two lobes with a cleft between. The cut-off level of density is 10% of the maximum (reprinted from Suck et al., 1981, courtesy of D. Suck and The National Academy of Sciences, USA).

name has persisted despite its hasty assignment and until the controversy surrounding the role of this protein in the control of contraction has been resolved, it would be premature to attempt a more apt alternative.

A. Crystals, Fibres and Tactoids

Bailey (1948) prepared plate-like crystals of tropomyosin in dilute salt solution but, in an accompanying paper, Astbury *et al.* (1948) recorded X-ray diffraction patterns not from these crystals but from thin air-dried films of tropomyosin fibres. The patterns showed the α-diagram characteristic of the keratin-myosin-epidermis-fibrinogen group of proteins (Perutz, 1951). In the electron microscope, Astbury *et al.* (1948) showed with gold shadowing that their tropomyosin preparation contained long fibrils 20–30 nm thick, which treatment with osmic acid indicated were cross-striated. The analysis of Bailey's crystals was not taken up for eleven years, when Hodge (1959) showed from sections stained with phosphotungstic acid the presence of a two-dimensional net structure with spacings of about 20 nm and 40 nm. Similar observations were made, at higher resolution, by Huxley (1963) on negatively stained fragments of crystals. The length of the tropomyosin molecule was measured as 40 nm by Rowe (1964) from shadowed preparations. Before shadowing, the protein was sprayed from a solution of high ionic strength containing a volatile buffer, a procedure designed to avoid polymerization. Similar observations were made by Peng *et al.* (1965) and Ooi and Fujime-Higashi (1971). Tsao *et al.* (1965) prepared spindle- or needle-shaped paracrystals by precipitation of tropomyosin near its isoelectric point (pH 5·1). They observed in specimens from many sources, stained with phosphotungstic acid, that the band pattern sometimes repeated at 20 nm or 80 nm instead of 40 nm. In some specimens the sub-banding with the 40 nm repeat occurred at intervals of one third of the repeat, a distance which has subsequently been favoured as the pitch of the tropomyosin coiled-coil (see Section V.C). The similarity between the paracrystals and the fibrils of tropomyosin, as studied by Astbury *et al.* (1948) and prepared in the absence of salt at neutral pH, was pointed out by Kung and Tsao (1965). Although the fibrils were less regular and much longer, shadowing revealed similar band patterns.

Cohen and Longley (1966) showed that a more reliable method of preparing the needle-shaped paracrystals, which they called tactoids, was the use of divalent cations at neutral or slightly alkaline pH. Negative staining of the tactoids revealed a symmerical pattern of sub-bands within each 40 or 80 nm repeat, indicating that the tropomyosin molecules (assumed polar) were packed antiparallel. The 80 nm repeat, observed in some tactoids of smooth muscle tropomyosin, was attributed to the formation of an end-to-end dimer.

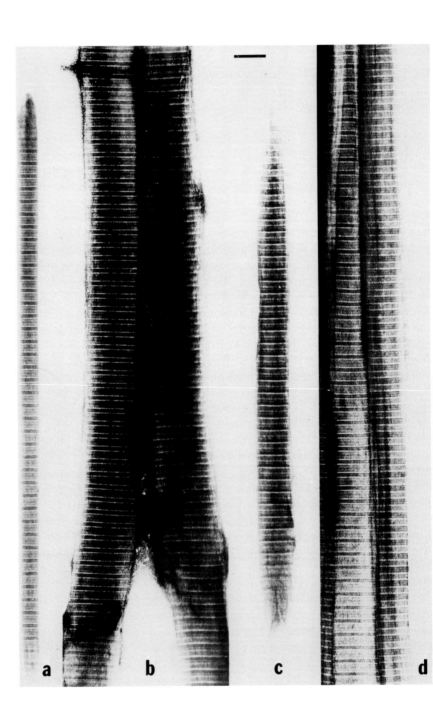

In an extended study of crystals and tactoids of rabbit skeletal muscle tropomyosin, Caspar *et al.* (1969) showed that the cations of magnesium, lead and barium all produced tactoids with a 39·5 nm band period (Fig. 12). The magnesium and barium tactoids had symmetrical patterns of sub-bands but those of the lead tactoids were polar. Some of the barium tactoids had a slightly longer repeat, 40·5 nm (Fig. 12d). The appearance of the lead tactoids confirmed that the tropomyosin molecule was polar, and the pattern of sub-banding within these tactoids had two light regions, one with a dominant narrow band and the other a spaced doublet with another narrow light line to one side. The distance between the centre of the doublet and the dominant narrow band was about 17·0 nm on one side and 22·5 nm on the other, spacings which corresponded to the separations of the cross-over points of the strands of the nets in the crystals (Fig. 13). The long-short pattern of cross-over points gave the nets a characteristic kite shape. Other smaller crystalline specimens showed rectangular or square nets with 40 nm or smaller spacings, and in some of these, analogous patterns of long-short internodal distances were observed. The strands generally had a width of about 4·0 nm and were sometimes seen to comprise of two filaments (Fig. 13). Kite-shaped and square nets were also observed by Higashi-Fujime and Ooi (1969) in sectioned and negatively stained crystals. In some specimens the transformation between a net and banded structure could be seen (Fig. 14), illustrating the relationship between filament packing in crystals and tactoids. Millward and Woods (1970) prepared tactoids of tropomyosin from vertebrate smooth and invertebrate muscles, and demonstrated similar band patterns to those observed for rabbit skeletal muscle tropomyosin. They also showed that rabbit tropomyosin, after dissociation into its subunits and reassembly, would form crystals and tactoids with the same appearances as those of the native protein.

At acid pH, other kinds of tactoids were observed, some having striations approximately perpendicular to the long axis and spaced about 4·5 nm (Cohen and Longley, 1966; Nonomura *et al.*, 1968) and others with more oblique striations spaced at 6–7 nm (Caspar *et al.*, 1969). Longley (1977) analysed the structure of the first type, prepared by precipitating tropomyosin with divalent cations in the pH range 4 to 6·5, and showed that the tactoids had a molecular packing related to that of the alkaline form. He constructed a model

---

Fig. 12. Tropomyosin tactoids formed with divalent cations, negatively stained with uranyl acetate. (a) Magnesium tactoid. (b) On the left, magnesium tactoid in register with lead tactoid on right. (c) Lead tactoid showing polar nature of the band pattern clearly. (d) Barium tactoids: on the left with an appearance indistinguishable from a magnesium tactoid, and on the right having a periodicity about 2·5% larger, and a distinctive band pattern. Scale marker represents 100 nm (reprinted from Caspar *et al.*, 1969, courtesy of C. Cohen and Academic Press Inc.).

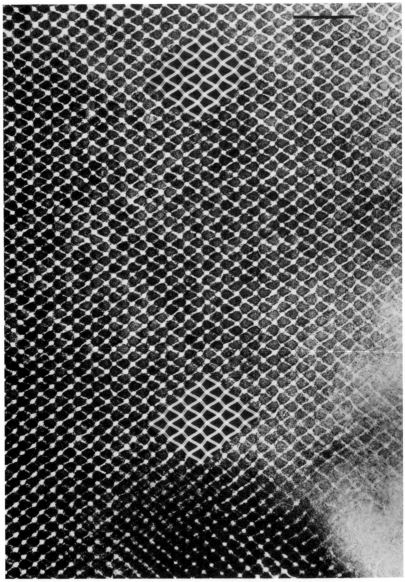

Fig. 13. *Trompe l'oeil* view of a tropomyosin crystal, negatively stained with uranyl acetate. The long-short pattern of cross-over points of the strands in the net give the repeating motif a characteristic kite shape. Two patches of the electron density map calculated from the X-ray diffraction pattern of the hydrated crystals are superimposed on the electron micrograph. Scale marker indicates 100 nm (reprinted from Caspar *et al.*, 1969, courtesy of C. Cohen and Academic Press Inc.).

1. Actin and Thin Filaments 25

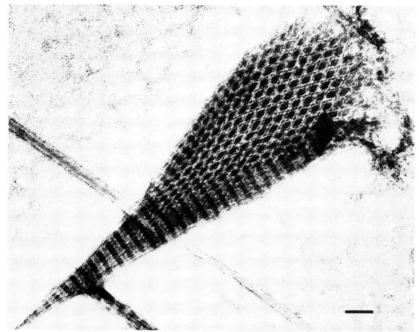

Fig. 14. Negatively-stained specimen of tropomyosin in which the net and banded structures merge into each other, illustrating the relationship between filament packing in crystals and tactoids. The net form is the "double diamond" lattice, as observed in the presence of TnT (cf Fig. 23), suggesting that the tropomyosin was contaminated with this component (Yamaguchi et al., 1974). Scale marker indicates 100 nm (reprinted from Higashi-Fujime and Ooi, 1969, courtesy of S. Higashi-Fujime and Société Française de Microscopie Electronique).

of overlapping sheets, in each of which the pair of prominent white, stain excluding lines seen every 40 nm along the alkaline tactoids (Fig. 12a, b) was used as the repeating motif. Instead of extending right across the tactoid, however, adjacent lateral segments of each line were displaced axially to give rise to an oblique lattice.

B. The Coiled-coil

The description of tropomyosin as two α-helical strands wound round each other was proposed by Crick (1953) to explain the X-ray diffraction patterns obtained by Astbury et al. (1948). Crick suggested that the non-polar hydrophobic groups, which accounted for about 2/7 of the constituent amino acids, would tend to occur on the inside of the molecule between the chains and that as a consequence there would be an alternation of groups of polar and non-polar residues, the non-polar occurring at an average interval of 3·5 residues. The determination of the sequence (Hodges *et al.*, 1972; Stone *et al.*, 1974) confirmed this type of distribution: a repeating pattern of seven residues in which hydrophobic groups occur at intervals of 3 and 4 residues alternately. The presence of two chains of molecular weight 33 500 in the molecule was shown by Woods (1967), but that they are often non-identical was shown by Cummins and Perry (1973), who distinguished two types α and β of differing electrophoretic mobility and cysteine content. From the sequence of rabbit skeletal α-tropomyosin, Hodges *et al.* (1972) showed how two parallel chains twisted slowly round each other would provide mutual contacts in which the hydrophobic residues of one chain were interwoven with those of the other, the "knobs in holes" packing proposed for the coiled-coil by Crick (1953). This arrangement necessitated the chains being in register or staggered by 7 or a multiple of 7 residues. The number of residues in each chain, 284, indicated a minimum molecular length of about 42·3 nm (284 × 0·149 nm, the axial repeat per residue of an α-helix), somewhat larger than the repeats observed in crystals and tactoids. This suggested that end-to-end aggregation of tropomyosin involved a short molecular overlap.

The negatively stained tactoids of tropomyosin show a detailed pattern of sub-banding. Caspar *et al.* (1969) recorded X-ray diffraction patterns of air-dried fibres formed from unstained tactoids and noted the presence of an axial reflection at 2·8 nm, indexing as the 14th order of the 39·6 nm repeat. Parry and Squire (1973) correlated this with both the positively and negatively stained band patterns, showing by convolution enhancement of the electron micrographs the presence of a pronounced 14-fold axial sub-repeat. In the amino-acid sequence, Parry (1974) and Stone *et al.* (1974) noted a repeat of the non-polar groups of 19·5 residues, and a less well-developed repeat of this value for the negatively and positively charged groups. A similar analysis by Stewart and McLachlan (1975) and McLachlan and Stewart (1976), however, involving Fourier transformation of the sequence, found the repeat of the negatively charged groups as the most significant, refining it to 19·73 residues. The repeat of the non-polar groups was considered a less well-developed one, which might arise as a consequence of the repeat of the acidic residues, since the two periods were out of phase. These authors noted that the 19·73 residue pattern would repeat 14·39 times in 284 residues, so that if the tropomyosin molecules, each consisting of the two chains in register, were joined end-to-end without introducing a break in the periodicity, there would be an apparent molecular length of 276 residues and an overlap of 8 residues. A similar result was obtained by Parry (1975b).

McLachlan and Stewart (1975) considered that the most favourable packing of the hydrophobic groups between the two chains would occur when they were in register. This was also indicated experimentally when it was

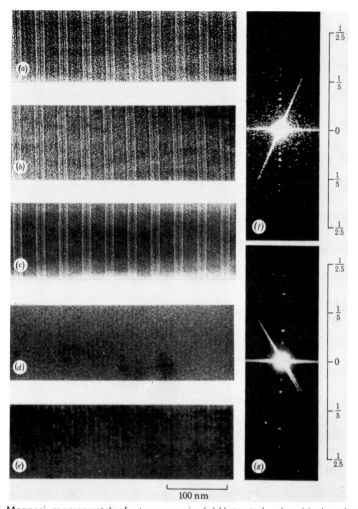

Fig. 15. Magnesium paracrystals of α-tropomyosin. (a) Untreated and positively stained with uranyl acetate. (b) Treated with methyl mercury in 6 M guanidinium chloride and positively stained with uranyl acetate. (c) Carboxymethylated prior to treatment with methyl mercury as in (b) and positively stained with uranyl acetate. The similarity of (b) and (c) to (a) indicates that the chemical treatments do not alter the molecular and paracrystal structure. (d) Carboxymethylated and treated with methyl mercury as in (c) and viewed unstained (that is, in the absence of uranyl acetate). (e) Treated with methyl mercury as in (b) and not carboxymethylated so that the cysteine residue was free to bind the mercury. Viewed unstained as in (d). There is an additional dark line at 40 nm intervals, which is seen more clearly if the micrograph is viewed at a glancing angle. (f) Optical diffraction pattern of (d). (g) Optical diffraction pattern of (e) (reprinted from Stewart, 1975b, courtesy of M. Stewart and The Royal Society).

shown that the cysteine residues could be cross-linked. α-tropomyosin has one cysteine in each chain, located 94 residues from the C-terminus (Sodek et al., 1972), and cross-linking obtained by oxidation to form a disulphide bridge was shown as intramolecular by the relative absence of oligomeric aggregation (Lehrer, 1975; Johnson and Smillie, 1975; Stewart, 1975a). When labelled with methyl mercury, the position of the cysteine could be related to the band patterns of positively or negatively stained tactoids. The presence of only a single mercury-stained band at intervals of 40 nm indicated that the cysteine residues were located close to the dyad axes relating adjacent antiparallel molecules (Stewart, 1975b; Fig. 15).

Knowledge of the sequence prompted several studies designed to determine the molecular packing and explain the band pattern in tactoids. Assuming that uranyl ions attached to the acidic residues, Stewart and McLachlan (1976) showed clearly the presence of the 14-fold repeat in positively-stained specimens (Fig. 16). The band pattern was matched when the overlap between the C-termini of adjacent antiparallel molecules was 175 residues and that between the N-termini was 107 to 122 residues. Hurwitz and Walton (1977) evaluated the charge interactions between adjacent molecules in the tactoids, and considered that the C-terminal overlap was 201 residues. Stewart (1981) reaffirmed a value of 176 ± 6 for this parameter, in a detailed analysis, and gave a value of 11 ± 5 for the number of residues in the head-to-tail, N to C overlap region of molecules with the same polarity. He showed that the intensity of the prominent white lines in the tactoid (Fig. 16) was reduced after digesting the tropomyosin with carboxypeptidase A, implying that, even in supposedly positively-stained material, the white lines resulted from the exclusion of residual negative stain at the head-to-tail molecular overlap. Comparison of predicted and observed patterns near the molecular ends suggested that the conformation of the head-to-tail overlap was not completely α-helical, and a model with a globular N-terminus and an extended C-terminus was proposed. Katayama and Nonomura (1979a, b) showed how the appearance of negatively-stained polar tactoids could be simulated when the amino acids were assigned a "bulkiness" parameter, defined as the ratio of side-chain volume to length. Photographic transparencies of the polar tactoids, when superimposed in antiparallel pairs with different axial translations, reproduced the various appearances of the symmetrical tactoids. The overlap deduced for the antiparallel molecules in the Mg^{2+} tactoid agreed with that given by Stewart and McLachlan (1976) for positively-stained specimens (Fig. 16).

C. Crystal Structure

After the publication of the analysis of nets and tactoids by Caspar et al.

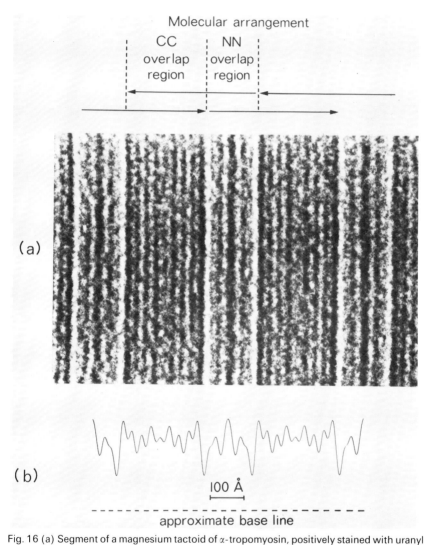

Fig. 16 (a) Segment of a magnesium tactoid of α-tropomyosin, positively stained with uranyl acetate. (b) Average filtered trace of the density projected onto the axis of the tactoid. In each repeat there are 14 bands, with 9 in the C-C overlap region and 5 in the N-N overlap region. The arrangement of the molecules is shown diagrammatically above the tactoid, where the arrowheads represent the C-terminus of the molecules (reprinted from McLachlan and Stewart, 1976, courtesy of M. Stewart and Academic Press Inc.).

(1969), electron microscope studies were concentrated mainly on the latter, but the nets, since they form true macroscopic crystals (albeit mostly water!) continued to be intensively analysed by X-ray diffraction. From electron micrographs of the nets, Caspar *et al.* (1969) constructed a model which they refined to fit the X-ray data corresponding to the *a*-axis projection of the crystals. This type of analysis was extended by Cohen *et al.* (1971). It was not possible to prepare isomorphous derivatives of the crystals, and structural analysis involved comparing crystals with different lattice parameters, with the restrictions that the length and mass distribution of the coiled-coil molecule remained invariant. From data extending to Bragg spacings of 2 nm, Phillips *et al.* (1979) determined a three-dimensional electron density map based on the solution of the three centric projections of the crystal. Phase origins were chosen for the projections that produced a three-dimensional filament path consistent with the $P2_12_12$ space group and the continuity of the filament. The phases of the reflections were refined so that the electron density in the solvent regions was uniform. The length of the filament, winding along a line parallel to the body diagonal of the unit cell, was measured as $41 \cdot 03 \pm 0 \cdot 14$ nm. Subtracting this length from that predicted for the whole molecule, 284 residues of coiled-coil, indicated a molecular overlap of $8 \cdot 6 \pm 1 \cdot 0$ residues. Location of the cysteine residue in mercury-labelled crystals established the probable positions of the molecular ends, which corresponded to regions of increased density in the map (Fig. 17). The shape of these regions suggested that they comprised intermeshed polypeptide chains rather than overlapped coiled-coils. An analysis of diffuse scattering observed in the X-ray diffraction patterns, and the presence of irregularities within the filament in the electron density map, implied localized flexible regions of the molecule, and possible departures from the stable α-helical conformation. The direction of the molecule and the positions of its ends were confirmed in an analysis of crystals of rabbit foetal skeletal tropomyosin, which contains a high proportion (70%) of β-subunits (Phillips *et al.*, 1980). Mercury labelling of the second cysteine present in β-tropomyosin showed that the position of residue 36 was in the position predicted in the molecular assignment as in Fig. 17.

D. Association with Actin

Tropomyosin was shown to be extracted together with actin from myofibrils at low ionic strength, and phase microscopy revealed the concomitant disappearance of the I-band (Perry and Corsi, 1958; Corsi and Perry, 1958; Huxley, 1960). Endo *et al.* (1966) showed by phase microscopy that fluorescent-labelled tropomyosin stained the I-bands of glycerinated muscle that had been partially digested with trypsin. In undigested muscle they showed that fluorescent-labelled antibody to tropomyosin stained the I-band,

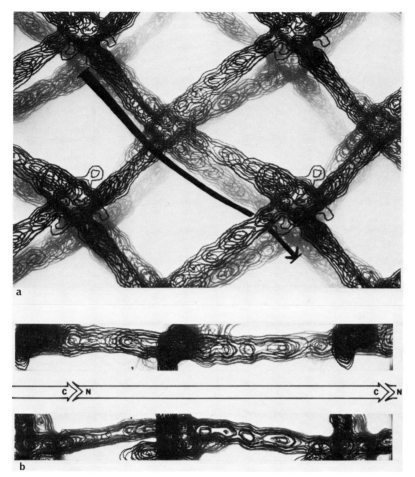

Fig. 17. Three-dimensional electron density map of tropomyosin obtained from X-ray diffraction analysis of crystals. (a) Several unit cells of the structure, with the proposed position and orientation of the molecule indicated by the arrow, tip of which represents the C-terminus. (b) A filament sectioned along its length. Top and bottom views are rotated by 90° about the filament axis (reprinted from Phillips et al., 1979, courtesy of G. N. Phillips, Jr and Macmillan Journals Ltd).

sometimes including the Z-region, and the centre of the A-band. Similar antibody results were obtained by Pepe (1966). Hanson and Lowy (1963, 1964) suggested that tropomyosin was located on the outside of the filament, running in the long-pitch helical grooves of actin. This was supported by the comparison of optical diffraction patterns of electron micrographs of paracrystals of pure and impure actin, that is, actin contaminated with

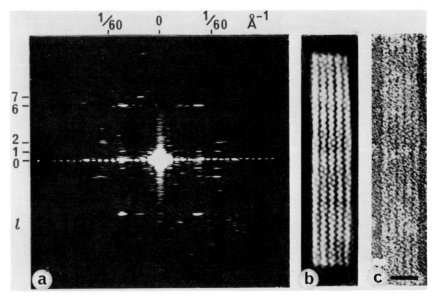

Fig. 18. (a) Optical diffraction pattern of the paracrystals shown in (c). The paracrystal was formed from actin plus tropomyosin:troponin, and negatively stained. The diffraction pattern shows a higher intensity of the second layer-line compared with corresponding patterns from pure actin paracrystals (Fig. 5). (b) Optically filtered image of (c). The filaments are further apart and have a different packing from those in pure actin paracrystals. Scale marker in (c) indicates 30 nm (reprinted from Hanson et al., 1972, courtesy of Cold Spring Harbor Laboratory).

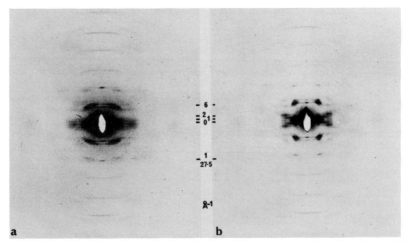

Fig. 19. X-ray diffraction patterns of oriented gels of filaments reconstituted from (a) purified actin and (b) actin prepared by KI extraction and therefore associated with tropomyosin:troponin. The second layer-line in (b) is much stronger than in (a) (reprinted from Hanson et al., 1972, courtesy of Cold Spring Harbor Laboratory).

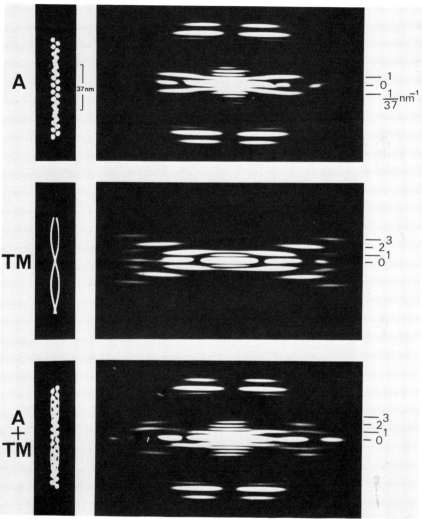

Fig. 20. Optical diffraction patterns obtained from computer-drawn models of actin, tropomyosin and actin+tropomyosin. Tropomyosin contributes to the low-order layer-lines only. Whereas actin alone gives rise to a diffraction pattern which is weak on the second layer-line, the intensity is increased by the presence of tropomyosin in the filament (reprinted from O'Brien et al., 1980, courtesy of Japan Scientific Societies Press, Tokyo).

tropomyosin (and troponin). O'Brien et al. (1971) and Hanson et al. (1972) showed that when tropomyosin was present the second layer-line in the diffraction pattern was intensified (Fig. 18), and Hanson et al. (1972) showed the same effect when X-ray diffraction patterns of oriented gels of filaments were compared (Fig. 19). An explanation of this effect is given in Fig. 20, which shows diffraction patterns of model structures. Tropomyosin, drawn as a smooth helix, gives rise to reflections on the low-order layer-lines only. Whereas the diffraction pattern from actin alone is very weak on the second layer-line, the intensity is increased when tropomyosin is present.

IV. TROPONIN

Until recently, troponin was not considered a universal component of the thin filament, since it had not been observed in significant amounts in some muscles, particularly those from molluscs (Lehman and Szent-Györgyi, 1975). This view has been modified, however, though the amount of troponin in the intact scallop muscle is not known (Goldberg and Lehman, 1978), and it is possible that the periodic attachment of this protein to actin and tropomyosin, as in Fig. 1, is a general feature of muscle.

A. Isolation and Components

Ebashi (1963) and Ebashi and Ebashi (1964) showed that a protein component resembling tropomyosin could be prepared which differed in that it was able to confer Ca^{2+}-sensitivity to actomyosin superprecipitation. The component had a larger sedimentation constant and a higher viscosity than tropomyosin, and Ebashi and Kodama (1965) showed that it could be separated into two proteins, one of high and one of low intrinsic viscosity. The one with high viscosity was tropomyosin and the other, a globular protein called troponin, was shown to increase dramatically the viscosity of F-actin in the presence of a relatively small amount of tropomyosin (Ebashi and Kodama, 1966).

The discovery that troponin itself had a complex structure opened a fascinating new field of protein investigation relevant to the thin filament. The first separation was into two components, troponin A, sensitive to Ca^{2+} and troponin B, which had an inhibitory effect on actomyosin interaction (Hartshorne and Mueller, 1968). Subsequently a spate of papers increased this number to three, TnC (troponin C), the Ca^{2+}-binding component, TnI (troponin I), responsible for inhibition and TnT (troponin T), which bound the complex to tropomyosin (Schaub and Perry, 1969, 1971; Hartshorne and Pyun, 1971; Ebashi *et al.*, 1971; Greaser and Gergely, 1971; Drabikowski *et al.*, 1971; Staprans *et al.*, 1972; Ebashi *et al.*, 1972; for review see Weber and Murray, 1973). There were some discrepancies in the molecular weights reported for the components, partly because there are actual differences according to the source muscle type. For rabbit skeletal muscle, however, the position was resolved by the determination of the amino-acid sequences, which gave TnC, 17 800 (Collins *et al.*, 1973; Collins, 1974); TnI, 20 900 (Wilkinson and Grand, 1975); TnT, 30 503 (Pearlstone *et al.*, 1976).

In the light microscope, fluorescent-labelled troponin was shown to stain the I-bands of glycerinated myofibrils that had been partially digested with trypsin (Endo *et al.*, 1966). Similar staining was achieved by these authors when the undigested muscle was treated with fluorescent-labelled antibody to troponin. All the I-band except the Z-line was stained. This pattern was also observed by Perry *et al.* (1972) with fluorescent-labelled antibody to troponin and to TnI, who noted that for the anti-troponin the gap between the fluorescent region and the Z-line was narrower than when anti-ThI was used.

B. Interaction with Tropomyosin

Nonomura *et al.* (1968) compared Mg^{2+}-induced tactoids of pure tropomyosin with those of tropomyosin-troponin. In negatively-stained specimens they noted the presence of a broad light band attributable to troponin and repeating at intervals of 37–39 nm along the paracrystals. This band was enhanced when the troponin was first conjugated with ferritin. Troponin was assigned a position about 13 nm from one end of the tropomyosin molecule, from observations of negatively-stained tropomyosin tactoids with and without troponin (Ebashi *et al.*, 1972; Ohtsuki, 1974). Some of the tactoids appeared severed or broken on the grid, revealing a fringe of fibrils at the ends, and the end of the fringe was assumed to mark the end of the tropomyosin molecules. This feature was related to the enhanced light band in the tactoids with troponin (Fig. 21). Cohen *et al.* (1972) considered that this band consisted of pairs of troponin units, which meant that although the distance from midway between the units to the end of the tropomyosin was 12·5 nm, this was an average value and that the troponin was bound either at about 15·0 or 10·0 nm from a molecular end.

Greaser *et al.* (1972) showed that TnT bound to pre-formed tropomyosin tactoids in the middle of the light band, as had been shown for whole troponin. At the ends of the tactoids a transition to a hexagonal net of dimension 40 nm was observed (Fig. 22), and comparison of the band and net patterns indicated that TnT bound to the end of the tropomyosin molecules, rather than a third the way along. When the two proteins were mixed before precipitation, the hexagonal net form predominated. Addition of TnT to pre-formed tactoids also produced a double-stranded net, sometimes transforming into tactoids. Higashi-Fujime and Ooi (1969) had observed very similar net and tactoid appearances, and the transformation between them (Fig. 14), in a preparation they considered free of troponin, but it seems likely that TnT was present (Yamaguchi *et al.*, 1974). Cohen *et al.* (1972) and Margossian and Cohen (1973) considered that the TnT was probably located at the acute vertex of the

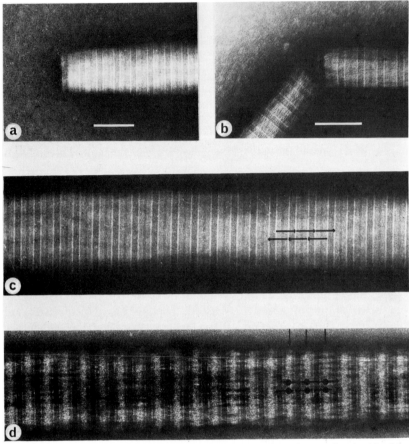

Fig. 21. (a) A tropomyosin tactoid, negatively stained, showing a fractured end, with a fringe of molecules. (b) A broken tactoid, each half showing a fringe. (c) Tropomyosin tactoid, with the proposed arrangement of molecules superimposed. The ends of the molecules are aligned with the edges of the dark bands of the tactoids, as suggested by the appearance of the fringes in (a) and (b). (d) Tactoid of tropomyosin plus troponin. Troponin is located in the middle of the wide band (indicated by arrows). Scale markers represent 100 nm; (c) and (d) are at the same magnification as (b) (reprinted from Ebashi et al., 1972, courtesy of I. Ohtsuki and Cold Spring Harbor Laboratory).

diamonds in the double-stranded net form (Fig. 23) and that TnT generated different cross-connections between tropomyosin filaments from those formed with whole troponin. In the presence of whole troponin, the net type of paracrystals were of the normal type with the kite-shaped repeating unit, and Cohen et al. (1972) showed that in this case the troponin attached at the middle of the long arm of the unit (Fig. 24). At lower resolution this pattern of

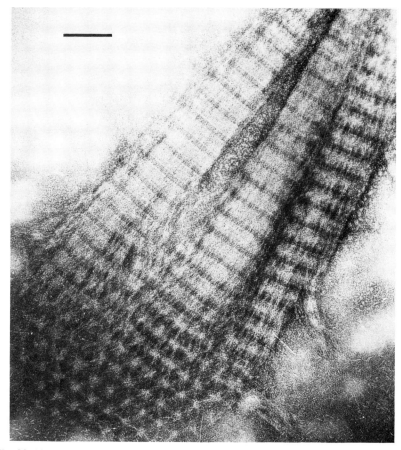

Fig. 22. Magnesium tactoid of tropomyosin to which TnT has been added. The end of the tactoid has transformed into a hexagonal net pattern. If the TnT binding periodicity is followed through the transformation region, it suggests that TnT is located at the intersection points of the hexagonal net pattern. Scale marker indicates 100 nm (reprinted from Greaser et al., 1972, courtesy of M. L. Greaser and Cold Spring Harbor Laboratory).

attachment would produce the 30 nm repeat of the striations observed by Higashi and Ooi (1968). Margossian and Cohen (1973) found that the diamond nets were not formed if TnC as well as TnT was mixed with tropomyosin, and that although tactoids were formed, they resembled those formed with whole troponin present rather than with TnT alone. If the TnC was first saturated with Ca^{2+}, the tactoids showed a marked increase in the contrast of the bright-staining band attributed to the bound components, suggesting that TnC bound more strongly to TnT in the presence of saturating Ca^{2+}, an observation consistent with the turbidity measurements of TnC:TnT

Fig. 23. Portion of a double-stranded ("double-diamond") net crystal of tropomyosin plus TnT. The superimposed drawing accentuates the molecular filaments and distinguishes the two types of cross connections. The dotted lines indicate an orthogonal pair of mirror lines of the plane group *cmm*. The lines correspond to two-fold axes of the crystal lattice seen in projection. Scale marker indicates 50 nm (reprinted from Cohen et al., 1972, courtesy of C. Cohen and Cold Spring Harbor Laboratory).

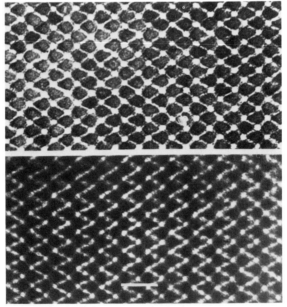

Fig. 24. Crystals of (a) tropomyosin and (b) tropomyosin:troponin, negatively stained with uranyl acetate. Comparison of (a) and (b) shows that the troponin binds at the middle of the long arm of the kite-shaped mesh. From the symmetry of the crystal, each strand of the net consists of a pair of oppositely directed molecular filaments. Correspondingly, the thickening identified with troponin consists of a pair of units related by a dyad. Scale marker indicates 50 nm (reprinted from Cohen et al., 1972, courtesy of C. Cohen and Cold Spring Harbor Laboratory).

mixtures at different Ca^{2+} levels made by Ebashi et al. (1972). This result depended on previous observations indicating that TnC did not bind to tropomyosin (Hartshorne et al., 1969; Greaser et al., 1972). In contrast, Yamaguchi and Greaser (1979), working with cardiac tropomyosin and troponin, found that although these proteins gave mostly the same crystal and tactoid patterns as those from skeletal muscle, the binding of TnC and TnT to tactoids depended on the absence rather than the presence of Ca^{2+}.

Yamaguchi et al. (1974) found that the kite-shaped as well as the double-diamond lattice could be formed in the presence of TnT and that the binding site was the same as that for whole troponin (Fig. 24). They considered that not only the double-diamond but also the square lattices observed by previous authors depended on the presence of TnT. When the tropomyosin was pure only the tactoids, the kite-shaped lattice and an additional net form with a finer mesh were observed. The fine-mesh crystals were formed when both chains of the tropomyosin molecule were α, and were later shown to comprise a double array of kite-shaped units (Greaser et al., 1977).

C. The 40 nm Stripe

The observation of a regular 40 nm cross-striation in the I-bands of myofibrils was made very early, at the same time as the myofilaments were discovered, and soon after the development of electron microscopy and its application to biological structure. Draper and Hodge (1949) showed the striation in myofibrils isolated from formalin-fixed toad muscle and stained with phosphomolybdic acid. A clearer demonstration followed a few years later with the development of the ultramicrotome for cutting thin sections of specimens embedded in plastics, and Hodge et al. (1954) showed regular cross-striations in both A- and I-bands of sarcomeres of rabbit muscle (Fig. 25). Their thin sections were fixed in buffered osmium tetroxide and stained (positively) with phosphotungstic acid. Careful measurements of the cross-striation repeat in the I-bands of thin sections of several muscle types gave corrected values between 39·5 and 41·7 nm, with a mean of 40·6 nm (Page and Huxley, 1963; Page, 1964). In order to limit filament shortening during fixation, the muscles were held at constant length, and a variety of fixatives were employed. The factors necessary to correct for filament shortening were found by comparing filament lengths in sectioned material with those in negatively-stained or metal-shadowed I-segments (the bundles of thin filaments still attached to the Z-line but separated from the A-filament assemblies).

It was noted that the cross-striation repeat was significantly greater than the cross-over repeat of the two actin strands and that this indicated the presence

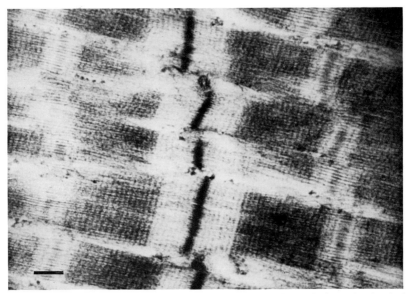

Fig. 25. Thin longitudinal section of rabbit muscle fixed in buffered OsO$_4$ and stained with phosphotungstic acid, showing the regular cross-striation present in all bands of the sarcomere. Scale bar indicates 200 nm (reprinted from Hodge et al., 1954, courtesy of A. J. Hodge).

of additional material (Hanson and Lowy, 1963, 1964; Huxley, 1963). The same conclusion was drawn from the X-ray diffraction pattern of living muscle, which showed a meridional reflection of spacing 38·5 nm, attributed to the presence of tropomyosin in association with other I-filament proteins (Huxley and Brown, 1967). The presence of this reflection was also noted in optical diffraction patterns of electron micrographs of sectioned muscle (O'Brien et al., 1971), and of negatively-stained isolated I-segments (Hanson et al., 1972; Fig. 26). Paracrystals formed from impure actin were shown to be striped (O'Brien et al., 1971; Hanson et al., 1972; Fig. 26) and a study of both negatively-stained and sectioned paracrystals showed that the stripes were due to the presence of troponin (Hanson, 1973). Spudich et al. (1972) measured the stripe repeat as approximately 38·5 nm in negatively-stained paracrystals formed from actin and the tropomyosin:troponin complex, and observed a corresponding meridional reflection in optical diffraction patterns. Similar results were obtained by Ohtsuki and Wakabayashi (1972) and by Hanson et al. (1972; Fig. 26). The results from electron microscopy and optical diffraction thus supported the assignment of a 38·5 nm meridional reflection, in X-ray diffraction patterns of living muscle, to the stripe repeat and to troponin. In the X-ray diffraction pattern of glycerinated muscle, the

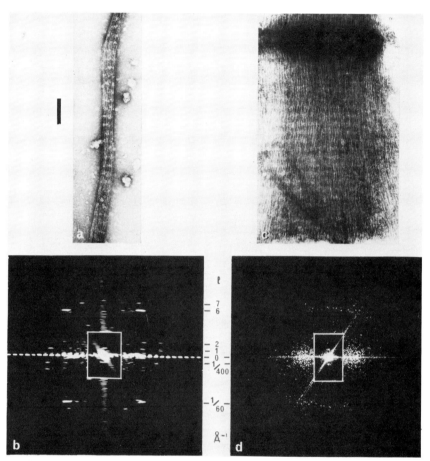

Fig. 26. Negatively-stained preparations of (a) paracrystal of filaments synthesized from impure actin (actin extracted by KI), and (c) part of an I-segment separated from a relaxed fibril of frog skeletal muscle. In each case the cross-striation is attributed to troponin. (b) And (d) are optical diffraction patterns of parts of the micrographs shown in (a) and (c), respectively. The central area of each pattern, which has been exposed less than the rest, shows a meridional first order reflection from the cross-striation, and (d) shows higher orders also. Scale marker indicates 100 nm, (a) and (c) same scale (reprinted from Hanson et al., 1972, courtesy of Cold Spring Harbor Laboratory).

contribution of troponin to the 38·5 nm meridional reflection was confirmed by the enhanced reflection intensity in the presence of antibody to TnC (Rome et al., 1973).

Ohtsuki et al. (1967) used antibody and ferritin-labelled antibody to troponin and demonstrated the 40 nm repeat in thin sections of glycerinated muscle and in negatively stained I-segments, counting 24 stripes on each side of the Z-line. Antibodies prepared to each of the three troponin components were shown to label I-segments by Ebashi et al. (1972), who noted that the stripes were broader in the case of anti-TnT, having a width of nearly 15·0 nm (Fig. 27). The leading and trailing edges of the first anti-TnT band were 25·0 nm from the end of the segment, the former coinciding with the site of attachment of the antibodies against the other two components and that against whole troponin. The authors suggested that whole troponin might not lead to the formation of a wide band if the antibody to TnT was sterically

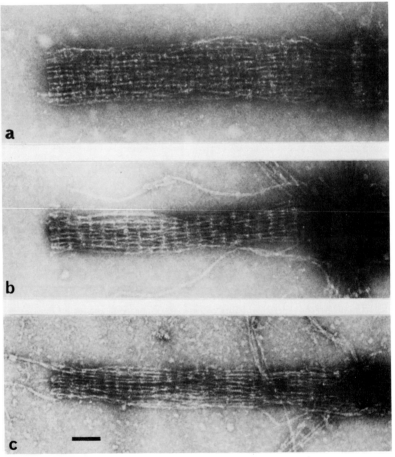

Fig. 27. I segments stained with (a) anti-TnT, (b) anti-TnC and (c) anti-TnI. The cross-striation is wider in (a) than in (b) or (c), having a width of nearly 15·0 nm. Scale marker indicates 100 nm (reprinted from Ebashi et al., 1972, courtesy of I. Ohtsuki and Cold Spring Harbor Laboratory. In the original paper the troponin components T, C and I are referred to by their older names, troponin I, troponin A and troponin II, respectively).

hindered from binding in its proper position by the presence of the antibodies to the other components. Ohtsuki (1975) noted that the broad bands formed by anti-TnT were sometimes split in two (Fig. 27a), and pointed out that the mid-point of the first band or doublet band was about 6 nm further from the end of the I-segment than the 27 nm deduced for the position of troponin from one end of the tropomyosin molecule, so that the tropomyosin might not extend to the very end of the I-filaments. In a later paper Ohtsuki (1979) digested TnT with chymotrypsin and prepared antibodies to the two fragments. One of these attached to I-segments at the position corresponding to that observed for antibodies to TnC and TnI and the other at a distance 13 nm further towards the Z-line. These positions were the same as those observed for the double stripe formed with antibody to whole TnT. Ohtsuki suggested that the TnT molecule was rod-shaped, about 10 nm long and associated in parallel with the coiled-coil structure of the tropomyosin filament, probably by helix-helix interactions, as had been suggested by Pearlstone *et al.* (1976) from the amino-acid sequence. The distribution of the two TnT fragments on tropomyosin proposed by Ohtsuki (1980) is shown in Fig. 28. Evidence that TnT might bind over an even more extended region of tropomyosin was given by Mak and Smillie (1981), who showed that the removal of residues at the C-terminal led to reduced affinity for the cyanogen bromide fragment of TnT which contained residues 1–151. In addition, Pato *et al.* (1981) showed that the binding of this fragment to tropomyosin enhanced the head-to-tail aggregation of the molecule.

Recent observations of shadowed preparations of TnT have revealed it as rod-shaped with a length 18.5 ± 2.5 nm and width about 2.0 nm (Flicker *et al.*, 1982), so that it is long enough to span about one half of the tropomyosin length and could extend through the head-to-tail overlap region of the tropomyosin strands in the thin filament. Molecules of whole troponin showed the presence of a globular region of diameter about 10 nm, with a rod-shaped tail portion attached. The dimensions of the tail corresponded closely with those of isolated TnT indicating that TnC and TnI made up the globular region. Complexes of tropomyosin with troponin were prepared, either intact from muscle or reconstituted from the purified proteins. The troponin bound at intervals of about 40 nm along narrow filaments of tropomyosin, with the globular head portion often appearing displaced from the filament. In the reconstituted assembly, two troponin molecules were sometimes seen at the same axial level along the tropomyosin filament, and Flicker *et al.* proposed that each polypeptide chain of the tropomyosin molecule provided a binding site for troponin. When tropomyosin was wound on the thin filament in muscle, however, only one binding site would be available.

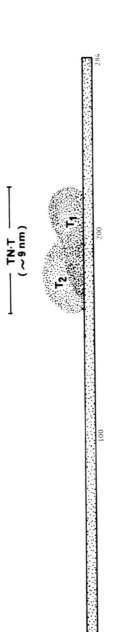

Fig. 28. The positions of the two chymotryptic fragments of TnT on tropomyosin. T_2 is in the position where TnI or TnC are located, as judged by antibody studies. T_1 is further towards the C-terminus of tropomyosin, and towards the Z-line in I-bands (reprinted from Ohtsuki et al., 1980, courtesy of I. Ohtsuki and Japan Scientific Societies Press).

V. ROLE OF TROPOMYOSIN AND TROPONIN IN REGULATION

Tropomyosin, as a rather rigid coiled-coil rod-like molecule, would be expected to confer stability to the actin filament when lying in the long-pitch helical grooves as in Fig. 1. Thus its physical properties immediately suggest one structural role in the thin filament. Its presence as end-to-end polymers on which troponin is periodically attached suggest a further role as a kind of molecular scaffolding for the calcium-sensitive protein. Since for a given length of filament the number of tropomyosin and troponin filaments are equal, but only one seventh of the number of actin subunits, this suggests a further role, that tropomyosin and troponin might act in concert, with the rod-like component extending the influence of the globular one.

A. Tropomyosin Movement

X-ray diffraction patterns from muscle at rest, contracting or in rigor showed changes in the intensities of the low-order layer-lines, according to the state of the muscle. Such patterns, recorded from vertebrate striated muscle (Huxley, 1970, 1971, 1972; Haselgrove, 1972) and from vertebrate smooth and molluscan muscles (Vibert *et al.*, 1972a, b) showed that the second layer-line, very weak when the muscle was at rest, was clearly visible during contraction or rigor. Since tropomyosin was the component identified as contributing significantly to this layer-line, its movement in the long-pitch helical grooves of actin could be invoked to explain the changes seen in the X-ray diffraction patterns, and to provide a model for the regulatory mechanism in which tropomyosin masked the myosin-combining sites of actin during the resting state of the muscle, but moved out of the way on activation (Hanson *et al.*, 1972). This model was developed in detail by Haselgrove (1972), Huxley (1972) and Parry and Squire (1973). They defined an azimuthal angle, ϕ, as that between the line joining the centre of an actin subunit to the helix axis and the line joining the centre of the tropomyosin strand to that axis. With $\phi = 45$–$50°$, the diffraction pattern calculated for the model gave a weak second layer-line, whereas for $\phi = 70$–$80°$ it was much stronger. For the X-ray diffraction patterns from vertebrate striated muscle, these angles could also explain the observed intensities of the third layer-line. At the lower angle it was proposed that the tropomyosin was located at the same position as that shown for myosin attachment by Moore *et al.* (1970) from three-dimensional reconstructions of thin filaments decorated with S-1 (myosin subfragment-1), a structure likely to be essentially similar to the actomyosin configuration in muscle in rigor. Assuming that this tropomyosin location corresponded to that in resting muscle, the model indicated that the change from rest to contraction was accompanied by a tropomyosin movement of about 1·5 nm towards the centre of the actin groove (Haselgrove, 1972).

B. Influence of Troponin and Ca^{2+}

The interpretation of the muscle X-ray diffraction patterns relied on the assumption that only the thin and not the thick filament components were contributing to the region of interest. Electron microscope observations of isolated thin filament preparations were designed to establish more directly the position of tropomyosin under different conditions. Spudich et al. (1972) prepared filaments from purified actin and the complex of tropomyosin with troponin (TmTn) in the presence of 0·2 mM Ca^{2+}. Three-dimensional reconstructions, calculated from electron micrographs of the filaments present in paracrystals resolved tropomyosin at a radius of about 3·2 nm and an angle $\phi = 60$–$70°$. The assignment of the tropomyosin peak in the reconstructions was confirmed by comparison of the results for actin + TmTn with those for actin alone (Fig. 29). The structure of the complex was very similar to that reported previously by Moore et al. (1970) from their reconstruction of a natural thin filament isolated directly from muscle. There was a variation in the density of the tropomyosin, with the portion in contact

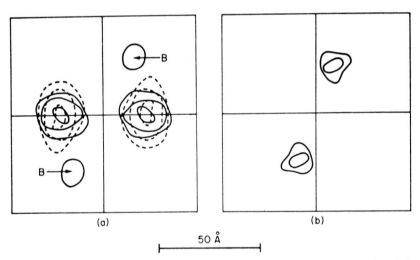

Fig. 29. (a) Helical projection of a filament of actin + tropomyosin (solid contours) with the projections of F-actin superimposed (dotted contours). The maps are the averages obtained from reconstructions of filaments within paracrystals. (b) The result of subtracting the actin contribution in (a) from the actin + tropomyosin, after scaling the densities at the peak of the actin monomer to be the same (reprinted from Spudich et al., 1972, courtesy of H. E. Huxley and Academic Press Inc.).

with the actin subunit having a higher level than that in between subunits. Spudich et al. (1972) considered that the negative stain might be more excluded in the region of contact than elsewhere and pointed out that in the two of their reconstructions where the variation of density was such that the tropomyosin appeared discontinuous, this effect could be eliminated by contouring to lower levels on the density maps. The paracrystals prepared by these authors showed the characterisic cross-striation at intervals of about 38·5 nm, attributable to troponin. Ohtsuki and Wakabayashi (1972) prepared similar paracrystals from actin + TmTn only when the amount of troponin used was several times that of tropomyosin. At lesser amounts paracrystals lacking stripes and resembling those of pure actin were observed. Gillis and O'Brien (1975) showed that pure actin-like paracrystals were observed when the Ca^{2+} concentration was below 10^{-6} M (Fig. 30). They confirmed the presence of the regulatory proteins in these paracrystals by SDS gel electrophoresis, and showed that if they were resuspended in a high Ca^{2+} medium, the reappearance of the stripes could be demonstrated. In the paracrystals formed at high Ca^{2+} ($> 10^{-5}$ M) the paracrystals were broad ribbon-like assemblies, but at low Ca^{2+} they were more compact, with the filaments closer together. In the corresponding optical diffraction patterns the second layer-line was strong at high Ca^{2+} and weak at low Ca^{2+} (Fig. 30), and measurements of the layer-line spacings showed the actin helical symmetry comprised 28/13 subunits per turn at high Ca^{2+} and 13/6 at low. Since electron microscopy involves dehydration and staining, so that the structures are observed in conditions far from physiological, Gillis and O'Brien recorded X-ray diffraction patterns from oriented gels of actin + TmTn, and showed a similar change of the second layer-line intensity according to the Ca^{2+} level, but the patterns were not sufficiently well defined for the parameters of the actin helix to be obtained.

The electron microscope results showed that a Ca^{2+} concentration in the same range as that controlling the onset of muscular activity marked the distinction between two filament structures, one with tropomyosin towards the centre of the actin groove and the other with tropomyosin towards the edge, integrated with the actin in some way so that the paracrystals resembled those of pure actin. These results were therefore consistent with the models for tropomyosin movement proposed to explain the X-ray diffraction patterns from muscle. Further direct evidence of this kind was provided by the analysis of partially reconstituted thin filaments, that is, filaments lacking the full complement of regulatory components. Wakabayashi et al. (1975) compared the structure of actin + tropomyosin, a filament composition considered not to inhibit actomyosin ATPase, with that of actin + tropomyosin + TnT + TnI, which is inhibitory. Three-dimensional reconstructions of electron micrographs of filaments present in paracrystals revealed a difference of tropomyo-

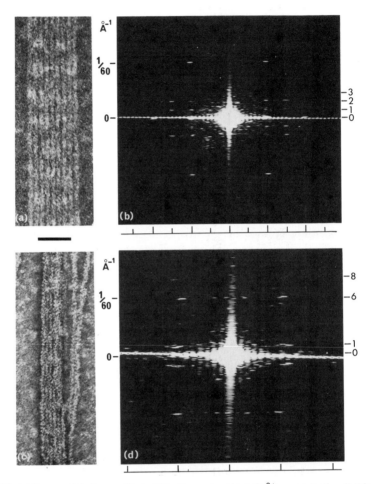

Fig. 30. (a) Paracrystal of reconstituted thin filaments at high Ca^{2+} concentration ($>10^{-5}$ M). (b) Optical diffraction pattern of (a). The second layer-line is strong, of about the same intensity as the first. The graduations underneath the pattern show the positions of the row-lines. (c) Paracrystal of reconstituted thin filaments at low Ca^{2+} concentration ($<10^{-6}$ M). The cross-over points of the long-pitch actin helices in adjacent filaments are aligned. (d) Optical diffraction pattern of (c). The second layer-line is no longer visible. The graduations underneath show the row-line positions, which are at orders of the inter-filament spacing, rather than twice this spacing as in (b). The filament packing and structure are substantially different, therefore, at the two Ca^{2+} levels. Scale marker indicates 50 nm and applies to both (a) and (c) (reprinted from Gillis and O'Brien, 1975, courtesy of Academic Press Inc.).

sin position in the two structures of approximately 20° in azimuthal angle and 1·0 nm in distance. Cylindrical sections of the reconstructions, superposed in radial projection, are shown in Fig. 31. The tropomyosin strands were more clearly resolved from the actin when at the higher azimuthal angle, in the actin + tropomyosin filaments, and followed a zigzag path. In the inhibitory filaments, the tropomyosin was closely associated with the actin subunits, and followed a smoother course. The actin subunits showed an asymmetric shape, described as "chick-shaped", and were joined along the left-handed genetic helix, though not along the long-pitch helices. The volumes of the reconstruc-

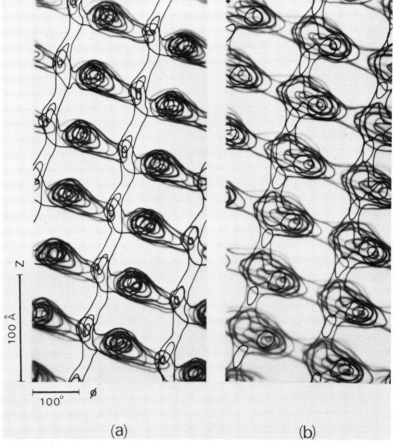

Fig. 31. Cylindrical sections of the three-dimensional reconstructions of (a) actin+tropomyosin and (b) actin+tropomyosin+TnT+TnI are shown superposed in radial projection. The sections are at radii 1·0, 1·5, 2·0 and 2·5 nm. Contour lines at small radii are drawn fainter while those at larger radii are thicker and sharper (reprinted from Wakabayashi et al., 1975, courtesy of T. Wakabayashi and Academic Press Inc.)

tions were about one third those expected from the molecular weights of the components, a reduction attributed to the effects of stain penetration into and lateral shrinkage of the filaments. O'Brien and Couch (1976) also calculated a three-dimensional reconstruction of actin + tropomyosin filaments, using a combination of optical and computer methods. They found a similar azimuthal angle of 70° for the tropomyosin to that observed by Wakabayashi *et al.*, but a larger radius, 4·0 nm compared to about 2·3 nm, and the strands followed a smooth rather than a zigzag path. An analysis with improved data confirmed these features, and showed that the actin subunit was elongated and comprised two unequal domains, tropomyosin being attached to the smaller one (O'Brien *et al.*, 1983; Fig. 32). The elongated shape of adjacent subunits along the genetic helix gave rise to the appearance of a V-shaped feature, also observed by Vibert and Craig (1982) in three-dimensional reconstructions of S1 decorated thin filaments, and so established the polarity of the filament with respect to the Z-line in muscle. The presence of two domains in the actin subunit was also indicated in the radial density distribution of actin + tropomyosin filaments calculated from the equatorial data in X-ray diffraction patterns of oriented gels of the filaments (O'Brien *et al.*, 1983).

O'Brien *et al.* (1975) showed by decoration with HMM (heavy meromyosin) that the filaments in paracrystals of reconstituted thin filaments (actin + TmTn) formed at high Ca^{2+} had alternating polarity across the paracrystal. The troponin stripe repeat was shown to be the same as the actin cross-over repeat, 38·4 nm, in these paracrystals, and a model for the filament packing was proposed in which the opposite polarity of adjacent filaments explained the rather wide stripes seen in these paracrystals compared with those observed in the I-bands of muscle sections. In the thin filament paracrystals at low Ca^{2+} the presence of 13/6 rather than 28/13 actin helical symmetry, with a consequently shorter cross-over repeat, meant that the troponin positions in adjacent filaments had no axial relationship and there were no stripes. Paracrystals of actin + tropomyosin were shown to have the same actin helical symmetry and antiparallel filament packing as those of fully reconstituted thin filaments at high Ca^{2+}, even though they lacked stripes. The observation of 28/13 actin helical symmetry in actin + tropomyosin paracrystals disagreed with the assignment of 13/6 symmetry by Wakabayashi *et al.* (1975) to their paracrystals of this constitution. Morris (1979) and O'Brien *et al.* (1980) observed that there were in fact two types, resembling in filament symmetry and packing the paracrystals formed from thin filaments at high and low Ca^{2+}. The two types were also formed from filaments composed of actin + tropomyosin + TnI, a complex which inhibits actomyosin ATPase. Observation of the structural effect of the TnI was clearer in the paracrystals having 28/13 helical symmetry. The stain-filled regions between the filaments were more pronounced in paracrystals containing TnI than in those of

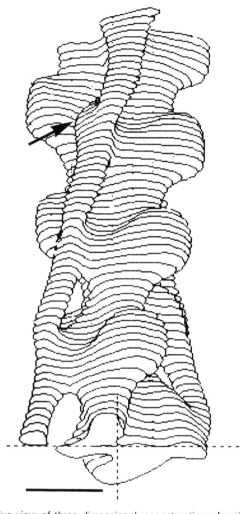

Fig. 32. Perspective view of three-dimensional reconstruction of actin+tropomyosin. The arrow marks the smaller domain of the actin monomer, to which tropomyosin is attached. The dotted cross marks the position of the filament axis. The elongated shape of the actin monomers gives rise to the appearance of a V-shaped formation pointing downwards. If the filament were in the muscle, the Z-line would be at the bottom of the Fig. Scale bar indicates 4 nm (reprinted from O'Brien et al., 1983, courtesy of Academic Press Inc.).

actin + tropomyosin alone (Fig. 33), an effect that could be explained by the movement of tropomyosin towards the edge of the actin groove (Fig. 34). In three-dimensional reconstructions of actin + tropomyosin + TnI filaments, O'Brien et al. (1983), observed the actin monomer to comprise two domains, as in actin + tropomyosin (Fig. 32). The tropomyosin was not clearly resolved but was tentatively assigned to a position in contact with the large actin domain, and thus towards the periphery of the filament, consistent with the model given in Fig. 34b.

C. Actin-tropomyosin Interaction

Longley (1975) suggested that, if the α-helical rods of the coiled-coil were regarded as smooth, the tropomyosin molecules in tactoids might be close-packed in sheets by staggering each double helix by one quarter of the

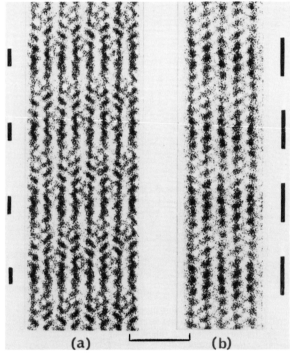

Fig. 33. Computer filtered images of electron micrographs of paracrystals of (a) actin + tropomyosin and (b) actin + tropomyosin + TnI (negative contrast). (a) Shows clearly the antiparallel packing of adjacent filaments. In (a) relatively little stain penetrates into gaps between the filaments, but in (b) the gaps are longer and more pronounced. The lines at the sides of the paracrystals mark the position of the gaps. Scale marker indicates 30 nm (reprinted from O'Brien et al., 1980, courtesy of Japan Scientific Societies Press).

pitch with respect to the next. Electron micrographs and optical diffraction patterns of the tactoids indicated this stagger was 2·8 nm, so that the pitch of the double helix was 11·4 nm. Half of this pitch would match the distance between actin subunits in the thin filament at the radius at which tropomyosin lay. Parry (1975a) considered, however, that the assembly of the tactoids might not be controlled solely from packing considerations, but might be dominated by the bridging effect of Mg^{2+} between acidic moieties of adjacent molecules. He suggested that the minimum energy state for a coiled-coil might occur when the pitch was about 14·0 nm, a value experimentally noted for several fibrous proteins. McLachlan and Stewart (1975) proposed that for a tropomyosin molecule in which the two chains were in register, there would be an integral number of n half-turns in 38·5 nm length of thin filament, relative to the twist of the actin long-pitch helical strands. If $n=7$ the molecule would present a similar orientation of its outer surface to all seven actins along its length (Fig. 35a). The twist of the actin strands is such that there is a rotation of approximately 180° every 38·5 nm in the opposite sense to the twist of the tropomyosin double helix. Allowance for this meant that tropomyosin in the thin filament required only six, not seven half-turns to match the seven actin subunits, giving a coiled-coil pitch of 13·7 nm, one third the tropomyosin molecular length (Fig. 35b), as also proposed by Parry (1975a). The requirement of making pseudo-equivalent actin-tropomyosin interactions along the long-pitch helical strand, with its introduction of a 180° twist every 38·5 nm, thus favoured a larger coiled-coil pitch than that proposed by Longley (1975). Refinement of a model of tropomyosin to fit the X-ray diffraction data from crystals also indicated a pitch of 13·7 nm (Phillips *et al.*, 1980; Phillips, G.N., unpublished observations).

Stewart and McLachlan (1975) and McLachlan and Stewart (1976) suggested the 14-fold pseudo repeat of the tropomyosin sequence might indicate the presence of two interleaved sets of seven binding sites to actin, one being used when tropomyosin was in the inhibitory position and the other when it was not. If in one position the narrow edge of the tropomyosin double helix pointed inward towards the axis of the actin filament for one set of sites, then it would point tangentially for the other (Fig. 35b). A quarter-roll of the tropomyosin about the helix axis would reverse the radial and tangential positions. This rolling would be sufficient to move the tropomyosin from towards the edge to towards the centre of the actin groove, through a distance of 1·0 to 1·5 nm. Similar ideas were proposed by Parry (1976), who suggested that in the inhibitory position, the tropomyosin coiled-coil might present its broad face at the region of actin contact, providing maximum coverage of the myosin binding site. A rolling from an applied torque was considered more easily transferred axially down the head-to-tail assembly of tropomyosin molecules than would be a local lateral translation, that is, a sliding of a

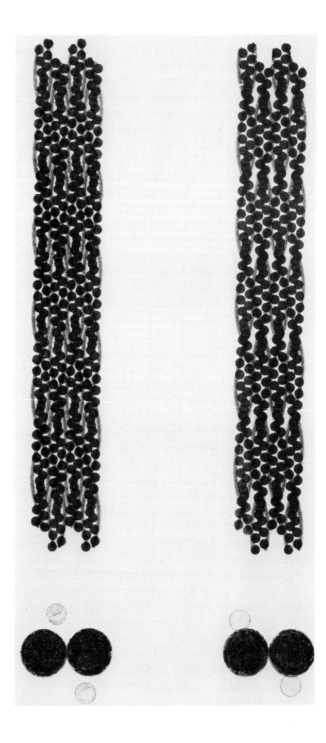

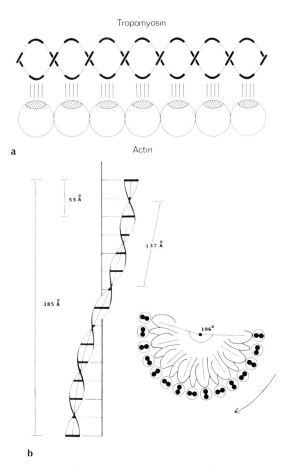

Fig. 35. (a) Proposed interaction of seven actin subunits with a tropomyosin molecule. The 28 negative zones on the tropomyosin molecule are indicated by the dense lines, and 7 of these are interacting with the actin strand. (b) The tropomyosin molecule as it would follow the actin strand in the filament, making 7 half-twists relative to the actin. The horizontal lines mark the positions of the negative zones as in (a). The view from above, along the helix axis, at the right shows how the cross-sections of alternate zones are twisted by a quarter turn relative to actin, so that as one set is tangential the other is radial. This view represents the projection of a series of 14 horizontal sections through one strand of the actin helix, stacked above one another to show the path of tropomyosin round the edge of the groove. The sketch of actin with two sites is not to be taken literally (reprinted from Stewart and McLachlan, 1975 and McLachlan and Stewart, 1976, courtesy of M. Stewart, Macmillan Journals Ltd and Academic Press Inc.).

Fig. 34. Computer-drawn models of paracrystals of actin+tropomyosin to simulate the change of structure induced when TnI is present. Left, the azimuthal position of tropomyosin is at 70° as illustrated in the projected view of a filament shown below the paracrystal, and right, this angle is 50°. At the smaller angle, the gaps between filaments are more pronounced, and these would produce the dark, stain-filled regions seen in electron micrographs of negatively-stained paracrystals of actin+tropomyosin+TnI (Fig. 33b) (reprinted from O'Brien et al., 1980, courtesy of Japan Scientific Societies Press).

portion of tropomyosin over an actin surface. Support for the presence of a seven-fold as well as a 14-fold periodicity in tropomyosin was given by Smillie *et al.* (1980) from a consideration of the distribution of the potential for forming an α-helix. The smoothed α-helix parameter as defined by Chou and Fasman (1974), averaged over stretches of 14 residues, had maxima and minima over intervals of close to 40 residues. This feature was well developed in the central and N-terminal half of the molecule, but became progressively less distinct towards the C-terminus.

D. Attachment Site of Myosin

Although there was considerable evidence for a tropomyosin movement associated with regulation, the proposal that in one position tropomyosin blocked the attachment of myosin heads to actin was called into question by the analysis of natural thin filaments in I-segments by Seymour and O'Brien (1980). In reconstructions of the S-1 decorated thin filament (Moore *et al.*, 1970) the tropomyosin was not resolved, and there remained the ambiguity of whether S-1 and tropomyosin were on the same or opposite sides of the actin groove. In I-segments the polarity of the filaments with respect to S-1 decoration was known, and three-dimensional reconstructions of electron micrographs of these filaments showed that tropomyosin was located on the opposite side of the actin groove to that previously assumed (Seymour and O'Brien, 1980; Fig. 36).

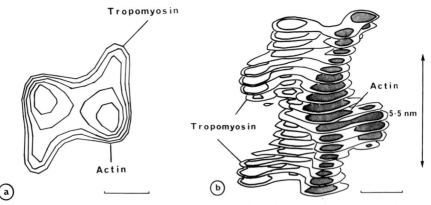

Fig. 36. (a) Helical projection of a thin filament within an I-segment, looking towards the Z-line (b) three-dimensional reconstruction of the filament, with the Z-line at the bottom. (a) And (b) were calculated from data averaged over several filaments. The different shadings indicate the long-pitch actin and tropomyosin strands on opposite sides of the helical axis. Scale bar represents 2 nm (reprinted from Seymour and O'Brien, 1980, courtesy of Macmillan Journals Ltd).

An improvement in the resolution in electron micrographs of thin filaments decorated with S-1, using minimal electron exposure of the specimens, led to a reassessment of the position of the myosin-binding site (Taylor and Amos, 1981). Three-dimensional reconstructions indicated that the site was located on the site of the actin groove opposite to that proposed by Moore *et al.* (1970) and thus on the same side as the tropomyosin position given by Seymour and O'Brien (1980). This result was also given by an analysis of decorated filaments prepared from scallop muscle proteins (Vibert and Craig, 1982). A further improvement in resolution was obtained by Amos *et al.* (1982) by selecting for analysis electron micrographs of decorated filaments whose diffraction patterns were compatible with X-ray diffraction patterns obtained from muscle infused with S-1. In this way, the specimens with the best negative staining were chosen. The reconstructions indicated that in addition to the main contact made by S-1 to an actin monomer, as previously observed, there was a second weaker one made to the adjacent monomer on the opposite strand. The S-1 thus reached around tropomyosin to form a second actin contact on the opposite side of the groove. Such a two-site interaction was consistent with the observations of Mornet *et al.* (1981) from chemical cross-linking studies. Amos *et al.* (1982) proposed a third interaction, between S-1 and tropomyosin, but their position for tropomyosin was uncertain as it was not resolved as a separate entity.

O'Brien *et al.* (1983) noted that the shape and orientation of the actin subunit in their reconstruction of actin + tropomyosin (Fig. 32) was very similar to the central part of the reconstructions of decorated filaments (Vibert and Craig, 1982; Taylor and Amos, 1982), but considered that the tropomyosin location relative to actin was more clearly established in their analysis. They suggested that the third interaction with S-1 proposed by Amos *et al.* (1982) involved actin as well as tropomyosin, and that the second interaction was at the site occupied by tropomyosin when TnI was present, that is, on the larger domain of the actin monomer. Movement of tropomyosin under the influence of TnI might thus be to a position where it could interfere with one of the myosin attachment sites.

VI. Z-DISC

An important, probably fundamental feature of contraction involving actin and myosin is the ability of these proteins to form bipolar filamentous arrays. In striated muscle, where these arrays are most highly organized, the bipolarity enables the mutual sliding of thick and thin filaments to take place in the same way in each half-sarcomere, with consequently even length changes throughout the muscle. Myosin has itself the property to form bipolar filaments (Huxley, 1963), which in the muscle are joined laterally near their centres by the M-line proteins to form large arrays. Actin has no such tendency, however, so the material at the Z-line, in the centre of the I-band, has the doubly important function of conferring bipolarity as well as structural integrity to the arrays. With this in mind it is not surprising that the Z-line, or more appropriately Z-disc, to convey its permeation throughout the muscle, has a complex structure which has puzzled and intrigued electron microscopists.

A. Z-disc in Insects and Vertebrates

The choice of letter followed the observation and naming of the *zwischenscheibe* (intermediate disc) by German microscopists of the nineteenth Century. Initial electron microscope studies by Hall *et al.* (1946) and Jakus and Hall (1947, 1949) showed the Z-disc clearly in myofibrils, either fixed with formalin and shadowed with chromium or stained with phosphotungstic acid. The presence of filaments linked to the Z-disc was observed by Rozsa *et al.* (1950), using similar techniques, and the diameter of each filament was noted as similar to that of fibrous actin. These features were confirmed in longitudinal sections of whole muscle (Hanson and Huxley, 1953; Huxley, 1953, 1957) and in negatively-stained I-segments (Huxley, 1963). Transverse sections of striated muscles showed that the number and arrangement of the thin filaments around the thick filaments were different in vertebrates and insects. This raised the question of how the respective Z-discs differed.

In transverse sections through the centre of the Z-disc of insect flight muscle from two *diptera*, Auber and Couteaux (1963) observed tubular structures, and proposed a model in which groups of three thin filaments from each half of the I-band connected with thinner filaments forming the walls of the tubes. Ashurst (1967) considered, however, from examination of the Z-disc of the flight muscle of the water bug *Lethocerus* that the thin filaments were not continuous across the Z-disc. Triangular groups of three filaments from each side of the Z-disc were interleaved to form the hexagonal lattice observed at the centre of the disc, and these filaments seemed to be cemented together by amorphous material. In the Z-disc of the honey-bee, Saide and Ullrick (1973) showed the presence of similar tubular structures to those of Auber and Couteaux (1963), but with a triangular cross-section and with an 8 nm thick wall which incorporated six thin filaments, three from each half of the I-band. The tubes were twisted along their length so that they rotated approximately 60° about the tube axis. The mutual orientation of the triangular groups of three filaments in the two half I-bands was thus 60° as in Ashurt's (1967) model. The Z-disc of the honey-bee was isolated by treatment with lactic acid, and in isolation presented a variety of appearances depending on the degree of expansion (Saide and Ullrick, 1974). Small discs showed little detail, but in swollen specimens a hexagonal lattice was observed. Sections of isolated discs showed projections about 130 nm in length extending outwards on either side, which were considered might represent insoluble stubs of thin filaments or "C"-filaments, which connect the thick filaments in these muscles to the Z-disc (Auber and Couteaux, 1963).

Analysis of vertebrate muscle also showed that the thin filaments were not continuous across the Z-disc. Knappeis and Carlsen (1962) showed in frog skeletal muscle that the thin filaments, which formed an approximately square lattice of spacing 20–25 nm in the vicinity of the Z-disc, were connected by a tetragonal arrangement of filaments within the disc. Each thin filament on one side of the Z-disc faced the centre of the space between four thin filaments on the opposite side. Similar observations were made by Franzini-Armstrong and Porter (1964) in sections through the Z-discs of several vertebrates. They considered, however, that the Z-filaments were part of a continuous membrane. In the transverse sections of rat and rabbit muscle obtained by Reedy (1964), which were thick enough to include the whole Z-disc together with the ends of the thin filaments, the pattern of Z-filament connections created a "basket-weave" appearance (see Fig. 38a). The group of four Z-filaments diverging from the end of the thin filament was shown by stereo-microscopy to connect with the neighbouring four filaments in the opposite sarcomere. Kelly (1967), from a study of the striated muscle of newts, considered that the Z-filaments might be formed by the two strands of actin diverging at the ends of the thin filaments, or by tropomyosin uncoiling from its thin filament position in the long-pitch helical grooves of actin. In the models proposed, the Z-filaments entered the disc and looped back, so that adjacent thin filaments were connected. The possibility that tropomyosin might form the filaments of the Z-disc had been mentioned earlier, by Huxley (1963), who noted the similarity of the lattice of the disc to that observed in tropomyosin crystals.

Landon (1970) made the significant observation that the type of lattice observed in transverse sections through the Z-disc depended on the method of fixation. He obtained the woven appearance described by Reedy (1964) when osmium was used as the primary fixative, but with glutaraldehyde fixation a smaller lattice of spacing 11 nm was seen (Fig. 37a). This was interpreted as the superposition of two square lattices, each of 22 nm spacing, with the corner of one square lying above the middle of the other. Longitudinal sections showed clearly the interdigitation of the ends of the thin filaments from adjacent sarcomeres (Fig. 37b).

Landon's results and their interpretation were confirmed by MacDonald and Engel (1971) and Kelly and Cahill (1972) who elaborated the looped model of Z-filament arrangement given by Kelly (1967). Greater complexity

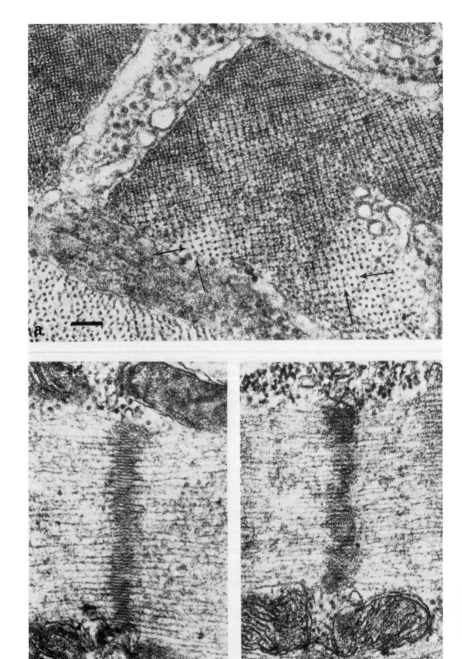

of Z-disc structure was observed by Rowe (1973) when red fibres of rat muscle were compared with intermediate and white ones, and models were proposed involving different numbers of interlocking hairpin loops. Katchburian *et al.* (1973) demonstrated the effect of tilting longitudinal sections on the Z-disc, but the model presented was basically similar to that proposed by Knappeis and Carlsen (1962) and did not involve looped Z-filaments. In fish muscle the Z-disc was shown to be relatively simple, and Franzini-Armstrong (1973) demonstrated that varying the fixative and sarcomere length did not alter its appearance. Again the proposed model involved direct linkage of thin filaments in adjacent sarcomeres across the Z-disc as proposed by Knappeis and Carlsen, without looped Z-filaments. In a comparative study of vertebrate Z-discs, Ullrick *et al.* (1977) disagreed with Landon's (1970) conclusion that the lattice depended on the fixative employed, as they found the woven appearance with either glutaraldehyde or osmium. They considered the woven appearance to be the most common, representing the fundamental lattice, and designed a model in which three curved Z-filaments diverged from the end of each thin filament (Fig. 38). The square and small-square lattices would then arise if the Z-filaments straightened out during fixation, the small-square lattice being seen only in thicker sections where the entire Z-disc was incorporated. The curved Z-filaments formed loops as described by previous authors, but did not interlock, which was considered consistent with observations that the Z-disc could split in the middle. This raised the question of why the sarcomere did not split during tension development, and the authors suggested there might be connecting filaments between the two halves of the Z-disc.

The Z-disc of cardiac muscle is generally similar to that of skeletal muscle. Special interest was attached, however, to the cardiac discs by the observation that in some sarcomeres the Z-regions extended for a considerable distance in the direction of the fibrillar axis, and that the anomalous discs were structurally related to the rod-like inclusions found in congenitally abnormal human skeletal muscle cells (Fawcett, 1968). Goldstein *et al.* (1977, 1979) showed that both normal and abnormal canine cardiac Z-discs contained 24 nm and 11 nm square lattices, and the explanation given for the smaller one as

Fig. 37. (a) A transverse section of rat striated muscle at a small angle to the plane of the Z-disc, showing the square array of thin filaments immediately adjacent to the disc (arrows) and the alignment of this array with alternate bands of the fine 11 nm lattice of the disc proper. (b) Longitidunal sections through the Z-disc. In the left-hand micrograph the interdigitation of the terminal portions of the thin filaments from adjacent sarcomeres is evident within the disc, with increase in their density throughout the zone of overlap, and extending for a short distance into the I-band. In the right-hand micrograph the orientation is such that the thin filaments from adjacent sarcomeres appear to be in longitudinal continuity. All specimens were fixed in glutaraldehyde and OsO$_4$. Scale bar in (a) represents 100 nm and applies also to (b) (reprinted from Landon, 1970, courtesy of D. N. Landon and Cambridge University Press).

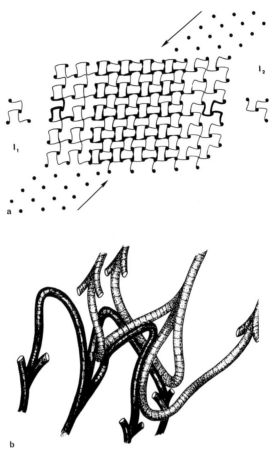

Fig. 38. (a) Model of vertebrate Z-disc, showing disposition of I-filaments as they enter the disc from adjacent sarcomeres, and the arrangement of Z-filaments arising from the ends of these filaments. In the middle of the disc, Z-filament arrays of opposite hand (I_1 and I_2) superimpose, forming a characteristic "basket-weave" pattern. (b) A drawing of the proposed structure of the Z-disc, in which at the end of each thin filament of the I-band, three curved filaments branch out, forming loops. The loops from the two sets of filaments from neighbouring sarcomeres do not interlock (reprinted from Ullrick et al., 1977, courtesy of W. C. Ullrick and Academic Press Inc.).

resulting from a superposition of two larger ones was the same as that given by Landon (1970). In longitudinal sections they found a periodicity of 38 nm in both types of disc, arising from Z-filaments inclined to and interconnecting the I-filaments. The normal Z-disc was considered to comprise three 38 nm-repeat units. In contrast to Ullrick et al. (1977), these authors proposed that the Z-filaments connected I-filaments from adjacent sarcomeres. The basket-weave as well as the square and small-square lattices was observed, and predominated in the thinnest sections, but it was not decided whether it arose from a re-arrangement or was merely a different view of the Z-filaments. In a study of slow skeletal muscle (rat soleus), Goldstein *et al.* (1982) favoured the former interpretation, and proposed a model for the interconversion of lattice types in which the axial filaments maintained the same diameter and relative position but the cross-connecting filaments changed in curvature and width.

B. α-actinin

Rasch *et al.* (1968) showed that urea extracted the Z-disc of rabbit skeletal muscle while leaving other components of the I- and A-bands intact. Stromer *et al.* (1969) achieved the extraction with a low ionic strength medium and showed that it was reversible. Busch *et al.* (1972) found that the Z-disc was removed by 1 mM Ca^{2+} in intact muscle but not in glycerinated muscle, and they identified a factor in the sarcoplasmic reticulum which was activated by Ca^{2+} and which removed Z-disc material containing tropomyosin and α-actinin, a protein first discovered by Ebashi *et al.* (1964) and Ebashi and Ebashi (1965), and considered to be associated with the Z-disc (Briskey *et al.*, 1967). Podlubnaya *et al.* (1975) observed that negatively stained α-actinin was rod-shaped, with length about 30·0 nm and width about 2·0 nm, and that it was able to form a regular array of cross-links between actin filaments. Suzuki *et al.* (1976) showed that the molecule contained two chains, with about 74% α-helix and a total weight of 200 000, and observed a considerably larger rod-shape in shadowed specimens, about 4·4 nm by 40–50 nm. Reddy *et al.* (1975) found that the Ca^{2+}-activated sarcoplasmic factor isolated by Busch *et al.* (1972) digested α-actinin specifically. Immunofluorescent studies on the Z-disc of chicken muscle, isolated by extraction with potassium iodide, showed that α-actinin was located in the main part of the disc. Also present in the isolated disc were actin, tropomyosin and a protein of molecular weight 50 000 which Reddy *et al.* called desmin and which was only present on the periphery of the disc.

In vertebrate smooth muscle the arrays of thin filaments are interspersed with concentrations of electron-dense material, called dense bodies, which may play a role similar to that of the Z-disc of striated muscle. α-actinin was identified as a component of the dense body by Geiger *et al.* (1981) by antibody labelling in frozen sections.

VII. ACTIN AND ASSOCIATED PROTEINS IN OTHER CELLS

Although muscle remains the most highly developed and specialized type of motile cell, which with its highly regular structure has been especially rewarding for analysis by electron microscopy and other structural techniques, other cells containing actin and myosin have been investigated for almost as long, and electron microscopy has played a crucial part in their characterization. Actin emerges as having a highly conserved amino-acid sequence in widely differing organisms, but has a considerable complexity of behaviour conferred by its ability to associate and interact with very many other cytoplasmic proteins, the study of which has become particularly intense in recent years.

A. Non-muscle Actin

One of the first to look for actomyosin-like proteins in undifferentiated tissue was Loewy (1952), who demonstrated that viscosity changes could be induced in the cytoplasm of slime mould by the addition of ATP. Contractile proteins were subsequently identified, mainly by their behaviour in the presence of ATP, in fibroblasts (Hoffmann-Berling, 1954), sarcoma cells (Hoffmann-Berling, 1956), and blood platelets (Bettex-Galland and Lüscher, 1959, 1961). The first use of electron microscopy was by Ts'o *et al.* (1957), who showed that shadowed extracts of the cytoplasm from slime mould comprised some globular material and rod-like structures of diameter between 6 and 8 nm with varying length up to 700 nm. Separation of the contractile proteins in blood platelets was achieved by Bettex-Galland *et al.* (1962) who described two components with actin- and myosin-like properties. Electron micrographs of thin sections of slime-mould cytoplasm showed the presence of aligned thread-like structures (Wohlfarth-Botterman, 1962, 1964a, b), with the same diameter as the shadowed filaments observed by Ts'o *et al.* (1957). Wolpert *et al.* (1964) and Nachmias (1964) made similar observations on amoeba cytoplasm, and Nagai and Rehbun (1966) in streaming *Nitella* cells.

The actin-like protein from slime mould was characterized in more detail by Hatano and Oosawa (1966a, b), who demonstrated its similarity to muscle actin. Their experiments involved G-F polymerization, measurements of sedimentation constant, amino-acid composition and ability to activate myosin ATPase. The G-F polymerization of this actin was followed in the electron microscope by Hatano *et al.* (1967), who obtained negatively-stained filaments of similar appearance to those observed by Hanson and Lowy (1963) for muscle actin. An actomyosin preparation from the same source was shown by Hatano and Tazawa (1968) to contain filaments resembling F-actin, in the presence of ATP and, in its absence, wider filaments of diameter about 10 nm resembling F-actin with granular particles attached to its surface. Arrowhead-like structures were not observed by these authors, but were seen by Senda *et al.* (1969) in a negatively-stained preparation from equine leucocytes. Ishikawa *et al.* (1969) made extensive use of the ability of HMM to decorate actin filaments *in situ*, showing firstly that this could be achieved in muscle cell cultures by glycerination and incubation with HMM, and then extending the technique to non-muscle tissues. Decorated filaments were shown in sectioned fibroblasts, chondrocytes, nerve cells and several types of epithelial cells. The specificity of the reaction was demonstrated by the lack of HMM binding with other cellular components such as microtubules or 10 nm filaments.

The work of Hatano and Tazawa (1968) on slime mould actomyosin was extended by Nachmias *et al.* (1970), who showed the presence of arrowhead structures in the natural extract and that, after treatment with ATP, the filaments could be decorated with muscle S-1 or HMM to form complexes of more regular appearance than that of the native actomyosin. Purified actin from amoeba was shown to form filaments that could be decorated with muscle HMM by Pollard *et al.* (1970), who also demonstrated *in situ* decoration. The same kind of study was carried out on the brush-border of intestinal epithelial cells by Tilney and Mooseker (1971). A further charactersitic of F-actin, the ability to form paracrystals in the presence of Mg^{2+}, was demonstrated for platelet actin by Spudich (1972). Similar paracrystals, formed at protein concentrations greater than 1 mg/ml, but without requiring divalent cations, were prepared from brain actin by Bray and Thomas (1976a, b), and optical diffraction patterns of electron micrographs of the paracrystals showed reflections characteristic of the actin helix.

B. Non-muscle Tropomyosin

A tropomyosin-like protein from human platelets was isolated by Cohen and Cohen (1972). It was described as a two-chain α-helical structure similar to muscle tropomyosin but with a smaller subunit, 30 000 molecular weight, and forming tactoids with a smaller axial period, 34·3 nm, and different band pattern. Fine *et al.* (1973) made similar observations on a preparation from nerve cells, and found that the molar ratio of actin to tropomyosin was 14·9 compared with 4·6 for actin to muscle tropomyosin. Fine and Blitz (1975) compared tropomyosins from several muscle types (striated and smooth) and non-muscle cells (brain, platelet and pancreas from calf, and fibroblasts from mouse), showing subunit molecular weights consistently of 35 000 for one class and 30 000 or the other, with correspondingly consistent tactoid repeats (40 nm vs 34·5 nm) and band patterns. Evidence for the association of actin and tropomyosin in bundles in a variety of non-muscle cells was provided by Lazarides (1975, 1976), from immunofluorescence studies.

Horse platelet tropomyosin was shown by Côté *et al.* (1978) from viscosity measurements to polymerize to a much smaller extent than muscle tropomyosin, and the end sequences of the two classes of protein were shown to be

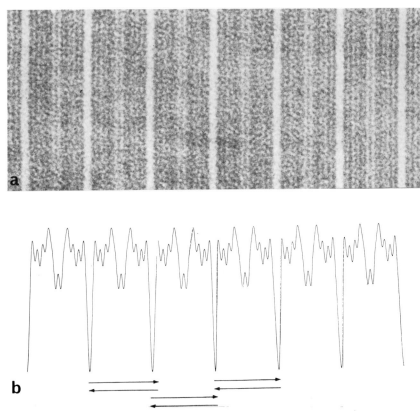

Fig. 39. (a) Magnesium-induced tactoids of platelet tropomyosin, negatively stained with uranyl acetate. The prominent white (stain-excluding) band repeats at intervals of 34·5 nm, and between this band the dark bands are spaced at about 2·8 nm intervals. (b) Computer-filtered trace of stain distribution projected onto the long axis of the tactoid, with the proposed molecular positions shown by arrows underneath (reprinted from der Terrossian *et al.*, 1981, courtesy of M. Stewart and Academic Press Inc.).

different. This tropomyosin and that from pig platelets were shown by der Terrossian *et al.* (1981) to have an amino-acid composition virtually identical except for the number of cysteine residues. These authors showed that cleavage at the cysteine residue of the pig tropomyosin produced a fragment of molecular weight 19 000, indicating that this residue was located about one third of the distance from a molecular end. The most prominent feature of the band pattern of tactoids, a stain-excluding band, contrasted with the pair of bands of this type seen in tactoids of muscle tropomyosin (compare Fig. 39 with Fig. 16). The tactoids fractured at the band, indicating that the molecular ends occurred here, without the extended overlap between adjacent anti-parallel molecules observed for muscle tropomyosin tactoids (Fig. 16). A sub-banding of 2·8 nm repeat was observed as for muscle tropomyosin, attributed to the repeat of negatively-charged groups. The position of the cysteine residue relative to the molecular ends was confirmed by labelling the paracrystals with a thiol-specific mercury compound. The molar ratio of G-actin to tropomyosin in the cell was estimated at about 14:1. Assuming that one tropomyosin molecule would bind to six G-actins, rather than the seven in muscle, less than 50% of the actin would be associated.

C. Troponin-like Proteins

A considerable amount of biochemical evidence has accumulated for the presence of Ca^{2+}-sensitive regulation of actomyosin in non-muscle cells. The most widely occurring protein responsible for this regulation is calmodulin, a Ca^{2+}-binding protein that has many similarities to troponin C. Calmodulin has been reviewed by Means and Dedman (1980) who point out that it is a poor antigen. This has made its localization within cells difficult to establish, and the association of calmodulin or other Ca^{2+}-binding proteins with actin or thin filaments has not been demonstrated by electron microscopy.

D. Thin-filament Assemblies

Although actin-containing filaments have been observed in a wide variety of cells, it is in relatively few of these that structures comparable in filament regularity and organization to the I-bands of muscle occur. One such structure that has been intensively studied is the microvillus of the brush border of intestinal cells. McNabb and Sandborn (1964) showed in electron micrographs of sectioned material the presence of parallel filaments, approximately 6 nm in diameter, running the whole length of the microvillus. The filaments were packed hexagonally with a spacing of 10–15 nm and in some places had a nodular appearance. The helical features of actin were recognized by Mukherjee and Staehelin (1971) in the filament arrays in their micrographs of sectioned microvilli (Fig. 40), and they also observed in freeze-fractured specimens numerous cross filaments connecting each assembly to the surrounding membrane. Tilney and Mooseker (1971) showed that actin could be isolated from the brush border and Mooseker (1974) showed that detergent-treated microvilli contained actomyosin. In the presence of Ca^{2+} and ATP the actin filaments moved into the terminal web, a filamentous network at the base of the microvillus. Decoration with S-1 by Mooseker and Tilney (1975) established that all actin filaments had the same polarity within the microvillus, with the arrowheads pointing away from the apical tip as if from a Z-line. After treatment with Mg^{2+}, these authors showed the cross

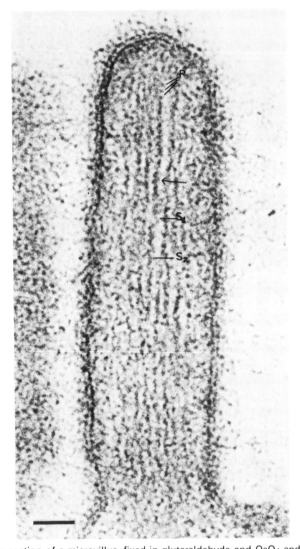

Fig. 40. Thin section of a microvillus, fixed in glutaraldehyde and OsO$_4$ and stained with uranyl acetate and lead. The filaments forming the core of the structure show features resembling the globular and double-helical aspects of F-actin. For example, at the region marked P, there are globular units with a centre-to-centre spacing of 6 nm, and the portion of filament marked with an arrow suggests a helical arrangement of subunits. The region between S$_1$ and S$_2$ shows less regular helical structure, which might be due to distortions introduced during sectioning. Scale marker indicates 50 nm (reprinted from Mukherjee and Staehelin, 1971, courtesy of T. M. Mukherjee and Cambridge University Press).

filaments connecting the actin arrays to the surrounding membrane were clearly visualized in thin sections and had a periodicity of 33 nm along the microvillus (Fig. 41). They also observed thick myosin-containing filaments in the region of the terminal web, and suggested a model for the contraction of the microvillus involving actomyosin interaction in this region.

Another example of ordered arrays of actin filaments is in the acrosomal process of the sperm of echinoderms, which can be up to 90 μm long and is formed very rapidly during fertilization of the egg. The reaction can be stimulated artificially, the "false discharge", by placing the sperm in alkaline sea water (Dan, 1952) or water that has contained eggs (Afzelius and Murray, 1957). Dan (1960) showed from sections that the acrosomal process was about 0·1 μm in diameter and contained fibrils arranged parallel to the outer membrane. Tilney *et al.* (1973) extended this analysis, showing that the fibrils were about 5 nm wide and could be decorated with muscle HMM, the resultant arrowheads pointing towards the main body of the sperm. In a water extract of an acetone powder of sperm, a major component of molecular weight corresponding to muscle actin was identified and shown to form a polymer which when negatively-stained resembled F-actin. When unreacted sperm were lysed with detergent and the extract analysed by centifrugation, at least 80% of the actin present was in the monomeric state, and sections of sperm showed a lack of organized structure in the acrosomal region. The authors therefore attributed the generation of the acrosomal process to the G- to F-actin transition. Myosin and actomyosin interaction would therefore play no part in this motile phenomenon.

In sperm of the horse-shoe crab, *Limulus*, filament bundles occur in both the unreacted and discharged states, and a transition between helical forms rather than a polymerization has been invoked to explain the formation of the process. In the unreacted state, Tilney (1975) showed that the bundle formed a coil around the base of the sperm. This unwound to form either a helical or straight process corresponding to the "false" and "true" discharged states. The unreacted and both types of discharged states could all be isolated, demonstrating their stability, though only the false discharge was obtained pure. Sections of each form showed that the constituent filaments were packed hexagonally, and in negatively-stained specimens the filaments could be seen to twist around each other in the coiled and helical forms but to run almost straight in the linear form with the cross-over points of the actin helices transversely aligned as in Mg^{2+}-induced paracrystals (Fig. 42). Associated with actin in the bundles were two proteins, one of molecular weight 95 000, thought to be α-actinin and present only in the coiled and false discharged states, and the other of molecular weight 55 000, present in the same molar concentration as actin. De Rosier *et al.* (1977) analysed the structure of the true discharge state in detail (Fig. 43a). They found that the actin filaments

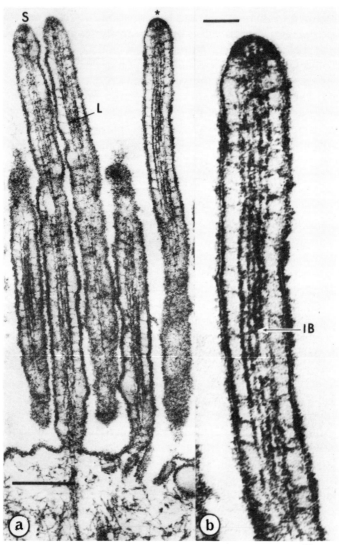

Fig. 41. (a) Thin section of an isolated brush border incubated in 15 mM Mg^{2+}. Cross-bridges connecting the filament arrays to the microvillus membrane are visible, and in some regions are periodic (33 nm). Lateral striations with the same periodicity are visible on the filament bundle in the microvillus marked S at its tip. Bridges of various lengths and diameters can be seen. An arrow (L) indicates the longest and thinnest bridge. (b) A higher magnification of the microvillus marked with a star at its tip in (a). In addition to membrane-filament cross-bridges, there are links (IB) between the filaments within the bundle. The scale bars in (a) and (b) represent 200 nm and 50 nm respectively (reprinted from Mooseker and Tilney, 1975, courtesy of L. G. Tilney and Rockefeller University Press).

had a symmetry comprising 28 subunits in 13 turns of the left-handed genetic helix, and showed how this symmetry could be accommodated within the hexagonal packing of the filaments if small distortions of the cross-links involving the 55 000 molecular weight protein were permitted. A three-dimensional reconstruction of the complex suggested that the 55 000 molecular weight protein occupied a binding site near that of myosin attachment to actin, which would account for Tilney's (1975) observation that HMM did not bind to the filaments of the false discharge. From a comparison of the actin helical symmetry in the coiled and true discharge states, De Rosier *et al.* (1980) measured a difference of twist of 0·23° per subunit between the two forms. They proposed that the change of twist occurring as the coil unwound generated a torsional force, which was converted to a linear force in manner analogous to the action of a simple spring wire. De Rosier *et al.* (1982) showed in freeze-etched material how the transformation from the coiled to the false discharge states involved a change in the supercoiling of the filaments from a right- to a left-handed sense. This change was considered to be produced by the effect of small molecules on subunit interactions, and an analogy was drawn with the effect of oxygen on haemoglobin.

A 55 000 molecular weight component was also identified associated with actin in extracts from unfertilized sea urchin eggs (Kane, 1975). De Rosier *et al.* (1977) analysed the structure of filament bundles formed from these extracts in parallel with their study of the acrosomal process. In the bundles from egg extracts the actin helical symmetry was 41 subunits in 19 turns, slightly different from that in the sperm discharge. A model was proposed in which filaments of this symmetry were packed in a hexagonal lattice and the 55 000 molecular weight protein was attached only at positions where cross-links between filaments could occur. This reproduced the distinctive transverse bands at intervals of 11 nm observed in micrographs of the bundles (Fig. 43b, c) and was consistent with the molar ratio of 55 000 molecular weight protein to actin being 0·21 as shown by Bryan and Kane (1977, 1978), who also redetermined the molecular weight as 58 000. One feature of the model was that the bands were not uniformly spaced but were either 4 or 5 actin subunits apart, a complete set of spacings between bands in one repeat of 41 units being described by the series: 5, 4, 5, 4, 5, 4, 5, 4 and 5. De Rosier and Censullo (1981) devised a computer search for this type of pattern in the electron micrographs, since it was not brought out by optical filtering. They verified the series and determined the number of bands in one true repeat along the bundle, from which the actin helical symmetry, with slightly more than 41/19 subunits per turn could be calculated to an accuracy of 1 in 20 000. The spacing of bands at intervals of 4 or 5 actin subunits was also noted by Spudich and Amos (1979) who analysed the structure of the bundles occurring *in situ* in the microvilli of fertilized sea urchin eggs. In other respects also the

structure of these bundles resembled that observed for the *in vitro* preparations, differing slightly in the actin helical symmetry present because of a small degree of supercoiling of the filaments. As indicated in Fig. 44, the crossbridge molecules might bind to adjacent *pairs* of actin monomers along each long-pitch helical strand, and this would automatically introduce an angle of just under 60° between successive cross-bridges, as is required for hexagonal packing of the filaments.

E. Actin-binding Proteins

A large number of proteins have been classified as binding to actin and thereby influencing its polymerization, aggregation and function in muscle and non-muscle cells, and the list is still growing rapidly. Some of these appear to be related to the proteins we have discussed, that is, tropomyosin, troponin, α-actinin and the 55 000 and 58 000 molecular-weight proteins, and many of them show similarities to each other, leading to the suggestion that they may be assigned to a limited number of functional groups, as discussed in a recent review by Weeds (1982).

Although the study of these proteins represents a large and fast-growing area of current research, the use of the electron microscope has so far provided relatively little detailed molecular information. Some actin-binding proteins are globular, others, particularly the ones that cross-link F-actin to form gels, are filamentous. The globular proteins often induce the formation of actin bundles, similar to those described in the previous section. Examples are the formation of paracrystals of F-actin by the presence of nerve growth factor (Castellani and O'Brien, 1981) and aldolase (Stewart *et al.*, 1980).

The electron microscope has revealed similarities in the filamentous group, and shadowed preparations have shown the presence of rod-like subunits, probably α-helical coiled-coils. Spectrin, a protein of molecular weight 460 000 which forms an important structural component of erythrocytes, was shown by Shotton *et al.* (1979), in shadowed preparations, to comprise two rod-like units joined to each other at each end. The dimer could associate into tetramers, the form responsible for cross-linking actin filaments (Ungewickell *et al.*, 1979). Filamin, isolated from vertebrate smooth muscle, and an actin-binding protein isolated from macrophages have similar molecular weights to that of the spectrin dimer. Tyler *et al.* (1980) showed the dimer of

---

Fig. 42. The acrosomal process in horseshoe-crab sperm, demembranated and negatively stained. (a) The coiled form as present in the unreacted state, (b) the false discharge, in which the process has a helical form, and (c) the true discharge, which is almost straight and in which the constituent actin filaments are mutually aligned across the structure, as in Mg^{2+}-induced paracrystals of F-actin. Scale bar indicates 100 nm (reprinted from Tilney, 1975, courtesy of L. Tilney and Rockefeller University Press).

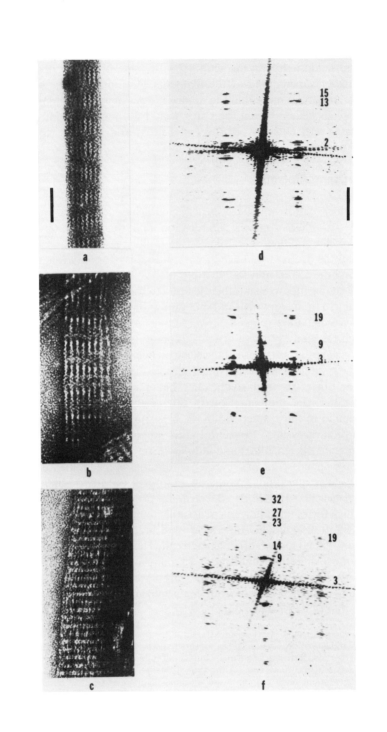

these molecules to be joined at one end only rather than forming a closed loop as in spectrin. This feature was also shown for filamin by Castellani et al. (1981; Fig. 45). The binding of the macrophage protein to F-actin was shown in negatively-stained and shadowed preparations by Hartwig and Stossell (1981). In order to inhibit gel formation, Castellani et al. (1981) prepared heavy merofilamin, with slightly less than one half the molecular weight of the intact molecule, and showed that it bound to F-actin without forming cross-links. Optical diffraction patterns of negatively-stained paracrystals showed an enhanced second layer-line intensity, suggesting that filamin might bind to actin in a similar way to tropomyosin, that is, in the long-pitch helical grooves.

VIII. SUMMARY AND DISCUSSION

A. Actin

The twin-stranded helical aspect of the actin filament was revealed in the electron microscope, in negative stain and by shadowing. This resolved the ambiguity of interpreting X-ray diffraction patterns of muscle, in which the reflections could arise either from a helical structure or from a net. The contribution of electron microscopy was further enhanced by the analysis of paracrystalline assemblies of F-actin, with the application of optical and computer methods for image processing. In this context, Moore et al. (1970) calculated a three-dimensional reconstruction of the actin subunit and filament (Fig. 7), and their results were confirmed by Spudich et al. (1972). The volume of the subunit was consistent with that expected from the amino-acid sequence. These studies were confined to the type of paracrystal in which the filaments run parallel to each other, first observed by Hanson (1967), and other types of paracrystal, particularly the net type, have not been studied in the same detail. Some of these other types are very interesting in their own right,

---

Fig. 43. (a) Negatively-stained acrosomal process from horseshoe-crab sperm in the true discharge state. (b) And (c) negatively stained bundles from sea urchin eggs, showing cross-striation at axial intervals of 11 nm. The bundle is viewed at a different angle in (c) and (b) so that the projected interfilament separation is different. The filaments of the bundle are hexagonally packed, and the projected interfilament spacing is $(3/2)^{\frac{1}{2}}$ times the actual spacing in (b) and 1/2 times this spacing in (c). The scale marker in (a) represents 50 nm and applies also to (b) and (c). (d) Optical diffraction pattern of (a). The indexing of the layer-lines is consistent with a helix having 28 subunits in 13 turns of the left-handed genetic helix (Fig. 6). (e) And (f) are optical diffraction patterns of (b) and (c), respectively, and are indexed as for a helix with 41 subunits in 19 turns of the genetic helix. The strong meridional reflection on the 9th layer-line arises from the axially repeating bands in the micrographs. The scale marker in (d) represents 0.1 nm^{-1} and applies also to (e) and (f) (reprinted from De Rosier et al., 1977, courtesy of D. J. De Rosier and Academic Press Inc.).

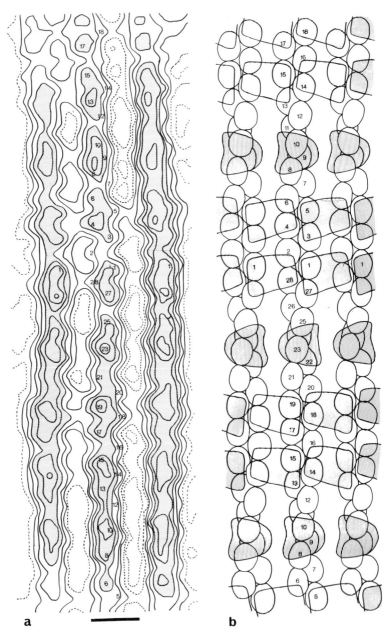

Fig. 44. (a) And (c) computationally filtered images of different regions of actin bundles from the microvillus of fertilized sea urchin eggs. Regions of higher protein density (i.e. absence of negative stain) are shaded. Densities thought to represent the positions of successive actin

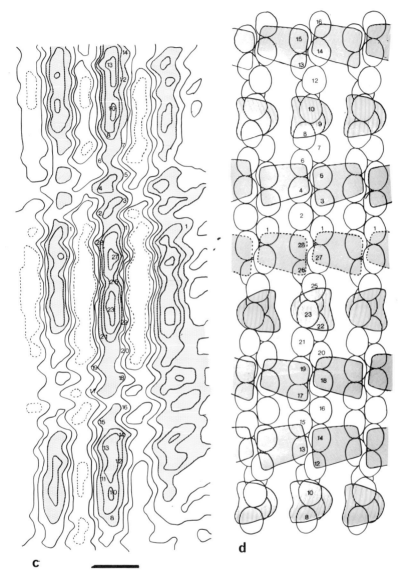

monomers have been numbered along the central filament. Each filament in the images probably arises from the superimposition of two or more filaments exactly in register in this view. (b) And (d) schematic drawings showing the probable arrangement of actin filaments and cross-bridges in (a) and (c) respectively. The interpretation of cross-bridge positions is based on density between filaments in (a) and (c), and on regions of high density along the filaments which cannot be accounted for as due to the superposition of actin monomers. The cross-bridges with broken outlines in (d) are those which differ in orientation from their counterparts in (b). Scale markers indicate 10 nm (reprinted, with modification, from Spudich and Amos, 1979, courtesy of L. A. Amos and Academic Press Inc.).

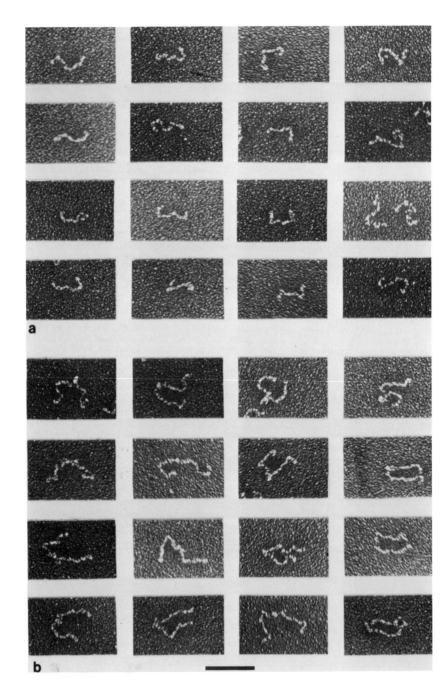

however, apart from demonstrating the polymorphism of F-actin aggregation, since they may be specifically induced by naturally occurring polyamines, such as spermine, known to induce cellular division (Oriol-Audit, 1978; Dickens and Oriol-Audit, 1980). Microinjection of spermine has been shown to induce cytokinesis of amoeba, a process during which oriented actin filaments are involved in the formation of the cleavage furrow (Gawlitta et al., 1981).

The derivation of a three-dimensional reconstruction of the actin subunit in the filament illustrated an advantage offered by electron microscopy over X-ray diffraction, where most of the attention was restricted to elucidating the helical parameters and comparing them with those of the thick filament. On the other hand, the determination of these parameters represented a fundamental step in understanding the respective contributions of the components of the sarcomere to the X-ray diagram, and investigating the changes taking place during contraction. The axial breadth of the layer-lines from actin in muscle X-ray patterns is much greater than that from the myosin filaments, although they derive from diffraction structures of comparable length. This and the observation of a variable cross-over repeat of the actin long-pitch helices in electron micrographs of single actin filaments, led Egelman et al. (1982) to propose the presence of a considerable degree of rotational disorder of the actin subunits. Azimuthal movement might be important in allowing actomyosin interaction over a range of angles in muscle, and providing greater flexibility of cross-linking of actin filaments by the proteins present in non-muscle cells.

X-ray diffraction analysis of single crystals of the actin monomer, combined with another protein to prevent filament formation, promised to extend dramatically our knowledge of the three-dimensional structure. The map of actin:DNase at 0·6 nm, although only a step in this direction, does show several interesting features (Suck et al., 1981; Fig. 11). The actin monomer is considerably elongated and divided into a small and large domain, with the ATP binding site assigned to the inter-domain cleft. Comparison with the earlier reconstructions of the actin monomer within the filament, from electron microscopy (Moore et al., 1970; Fig. 7), suggested that the long dimension was oriented parallel to the filament axis. More recent analysis, in which the two-zone as well as the elongated shape of the monomer is revealed, shows that an orientation roughly perpendicular to the filament axis is present (O'Brien et al., 1983; Fig. 32). This radical re-assessment may be due to

---

Fig. 45. (a) Shadowed filamin molecules showing typical curvilinear configurations. (b) Filamin molecules with twice the length of those shown in (a). They consist of two units, lying side by side or joined end to end. In some regions, the unit seems to comprise two strands. Scale marker indicates 100 nm (reprinted from Castellani et al., 1981, courtesy of Chapman and Hall Ltd).

increased resolution resulting from better paracrystal preparation and staining, but Egelman and De Rosier (1983) have suggested that in the Hanson-type actin paracrystals, on which the earlier work was carried out, the filaments are superimposed with alternating polarity so that the image presented is a composite. The later analysis (O'Brien et al., 1982) used paracrystals with defined filament polarity, which were probably a monolayer and which gave results consistent with those obtained for isolated filaments decorated with S-1 (Taylor and Amos, 1982, Vibert and Craig, 1982). In addition, the division of the actin monomer into two domains was indicated by a calculation of the radial density distribution of F-actin from the equatorial data of the X-ray diffraction pattern of oriented filaments (O'Brien et al., 1982). This is consistent with the long axis of the monomer oriented roughly perpendicular to the filament axis. Egelman and De Rosier (1983) have also deduced this orientation in a model in which the two monomer domains are of equal size and which is designed to match the appearance of electron micrographs of single actin filaments.

Another approach to the actin subunit structure, the analysis of the microcrystals formed in the presence of lanthanide ions, aimed to increase the resolution obtainable by electron microscopy by examining larger arrays than those formed from F-actin. The most visually attractive form of microcrystal is the tube, which varies in diameter according to the ionic radius of the lanthanide ion used (dos Remedios et al., 1980). The elongated actin monomers in the tube are arranged in pairs (Dickens, 1978; Fig. 9), the distance across each pair, 6·6 nm, being comparable to the F-actin width. Image analysis of the microcrystals has shown the two-zone nature of the monomer (Aebi et al., 1980; Fig. 10), though with some variability (Aebi et al., 1981). Further analysis of microcrystals may increase the resolution if the images are formed at low electron fluxes. The use of counterion rather than a whole protein molecule to inhibit filament formation may also prove an advantage, since the attribution of protein density to actin is unambiguous. In this respect, the observation that G-actin could be crystallized in the presence of polyethylene glycol was promising, but the crystals were very small and took a long while to grow (Oriol et al., 1977).

The overall diameter of F-actin observed by different authors varies considerably. In negative stain, Hanson and Lowy (1963) quoted 8·0 nm as an average, with a range 7·3 to 8·6 nm, for filaments recorded over holes in the carbon support film, but Huxley (1963) gave a lower range, 6·0 to 7·0 nm, where the filaments were not over holes. In the three-dimensional reconstruction of actin+tropomyosin calculated by O'Brien et al. (1983) from paracrystals (not over holes) the actin diameter is about 9·0 nm (Fig. 32), but the radial distribution calculated by these authors from X-ray data indicates a diameter about 2·0 nm less. This is in agreement with an earlier X-ray analysis carried out by Lednev and Tumanyan (1974). A larger diameter, 9·5 nm, has been given by Heuser and Kirschner (1980) for shadowed filaments in freeze-dried cells, though this figure does not allow for the thickness of the metal coating.

B. Tropomyosin

Tropomyosin has been extensively studied. Its ability to form many different types of aggregates offered considerable scope for electron microscopy, and the proposal that it was a coiled-coil (Crick, 1953) suggested a regularity of structure that should be revealed by appropriate staining methods. It is surprising that the macroscopic crystals observed by the protein's discoverer, Bailey (1948), were not analysed by X-ray diffraction for a decade, a period which saw the development of protein crystallography and the solution of the structures of myoglobin and haemoglobin. The contemporary X-ray analysis of Astbury *et al.* (1948) was confined to a fibrous preparation.

The determination of the sequence of α-tropomyosin (Hodges *et al.*, 1972) caused a flurry of interest and several groups tried to match the staining patterns observed in the microscope with that expected from the distribution of charged residues along the molecule. Consideration of the hydrophobic interactions between the two chains of the coiled-coil indicated an unstaggered structure (McLachlan and Stewart, 1975), and chemical cross-linking of the cysteine residues confirmed this feature experimentally (Lehrer, 1975, Johnson and Smillie, 1975; Stewart, 1975). The idea that a small stagger might be important in the end-to-end overlap region in tropomyosin polymers was therefore abandoned, but several studies indicated there was something different about the ends. The amino-acid sequence at the ends did not contain such a high proportion of α-helix-forming residues, and observations of the staining pattern of tactoids (Stewart, 1981) and of the electron density derived from X-ray analysis (Phillips *et al.*, 1979; Fig. 17) indicated the presence of a globular overlap region.

The pitch of the tropomyosin coiled-coil has not been determined unequivocally. Given the well-defined 14-fold periodicity of the acidic residues in the molecule (McLachlan and Stewart, 1976), it was reasonable to relate this to the seven actin subunits that would potentially be in contact with one tropomyosin molecule in the thin filament. Pseudo-equivalent contacts are made when the coiled-coil pitch matches the separation of actin subunits along one strand (Fig. 35). The pitch so derived, 13·7 nm, has not been shown experimentally in tropomyosin tactoids, but has been claimed to give the best fit to the chain density seen in the tropomyosin crystals, analysed by X-ray diffraction (Phillips *et al.*, 1980).

Tropomyosin prepared from non-muscle sources seems very similar in its basic attributes to the muscle protein, but there are significant and intriguing differences that cast doubt on how far the role of tropomyosin in muscle has general application. In the non-muscle tropomyosin most extensively studied, from platelets (der Terrossian *et al.*, 1981), there is probably a similar repeat of negatively charged groups, so that the coiled-coil pitch is likely to be the same, and the position of the cysteine residue, about one third of the way along the molecule, is also similar. The length of the molecule is shorter, however, as in other non-muscle tropomyosins, and there is insufficient protein present in the cells to combine with even half of the actin, if a model for association like that in muscle is assumed. Non-muscle tropomyosin does not readily polymerize, and its end sequences are different from those in muscle tropomyosin. It will be interesting when the full sequence is known, so that the deletion or deletions responsible for the shorter molecular length can be located. The lack of polymerizability indicates that tropomyosin in non-muscle cells cannot exert a co-operative effect on F-actin analogous to the stimulation of actomyosin ATPase observed with muscle proteins (Bremel and Weber, 1972).

C. Troponin

The discovery of troponin came much later than that of tropomyosin, about 15 years, but in the interim the numerous observations of the 40 nm stripe in electron micrographs of I-bands indicated the presence of an additional component. It was possible to speculate, however, that tropomyosin might lie perpendicular to as well as along the actin filaments.

Some resolution of the stripe was soon achieved when the troponin components were separated, as it was noted by Ebashi *et al.* (1972) that with antibody to TnT, the stripes were broader than for antibodies to the other components. Ohtsuki (1975) noted that the anti-TnT stripes were sometimes split into two, and went on to exploit this division of antigenic sites by preparing antibodies to two chymotryptic fragments of TnT (Ohtsuki, 1979; Fig. 28). Studies of the amino-acid sequence of TnT and of the affinity of various cyanogen bromide fragments for different regions of tropomyosin, have indicated that the molecule has a globular part which is located near the cysteine 190 of tropomyosin, and an α-helical part that extends towards the C-terminus (Mak and Smillie, 1981). The α-helical region may lie in close contact with the tropomyosin forming a three-stranded coiled-coil (Nagano *et al.*, 1980). There is no strong evidence, however, for a region of tropomyosin with a sequence complementory to that of TnT. In particular, the well-developed periodicity of 8·7 residues in both the acidic and basic residues observed over a considerable region of TnT (residues 71–151) does not have a convincing match in tropomyosin (Parry, 1981).

The decoration of tropomyosin paracrystals with troponin or the formation of paracrystals from the two proteins provided a convenient means of studying interactions between troponin and tropomyosin and between the troponin components. The determination of the location of troponin as about one third the way along the tropomyosin molecule was possible in both tactoids and net forms, though for the latter X-ray diffraction results were necessary to supplement the microscopy (Phillips *et al.*, 1980). With TnT rather than whole troponin, hexagonal and "double-diamond" lattices were observed in addition to the net with the kite-shaped unit, characteristic of tropomyosin alone (Cohen *et al.*, 1972; Fig. 23; Greaser *et al.*, 1972; Fig. 22). The conclusion that TnT is bound to the end of, rather than one third the way along, the tropomyosin molecule (Greaser *et al.*, 1972; Yamaguchi *et al.*, 1974) has not found confirmation elsewhere. In the kite-shaped lattice, TnT causes a thickening of the strands near the middle of the long arm of the unit (Yamaguchi *et al.*, 1974) and this must correspond to the globular region of the TnT molecule, binding near the cysteine 190 of tropomyosin. In shadowed preparations of isolated TnT, the globular region has not been observed and the TnT molecule has an appearance similar to that of tropomyosin, but about half as long (Flicker *et al.*, 1982). It seems likely that in combination with the other troponin components and with tropomyosin, the TnT is folded to some degree, and the extended conformation observed in isolation does not necessarily mean that TnT crosses the overlap region of the tropomyosin molecules in the thin filament. An alternative speculation, making use of the large TnT length, is that the molecule may make contact with its counterpart on the opposite tropomyosin strand. This would explain why the two troponin complexes at each 38 nm interval are at nearly the same axial level.

When TnC as well as TnT was present, tropomyosin tactoids only were formed, not nets. With skeletal muscle proteins, the amount of TnC bounds was higher when it was saturated with Ca^{2+} (Margossian and Cohen, 1973), but for cardiac proteins the binding of TnC depended on the lack of Ca^{2+} (Yamaguchi and Greaser, 1979).

D. Role of Tropomyosin and Troponin in Regulation

Since the thin filament is a dynamic element, taking part in the control as well as the production of muscular tension, the use of electron microscopy takes on an added dimension if different states relevant to these functions can be observed. F-actin seems to vary very little under different conditions, and its structure with S-1 attached, as in muscle in rigor, is evidently essentially the same as when it is alone or in combination with the regulatory proteins. The main thin filament change, at least at low resolution, when muscle is activated or isolated thin filaments are treated with Ca^{2+}, has been attributed to an azimuthal movement of tropomyosin in the long-pitch helical grooves of actin.

The tropomyosin movement was proposed to explain the changes of the low-order layer-lines of X-ray diffraction patterns of muscle when activated (Hanson et al., 1972; Haselgrove, 1972; Huxley, 1972; Parry and Squire, 1973). The movement provided a simple model for the control of contraction, in which tropomyosin could either block or allow the attachment of myosin heads to actin (the "steric-blocking" mechanism). This relatively crude mechanism was attractive partly because tropomyosin makes only quasi-equivalent bonds with actin along the long-pitch helical strands, so that a more intimate mechanism involving the two proteins seemed unlikely. The nature of the quasi-equivalent bonding was given a surer footing by the analysis of the tropomyosin sequence and the proposal of coiled-coil pitch of 13·7 nm to match the actin subunit repeat along each strand (Parry 1975; McLachlan and Stewart, 1976; Fig. 35). More recent studies, however, of tropomyosin crystals by X-ray diffraction, indicate that the molecule has a tendency to vibrate in a direction perpendicular to its length and that there are local regions of "kinking" particularly near the C-terminal region (Phillips et al., 1980). At any one time, the binding to F-actin is unlikely therefore to incorporate quasi-equivalent interactions with all the seven actin subunits per molecular length.

Several analyses of isolated thin filament preparations showed that tropomyosin could be detected in different azimuthal positions. A change of Ca^{2+} level was shown to induce a change of the type of paracrystal formed of reconstituted thin filaments, associated with a tropomyosin change of position (Gillis and O'Brien, 1975; Fig. 30). In the absence of TnC, a Ca^{2+}-independent change was observed, induced by TnI+TnT (Wakabayashi et al., 1975; Fig. 31) or by TnI on its own (O'Brien et al., 1980; Fig. 33). In view of the accumulating evidence for an elongated shape for TnT, with one end more globular and the other containing a large α-helical region, features which indicate a highly specialized structure, the observation that TnI alone can determine the tropomyosin position is of considerable interest. TnI in asociation with calmodulin is able to regulate actomyosin ATPase (Castellani et al., 1980), indicating that TnT may only be necessary in a more highly developed control mechanism.

It seemed important to establish whether the tropomyosin location was on the same or the opposite side of the long-pitch actin helical groove to the site of myosin attachment (Seymour and O'Brien, 1980; Fig. 36), to test one of the underlying assumptions of the steric-blocking mechanism of regulation. The discovery that the tropomyosin was on the opposite side to the S-1 attachment site previously proposed by Moore et al. (1970), led to a re-examination of the structure of decorated thin filaments by several groups, with the result that the

S-1 site was also moved to the other side. This, however, has now been shown to be only the major site of intraction, with a second weaker contact occurring between the S-1 and the adjacent actin subunit on the opposite strand (Amos et al., 1982). The possibility that tropomyosin may in some position block myosin attachment is thus very likely, but it would be at only one of two or more interaction sites, so that it would modulate rather than interrupt the attachment.

There is irony in the situation where, after it had been proposed that tropomyosin was on the opposite side of the actin groove to that previously suggested, the precise location proved less important for the validity of the steric-blocking mechanism of regulation because of the distributed nature of the actomyosin binding site. There is added piquancy in that the most recent analysis of the structure of actin + tropomyosin (O'Brien et al., 1983; Fig. 32) indicates that the tropomyosin is back on the original side! With the knowledge of this structure and of the two-domain, elongated nature of the actin subunit, the three-dimensional reconstruction of natural thin filaments in I-segments calculated by Seymour and O'Brien (1980, Fig. 36) may be re-interpreted. It seems clear that this reconstruction shows only the actin, with the position of tropomyosin unresolved. The density peak attributed to tropomyosin is probably the outer domain of the actin subunit, which would explain why this peak is not part of a continuous strand of density running along the long-pitch helices (compare Figs 32 and 36). A similar interpretation can be made of the reconstruction calculated by Spudich et al. (1972) from reconstituted thin filaments in paracrystals. In their Plate IX, the peak marked "B" which was assigned to tropomyosin is probably the outer domain of the actin subunit, and the weaker peak, marked "A", which was discounted for several reasons, is near the position of tropomyosin as deduced by O'Brien et al. (1983; Fig. 32).

The analysis of Seymour and O'Brien (1980) still represents the most detailed determination of the structure of natural thin filaments, and it stimulated the careful re-examination of the structure of S-1 decorated filaments. The positions of at least the main binding site of S-1 to actin and that of tropomyosin are unlikely to be significantly revised now. Movement of tropomyosin associated with regulation may take place between the two domains of the actin monomer. O'Brien et al. (1983) proposed that, under the influence of TnI, tropomyosin moved from making contact with the smaller to the larger domain, and then occupied the position identified as the second S-1 attachment site. The experiments of Chalovich and Eisenberg (1982) have shown that the binding of S-1 to thin filaments in the presence of ATP is virtually unaffected by the presence or absence of Ca^{2+}, a result which indicates that regulation by tropomyosin-troponin does not involve steric hindrance of myosin binding to actin. Possibly the blocking of a structurally weak contact such as the second S-1 attachment site may affect the overall configuration of attachment without appreciably changing the binding constant.

E. Z-disc

In the I-bands of striated muscle, two important features that have not found ready explanations are the uniform length of the filaments and the way in which the Z-disc provides a reversal of filament polarity. A molecular interpretation of filament length control was put forward by Huxley and Brown (1967) in which tropomyosin, with a slightly different axial repeat from that of the actin cross-over, made contacts with actin that varied in a vernier manner. The filament terminated when the tropomyosin-actin contact became too unfavourable. More recent studies, however, have favoured an exact match between the tropomyosin molecular repeat and the repeat of seven actin subunits along the long-pitch strand. Ohtsuki (1975) suggested from the positions of the bands formed by antibody to TnT that tropomyosin did not extend to the extreme end of the I-filaments, so there may be some specialized structure, possibly involving an additional component, in this region.

In the case of the Z-disc, the presence of a specialized structure is evident, but the problem of interpreting it is complicated. Comparison of electron microscope observations of the Z-discs of insect flight and vertebrate striated muscles show a clear distinction, as would be expected from the different number and distribution of the I- with respect to the A-filaments. In insect flight muscle, where the I-bands are extremely short, the thin filaments entering the Z-disc are constrained in the hexagonal arrangement with which they surround the thick filaments. As might be expected from the requirement of polarity reversal, the filaments are not continuous across the Z-disc, but the arrays from each half of the I-band are interleaved to form a hexagonal lattice (Ashurst, 1967). In vertebrate striated muscle the thin filaments in the neighbourhood of the Z-disc are not influenced by the A-band geometry, and are arranged on a square lattice, presumably dictated by the Z-disc cross-connections.

The difficulties of a detailed interpretation of the vertebrate Z-disc lattice have been increased by the controversy over the effects of fixation. Landon (1970) and several subsequent authors considered that the "basket-weave" appearance of the Z-disc in transverse section (Fig. 38a), was obtained only when osmium rather than glutaraldehyde was used as the primary fixative, but Ullrick *et al.* (1977) concluded that this appearance was the general one, and that the square and small-square lattices also observed (Fig. 37a) were derived from it. If the possibility of fixation artefacts is allowed, then the rather elaborate models for the Z-filament lattice proposed by several authors may represent over-interpretation of the data. Many of the models involved the extension of the thin filaments into the Z-disc and a doubling back to connect with the adjacent filament tip, that is, the presence of hairpin loops, but there were differences of opinion as to whether the loops from adjacent half I-bands interlocked. The idea that the loops could comprise actin strands bending around (Kelly, 1967) is improbable, as actin has shown no such verstility in isolated preparations. This proposal was made, however, before the characterization of α-actinin, and its assignment as an important constituent of the Z-disc. It seems likely that the thin filaments maintain their integrity as they enter and form part of the disc structure, since where the disc is observed thicker, as in cardiac muscle, the lattice repeat in the direction parallel to the fibrillar axis is 38 nm, that is, the same as the troponin-tropomyosin period (Goldstein *et al.*, 1979).

The isolation of Z-disc of the honey-bee (Saide and Ullrick, 1974) provided a promising alternative preparation for electron microscopy. Possibly, improvements in technique will enable this approach to be made for the vertebrate system.

F. Actin and Associated Proteins in Non-muscle Cells

Actin filaments and bundles have been observed in very many cells, and there is considerable interest in evaluating how far the mechanism of force production by actomyosin interaction occurring in muscle can find a parallel elsewhere. The regular arrays observed in the microvilli of the brush border were shown by Mooseker and Tilney (1975) to comprise actin filaments of uniform polarity. They proposed a model in which the thick myosin-containing filaments observed at the base of the microvilli would link adjacent actin arrays and produce their relative motion.

In other motile systems, actin filament arrays are observed but myosin is absent, and the resemblance to muscle contraction is lost. The acrosomal process of the sperm of echinoderms forms very rapidly during fertilization of the egg, but before formation there is no evidence for the presence of F-actin in the acrosomal region, and lysates of the cells contain mostly G-actin. The formation of the process has therefore been attributed to the rapid polymerization of actin (Tilney *et al.*, 1973). In the acrosomal region of the sperm of horseshoe crab F-actin bundles are present before as well as after the formation of the process (Tilney, 1975), and a change of the twist of the actin subunits within each filament has been proposed to provide the force for the extension (De Rosier *et al.*, 1980). It is difficult to prove, however, that the change of twist actually produces rather than just accompanies the formation of the process. Another feature that required clarification is the role of the

55 000 molecular weight protein, which is associated with actin in an equimolar ratio. The means for triggering the event are also unknown, and Ca^{2+} is not responsible. For the rapid polymerization of actin producing the process of the echinoderm sperm Tilney *et al.* (1978) have suggested a transient proton flux.

A 55 000 molecular weight protein (later 58 000) has also been shown to be present in bundles of actin filaments in sea urchin eggs. The preparations were either from extracts of unfertilized eggs (De Rosier *et al.*, 1977) or *in situ* in the microvilli of fertilized eggs (Spudich and Amos, 1979). In these assemblies the 55 000 molecular weight protein has only about 1/5 the actin molar concentration, and it forms cross-links between the filaments at intervals of 11 nm along the bundles (Figs 43, 44). Whether or not this protein has a dynamic as well as a structural role in the function of microvilli is not resolved.

The spectrum of actin-binding proteins is wide, but a classification may be attempted (Weeds, 1982). As with muscle, the filamentous components such as filamin and spectrin have interesting structural features that are fairly easily resolved in the electron microscope. More detailed analysis will be possible if crystals or paracrystals are prepared, and sequences are determined. For the globular components, the situation may be similar to that for troponin, where the structure and function are assessed mainly in conjunction with filamentous assemblies.

For a protein of highly conserved sequence and relatively invariant structure, F-actin shows remarkable versatility, endowed through its ability to associate and interact with many other proteins. It seems extraordinary that it can play such diverse roles in contractility and movement as in its interaction with myosin in muscle, its rapid polymerization in the formation of the acrosomal process of echinoderm sperm, and its change of helical symmetry and supercoiling in the formation of the acrosomal process of horseshoe-crab sperm. These roles may be found to be more closely related than is apparent when further structural details are elucidated, and, particularly in studies of muscle and the myosin cross-bridge cycle, much store is set on the advancement of X-ray diffraction studies to obtain atomic resolution. The study of structure at the tertiary and quarternary levels will be of no less importance, however, than the knowledge of the atomic co-ordinates of individual components, and it is here that electron microscopy will continue to play a key role for many years to come.

IX. ACKNOWLEDGEMENT

We are very grateful to Dr P. M. Bennett for a critical reading of the manuscript.

X. REFERENCES

Aebi, U., Fowler, W. E., Isenberg, G. and Pollard, T. D. (1981). *J. Cell Biol.* **91**, 340–351.
Aebi, U., Smith, P. R., Isenberg, G. and Pollard, T. D. (1980). *Nature* **288**, 296–298.
Afzelius, B. A. and Murray, A. (1957). *Exp. Cell Res.* **12**, 325–337.
Amos, L. A., Huxley, H. E., Holmes, K. C., Goody, R. S. and Taylor, K. A. (1982). *Nature* **299**, 467–469.
Ashurst, D. E. (1967). *J. Mol. Biol.* **27**, 385–389.
Astbury, W. T., Perry, S. V., Reed, R. and Spark, L. C. (1947). *Biochim. Biophys. Acta* **1**, 379–392.
Astbury, W. T., Reed, R. and Spark, L. C. (1948). *Biochem. J.* **43**, 282–287.
Auber, J. and Couteaux, R. (1963). *J. Microscopie* **2**, 309–324.
Bailey, K. (1946). *Nature* **157**, 368–369.
Bailey, K. (1948). *Biochem. J.* **43**, 271–279.
Barden, J., Tulloch, P.A. and dos Remedios, C. G. (1981). *J. Biochem.* **90**, 287–290.
Bettex-Galland, M. and Lüscher, E. F. (1959). *Nature* **184**, 276–277.
Bettex-Galland, M. and Lüscher, E. F. (1961). *Biochim. Biophys. Acta* **49**, 536–547.
Bettex-Galland, M., Portzehl, H. and Lüscher, E. F. (1962). *Nature* **193**, 777–778.
Bray, D. and Thomas, C. (1976a) *In* "Cold Spring Harbor Conference on Cell Proliferation 3, Cell Motility", Book B, pp. 461–473.
Bray, D. and Thomas, C. (1976b). *J. Mol. Biol.* **105**, 527–544.
Bremel, R. D. and Weber, A. (1972). *Nature New Biol.* **238**, 97–101.
Briskey, E. J., Seraydarian, K. and Mommaerts, W. F. H. M. (1967). *Biochim. Biophys. Acta* **133**, 424–434.
Bryan, J. and Kane, R. E. (1977). *J. Cell Biol.* **75**, 268a.
Bryan, J. and Kane, R. E. (1978). *J. Mol. Biol.* **125**, 207–224.
Busch, W. A., Stromer, M. H., Goll, D. E. and Suzuki, A. (1972). *J. Cell Biol.* **52**, 367–381.
Carlsson, L., Nyström, L. E., Lindberg, U., Kannan, K. K., Cid-Dresdner, H., Lövgren, S. and Jörnvall, H. (1976). *J. Mol. Biol.* **105**, 353–366.
Carlsson, L., Nyström, L. E., Sundkvist, I., Markey, F. and Lindberg, U. (1977). *J. Mol. Biol.* **115**, 465–483.
Caspar, D. L. D., Cohen, C. and Longley, W. (1969). *J. Mol. Biol.* **41**, 87–107.
Castellani, L., Morris, E. P. and O'Brien, E. J. (1980). *Biochem. Biophys. Res. Commun.* **96**, 558–565.
Castellani, L. and O'Brien, E. J. (1981). *J. Mol. Biol.* **147**, 205–213.
Castellani, L., Offer, G., Elliott, A. and O'Brien, E. J. (1981). *J. Muscle Res. Cell Motil.* **2**, 193–202.
Chalovich, J.M. and Eisenberg, E. (1982), *J. Biol. Chem.* **257**, 2432–2437.
Chou, P. Y. and Fasman, G. D. (1974). *Biochemistry* **13**, 211–222.
Cohen, C., Caspar, D. L. D., Johnson, J. P., Nauss, K., Margossian, S. S. and Parry, D. D. A. D. (1972). *Cold Spring Harb. Symp. Quant. Biol.* **37**, 287–297.
Cohen, C., Caspar, D. L. D, Parry, D. A. D. and Lucas, R. M. (1971). *Cold Spring Harb. Symp. Quant. Biol.* **36**, 205–216.
Cohen, C. and Hanson, J. (1956). *Biochim. Biophys. Acta* **21**, 177–178.
Cohen, C. and Longley, W. (1966). *Science* **152**, 794–796.
Cohen, I. and Cohen, C. (1972). *J. Mol. Biol.* **68**, 383–387.
Collins, J. H. (1974). *Biochem. Biophys. Res. Commun.* **58**, 301–308.

Collins, J. H. and Elzinga, M. (1975). *J. Biol. Chem.* **250**, 5912–5920.
Collins, J. H., Potter, J. D., Horn, M. J., Wilshire, G. and Jackman, N. (1973). *FEBS Lett.* **36**, 268–272.
Corsi, A. and Perry, S. V. (1958). *Biochem. J.* **68**, 12–17.
Côté, G., Lewis, W. G. and Smillie, L. B. (1978). *FEBS Lett.* **91**, 237–241.
Crick, F. H. C. (1953). *Acta Cryst.* **6**, 689–697.
Cummins, P. and Perry, S. V. (1973). *Biochem. J.* **133**, 765–777.
Dan, J. C. (1952). *Biol. Bull.* **103**, 54–66.
Dan, J. C. (1960). *Exp. Cell Res.* **19**, 13–28.
Depue, R. H. and Rice, R. V. (1965). *J. Mol. Biol.* **12**, 302–303.
De Rosier, D. J. and Censullo, R. (1981). *J. Mol. Biol.* **146**, 77–99.
De Rosier, D. J. and Klug, A. (1968). *Nature* **217**, 130–134.
De Rosier, D., Mandelkow, E., Silliman, A., Tilney, L. and Kane, R. (1977). *J. Mol. Biol.* **93**, 324–337.
DeRosier, D. J., Tilney, L. G., Bonder, E. M. and Frankl, P. (1982). *J. Cell Biol.* **93**, 324–337.
De Rosier, D., Tilney, L. and Flicker, P. (1980). *J. Mol. Biol.* **137**, 375–389.
der Terrossian, E., Fuller, S. D., Stewart, M. and Weeds, A. G. (1981). *J. Mol. Biol.* **153**, 147–167.
Dickens, M. J. (1978). *Proc. Roy. Microsc. Soc.* **13**, 80–81.
Dickens, M. J. and Oriol-Audit, C. (1980). *J. Muscle Res. Cell Motil.* **1**, 489.
dos Remedios, C. G. and Barden, J. A. (1977). *Biochem. Biophys. Res. Commun.* **77**, 1339–1346.
dos Remedios, C. G., Barden, J. A. and Valois, A. A. (1980). *Biochim. Biophys. Acta* **624**, 174–186.
dos Remedios, C. G. and Dickens, M. J. (1978). *Nature* **276**, 731–733.
Drabikowski, W., Dabrowska, R. and Barylko, B. (1971). *FEBS Lett.* **12**, 148–152.
Draper, M. H. and Hodge, A. J. (1949). *Aust. J. Exp. Biol. Med. Sci.* **27**, 465–505.
Ebashi, S. (1963). *Nature* **200**, 1010.
Ebashi, S. and Ebashi, F. (1964). *J. Biochem.* **55**, 604–613.
Ebashi, S. and Ebashi, F. (1965). *J. Biochem.* **58**, 7–12.
Ebashi, S., Ebashi, F. and Maruyama, K. (1964). *Nature* **203**, 645–646.
Ebashi, S., Endo, M. and Ohtsuki, I. (1969). *Quart. Rev. Biophys.* **2**, 351–384.
Ebashi, S. and Kodama, A. (1965). *J. Biochem.* **58**, 107–108.
Ebashi, S. and Kodama, A. (1966). *J. Biochem.* **59**, 425–426.
Ebashi, S., Ohtsuki, I. and Mihashi, K. (1972). *Cold Spring Harb. Symp. Quant. Biol.* **37**, 215–223.
Ebashi, S., Wakabayashi, T. and Ebashi, F. (1971). *J. Biochem.* **69**, 441–445.
Egelman, E. H., Francis, N. and De Rosier, D. J. (1982). *Nature* **298**, 131–135.
Egelman, E. H. and De Rosier, D. J. (1983). In "Actin—Its Structure and Function in Muscle and Non-Muscle Cells" (J. Barden and C. dos Remedios, eds), Academic Press, Australia.
Elzinga, M. and Collins, J. H. (1972). *Cold Spring Harb. Symp. Quant. Biol.* **37**, 1–7.
Endo, M., Nonomura, Y., Masaki, T., Ohtsuki, I. and Ebashi, S. (1966). *J. Biochem.* **60**, 605–608.
Fawcett, D. W. (1968). *J. Cell Biol.* **36**, 266–270.
Fine, R. E. and Blitz, A. L. (1975). *J. Mol. Biol.* **95**, 447–454.
Fine, R. E., Blitz, A. L., Hitchcock, S. E. and Kaminer, B. (1973). *Nature New Biol.* **245**, 182–186.
Flicker, P. F., Phillips, G. N., Jr and Cohen, C. (1982). *J. Mol. Biol.* **162**, 495–501.

Fowler, W. E. and Aebi, U. (1982). *J. Cell Biol.* **93**, 452–458.
Franzini-Armstrong, C. (1973). *J. Cell Biol.* **58**, 630–642.
Franzini-Armstrong, C. and Porter, K. R. (1964). *Zeit. Zellforsch. Mikrosk. Anat.* **61**, 661–672.
Gawlitta, W., Stockem, W. and Weber, K. (1981). *Cell Tissue Res.* **215**, 249–261.
Geiger, B., Dutton, A. H., Tokuyasu, K. T. and Singer, S. J. (1981). *J. Cell Biol.* **91**, 614–628.
Gillis, J. M. and O'Brien, E. J. (1975). *J. Mol. Biol.* **99**, 445–459.
Goldberg, A. and Lehman, W. (1978). *Biochem. J.* **171**, 413–418.
Goldstein, M. A., Schroeter, J. P. and Sass, R. L. (1977). *J. Cell Biol.* **75**, 818–836.
Goldstein, M. A., Schroeter, J. P. and Sass, R. L. (1979). *J. Cell Biol.* **83**, 187–204.
Goldstein, M. A., Schroeter, J. P. and Sass, R. L. (1982). *J. Muscle Res. Cell Motil.* **3**, 333–348.
Greaser, M. L. and Gergeley, J. (1971). *J. Biol. Chem.* **246**, 4226–4233.
Greaser, M. L., Yamaguchi, M., Brekke, C., Potter, J. and Gergeley, J. (1972). *Cold Spring Harb. Symp. Quant. Biol.* **37**, 235–244.
Greaser, M. L., Yamaguchi, M. and Vanderkooi, G. (1977). *J. Mol. Biol.* **116**, 883–890.
Hall, C. E., Jakus, M. A. and Schmitt, F. O. (1946). *Biol. Bull.* **90**, 32–50.
Hanson, J. (1967). *Nature* **213**, 353–356.
Hanson, J. (1968). *In* "Symposium on Muscle" (E. Ernst and F. B. Straub, eds), p. 93. Akademiai Kiadó, Budapest.
Hanson, J. (1973). *Proc. Roy. Soc. Lond. B.* **183**, 39–58.
Hanson, J. and Huxley, H. E. (1953). *Nature* **172**, 530–532.
Hanson, J. and Lowy, J. (1963). *J. Mol. Biol.* **6**, 46–60.
Hanson, J. and Lowy, J. (1964). *Proc. Roy. Soc. Lond. B.* **160**, 449–458.
Hanson, J., Lednev, V., O'Brien, E. J. and Bennett, P. M. (1972). *Cold Spring Harb. Symp. Quant. Biol.* **37**, 311–318.
Hartshorne, D. J. and Mueller, H. (1968). *Biochem. Biophys. Res. Commun.* **31**, 647–653.
Hartshorne, D. J. and Pyun, H. Y. (1971). *Biochem. Biophys. Acta* **229**, 698–711.
Hartshorne, D. J., Theiner, M. and Mueller, H. (1969). *Biochem. Biophys. Acta* **175**, 320–330.
Hartwig, J. H. and Stossel, T. P. (1981). *J. Mol. Biol.* **145**, 563–581.
Haselgrove, J. C. (1972). *Cold Spring Harb. Symp. Quant. Biol.* **37**, 341–352.
Hatano, S. and Oosawa, F. (1966a). *Biochim. Biophys. Acta* **127**, 488–497.
Hatano, S. and Oosawa, F. (1966b). *J. Cell Physiol.* **68**, 197–202.
Hatano, S. and Tazawa, M. (1968). *Biochim. Biophys. Acta* **154**, 507–519.
Hatano, S., Totsuka, T. and Oosawa, F. (1967). *Biochim. Biophys. Acta* **140**, 109–122.
Heuser, J. E. and Kirschner, M. W. (1980). *J. Cell Biol.* **86**, 212–234.
Higashi, S. and Ooi, T. (1968). *J. Mol. Biol.* **34**, 699–701.
Higashi-Fujime, S. and Ooi, T. (1969). *J. Microscopie* **8**, 535–548.
Hodge, A. J. (1959). *Rev. Mod. Phys.* **31**, 409–425.
Hodge, A. J., Huxley, H. E. and Spiro, D. (1954). *J. Exp. Med.* **99**, 201–206.
Hodges, R. S., Sodek, J., Smillie, L. B. and Jurasek, L. (1972). *Cold Spring Harb. Symp. Quant. Biol.* **37**, 289–310.
Hoffman-Berling, H. (1954). *Biochim. Biophys. Acta* **15**, 226–236.
Hoffman-Berling, H. (1956). *Biochim. Biophys. Acta* **19**, 453–463.
Hurwitz, F. I. and Walton, A. G. (1977). *Biochim. Biophys. Acta* **491**, 515–522.
Huxley, H.E. (1953). *Biochim. Biophys. Acta* **12**, 387–394.
Huxley, H. E. (1957). *J. Biophys. Biochem. Cytol.* **3**, 631–648.

Huxley, H. E. (1960). In "The Cell" (J. Brachet and A. E. Mirsky, eds), Vol. 4, pp. 365-481. Academic Press, New York and London.
Huxley, H. E. (1963). J. Mol. Biol. **7**, 281-308.
Huxley, H. E. and Brown, W. (1967) J. Mol. Biol. **30**, 383-434.
Huxley, H. E. (1970). In "Eighth International Congress Biochemistry", p. 23, Interlaken.
Huxley, H. E. (1971). Biochem. J. **125**, 85P.
Huxley, H.E. (1972). Cold Spring Harb. Symp. Quant. Biol. **37**, 361-376.
Ishikawa, H., Bischoff, R. and Holtzer, H. (1969). J. Cell Biol. **43**, 312-328.
Jakus, M. A. and Hall, C. E. (1947). J. Biol. Chem. **167**, 705.
Jakus, M.A. and Hall, C. E. (1949). Proc. 6th Int. Congress of Exp. Cytol., 1947. Exp. Cell Res. (Suppl. 1), 262-266.
Johnson, P. and Smillie, L. B. (1975). Biochim. Biophys. Res. Commun. **64**, 1316-1322.
Kane, R. E. (1975). J. Cell Biol. **66**, 305-315.
Katayama, E. and Nonomura, Y. (1979a). J. Biochem. **86**, 1495-1509.
Katayama, E. and Nonomura, Y. (1979b). J. Biochem. **86**, 1511-1522.
Katchburian, A., Burgess, A. M. C. and Johnson, F. R. (1973). Experientia **29**, 1020-1022.
Kawamura, M. and Maruyama, K. (1970). J. Biochem. **68**, 885-899.
Kelly, D. E. (1967). J. Cell Biol. **34**, 827-840.
Kelly, D. E. and Cahill, M. A. (1972). Anat. Rec. **172**, 623-642.
Klug, A. and Berger, J. E. (1964). J. Mol. Biol. **10**, 565-569.
Knappeis, C. G. and Carlsen, F. (1962). J. Cell Biol. **13**, 323-335.
Kung, T. H. and Tsao, T. C. (1965). Scientia Sinica **14**, 1383-1385.
Landon, D. N. (1970). J. Cell Sci. **6**, 257-276.
Lazarides, E. (1975). J. Cell Biol. **65**, 549-561.
Lazarides, E. (1976). J. Supramol. Struct. **5**, 531-563.
Lednev, V. V. and Tumanyan, V. G. (1974). Biophysics **19**, 1030-1033.
Lehrer, S. S. (1975). Proc. Natn. Acad. Sci. USA **72**, 3377-3381.
Lehman, W. and Szent-Györgyi, A. G. (1975). J. Gen. Physiol. **66**, 1-30.
Loewy, A. G. (1952). J. Cell Comp. Physiol. **40**, 127-156.
Longley, W. (1975). Nature **253**, 126-127.
Longley, W. (1977). J. Mol. Biol. **115**, 381-387.
Lowy, J. and Vibert, P. J. (1967). Nature **215**, 1254-1255.
MacDonald, R. D. and Engel, A. G. (1971). J. Cell Biol. **48**, 431-437.
Mak, A. S. and Smillie, L. B. (1981). J. Mol. Biol. **149**, 541-550.
Mannherz, H. G., Brehme, H. and Lamp, U. (1975). Eur. J. Biochem. **60**, 109-116.
Mannherz, H. G., Kabsch, W. and Leberman, R. (1977). FEBS Lett. **73**, 141-143.
Margossian, S. S. and Cohen, C. (1973). J. Mol. Biol. **81**, 409-413.
McLachlan, A. D. and Stewart, M. (1975). J. Mol. Biol. **98**, 293-304.
McLachlan, A. D. and Stewart, M. (1976). J. Mol. Biol. **103**, 271-298.
McNabb, J. D. and Sandborn, E. (1964). J. Cell Biol. **22**, 701-704.
Means, A. R. and Dedman, J. R. (1980). Nature **285**, 73-77.
Miller, A. and Tregear, R. T. (1972). J. Mol. Biol. **70**, 85-104.
Millman, B. M., Elliott, G. F. and Lowy, J. (1967). Nature **213**, 356-358.
Millward, G. R. and Woods, E. F. (1970). J. Mol. Biol. **52**, 585-588.
Moore, P. B., Huxley, H. E. and De Rosier, D. J. (1970). J. Mol. Biol. **50**, 279-295.
Mooseker, M. S. (1974). J. Cell Biol. **63**, 231a.
Mooseker, M. S. and Tilney, L. G. (1975). J. Cell Biol. **67**, 725-743.

Mornet, D., Bertrand, R., Pantel, P., Audemard, E. and Kassab, R. (1981). *Nature* **292**, 301–306.
Morris, E. P. (1979). PhD Thesis, University of London.
Mukherjee, T. M. and Staehelin, L. A. (1971). *J. Cell Sci.* **8**, 573–599.
Nachmias, V. T. (1964). *J. Cell Biol.* **23**, 183–188.
Nachmias, V. T., Huxley, H. E. and Kessler, D. (1970). *J. Mol. Biol.* **50**, 83–90.
Nagai, R. and Rehbun, L. I. (1966). *J. Ultrastr. Res.* **14**, 571–589.
Nagano, K., Miyamoto, S., Matsumura, M. and Ohtsuki, I. (1980). *J. Mol. Biol.* **141**, 212–222.
Nonomura, Y., Drabikowski, W. and Ebashi, S. (1968). *J. Biochem.* **64**, 419–422.
O'Brien, E. J., Bennett, P. M. and Hanson, J. (1971). *Phil. Trans. Roy. Soc. Lond. Ser. B.* **261**, 201–208.
O'Brien, E. J. and Couch, J. (1976). In "Proc. Sixth European Congress on Electron Microscopy", Vol. 2, pp. 153–154.
O'Brien, E. J., Couch, J., Johnson, G. R. P. and Morris, E. P. (1983). In "Actin—Its Structure and Function in Muscle and Non-Muscle Cells" (J. Barden and C. dos Remedios, eds), pp. 1–15. Academic Press, Australia.
O'Brien, E. J., Gillis, J. M. and Couch, J. (1975). *J. Mol. Biol.* **99**, 461–475.
O'Brien, E. J., Morris, E.P., Seymour, J. and Couch, J. (1980). In "Muscle Contraction: Its Regulatory Mechanisms" (S. Ebashi, K. Maruyama and M. Endo, eds), pp. 147–164, Japan Sci. Soc. Press, Tokyo/Springer-Verlag, Berlin.
Offer, G., Baker, H. and Baker, L. (1972). *J. Mol. Biol.* **66**, 435–444.
Ohtsuki, I. (1974). *J. Biochem.* **75**, 753–765.
Ohtsuki, I. (1975). *J. Biochem.* **77**, 633–639.
Ohtsuki, I. (1979). *J. Biochem.* **86**, 491–497.
Ohtsuki, I. (1980). In "Muscle Contraction: Its Regulatory Mechanisms" (S. Ebashi, K. Maruyama and M. Endo, eds), pp. 237–249, Japan Sci. Soc. Pres, Tokyo/Springer-Verlag, Berlin.
Ohtsuki, I. Masaki, T., Nonomura, Y. and Ebashi, S. (1967). *J. Biochem.* **61**, 817–819.
Ohtsuki, I. and Wakabayashi, T. (1972). *J. Biochem.* **72**, 369–377.
Ooi, T. and Fujime-Higashi, S. (1971). *Adv. Biophys.* **2**, 113–153.
Oriol-Audit, C. (1978). *Eur. J. Biochem.* **87**, 371–376.
Oriol, C., Dubord, C. and Landon, F. (1977). *FEBS Lett.* **73**, 89–91.
Page, S. (1964). *Proc. Roy. Soc. Lond. B* **160**, 460–466.
Page, S. and Huxley, H. E. (1963) *J. Cell Biol.* **19**, 369–390.
Parry, D. A. D. (1974). *Biochem. Biophys. Res. Commun.* **57**, 216–224.
Parry, D. A. D. (1975a). *Nature* **256**, 346–347.
Parry, D. A. D. (1975b). *J. Mol. Biol.* **98**, 519–535.
Parry, D. A. D. (1976). *Biochem. Biophys. Res. Commun.* **68**, 323–328.
Parry, D. A. D. (1981). *J. Mol. Biol.* **146**, 259–263.
Parry, D. A. D. and Squire, J. M. (1973). *J. Mol. Biol.* **75**, 33–55.
Pato, M. D., Mak, A. S. and Smillie, L. B. (1981). *J. Biol. Chem.* **256**, 602–607.
Pearlstone, J. R., Carpenter, M. R., Johnson, P. and Smillie, L. B. (1976). *Proc. Natn. Acad. Sci. USA* **73**, 1902–1906.
Peng, C. M., Kung, T. H., Hsiung, L. M. and Tsao, T. C. (1965). *Scientia Sinica* **14**, 219–228.
Pepe, F. A. (1966). *J. Cell Biol.* **28**, 505–525.
Perry, S. V., Cole, H. A., Head, J. F. and Wilson, F. J. (1972). *Cold Spring Harb. Symp. Quant. Biol.* **37**, 251–262.
Perry, S. V. and Corsi, A. (1958). *Biochem. J.* **68**, 5–12.

Perutz, M. F. (1951). *Nature* **167**, 1053–1054.
Phillips, G. N. Jr, Fillers, J. P. and Cohen, C. (1980). *Biophys. J.* **32**, 485–500.
Phillips, G. N. Jr., Lattman, E. E., Cummins, P., Lee, K. Y. and Cohen, C. (1979). *Nature* **278**, 413–417.
Podlubnaya, Z. A., Tskhovrebova, L. A., Zaalishvili, M. M. and Stefanenko, G. A. (1975). *J. Mol. Biol.* **92**, 357–359.
Pollard, T. D., Shelton, E., Weihing, R. R. and Korn, E. D. (1970). *J. Mol. Biol.* **50**, 91–97.
Rasch, J. E., Shay, J. W. and Biesele, J. J. (1968). *J. Ultrastr. Res.* **24**, 181–189.
Reddy, M. K., Etlinger, J. D., Rabinowitz, M. and Fischman, D. A. (1975). *J. Biol. Chem.* **250**, 4278–4284.
Reedy, M. K. (1964). *Proc. Roy. Soc. Lond. B.* **160**, 458–460.
Rome, E. M., Hirabayashi, T. and Perry, S. V. (1973). *Nature New Biol.* **244**, 154–155.
Rowe, A. (1964). *Proc. Roy. Soc. Lond. B.* **160**, 437–441.
Rowe, R. W. D. (1973). *J. Cell Biol.* **57**, 261–277.
Rozsa, G., Szent-Györgyi, A. and Wyckoff, R. W. G. (1949). *Biochim. Biophys. Acta* **3**, 561–569.
Rozsa, G., Szent-Györgyi, A. and Wyckoff, R. W. G. (1950). *Exp. Cell. Res.* **1**, 194–205.
Saide, J. D. and Ullrick, W. C. (1973). *J. Mol. Biol.* **79**, 329–337.
Saide, J. D. and Ullrick, W. C. (1974). *J. Mol. Biol.* **87**, 671–683.
Schaub, M. C. and Perry, S. V. (1969). *Biochem. J.* **115**, 993–1004.
Schaub, M. C. and Perry, S. V. (1971). *Biochem. J.* **123**, 367–377.
Selby, C. C. and Bear, R. S. (1956). *J. Biophys. Biochem. Cytol.* **2**, 71–85.
Senda, N., Shibata, N., Tatsumi, N., Kondo, K. and Hamada, K. (1969). *Biochim. Biophys. Acta* **181**, 191–200.
Seymour, J. and O'Brien, E. J. (1980). *Nature* **283**, 680–682.
Shotton, D. M., Burke, B. E. and Branton, D. (1979). *J. Mol. Biol.* **131**, 303–329.
Smillie, L. B., Pato, M. D., Pearlstone, J. R. and Mak, A. S. (1980). *J. Mol. Biol.* **136**, 199–202.
Sodek, J., Hodges, R. S., Smillie, L. B. and Jurasek, L. (1972). *Proc. Natn. Acad. Sci. USA* **69**, 3800–3804.
Spudich, J. A. (1972). *Cold Spring Harb. Symp. Quant. Biol.* **37**, 585–593.
Spudich, J. A. and Amos, L. A. (1979). *J. Mol. Biol.* **129**, 319–331.
Spudich, J. A., Huxley, H. E. and Finch, J. T. (1972). *J. Mol. Biol.* **72**, 619–632.
Staprans, I., Takahashi, H., Russell, M. P. and Watanabe, S. (1972). *J. Biochem.* **72**, 723–735.
Stewart, M. (1975a). *FEBS Lett.* **53**, 5–7.
Stewart, M. (1975b). *Proc. Roy. Soc. Lond. B.* **190**, 257–266.
Stewart, M. (1981). *J. Mol. Biol.* **148**, 411–425.
Stewart, M. and McLachlan, A. D. (1975). *Nature* **257**, 331–333.
Stewart, M. and McLachlan, A. D. (1976). *J. Mol. Biol.* **103**, 251–269.
Stewart, M., Morton, D. J. and Clarke, F. M. (1980). *Biochem. J.* **186**, 99–104.
Stone, D., Sodek, J., Johnson, P. and Smillie, L. B. (1974). *Proc. IX FEBS Meeting (Budapest)* **31**, 125–136.
Strzelecka-Golaszewska, H., Pròchniewicz, E. and Drabikowski, W. (1978). *Eur. J. Biochem.* **88**, 219–227.
Stromer, M. H., Hartshorne, D. J., Mueller, H. and Rice, R. V. (1969). *J. Cell Biol.* **40**, 167–178.

Suck, D., Kabsch, W. and Mannherz, H. G. (1981). *Proc. Natn. Acad. Sci. USA* **78**, 4319–4323.
Sugino, H., Sakabe, N., Sakabe,K., Hatano, S., Oosawa, F., Mikawa, T. and Ebashi, S. (1979). *J. Biochem.* **86**, 257–260.
Suzuki, A., Goll, D. E., Singh, I., Allen, R. E., Robson, R. M. and Stromer, M. H. (1976). *J. Biol. Chem.* **251**, 6860–6870.
Taylor, K. A. and Amos, L. A. (1981). *J. Mol. Biol.* **147**, 297–324.
Taylor, K. A. and Amos, L. A. (1983). *In* "Actin—Its Structure and Function in Muscle and Non-Muscle Cells" (J. Barden and C. dos Remedios, eds), Academic Press, Australia.
Tilney, L. (1975). *J. Cell Biol.* **64**, 289–310.
Tilney, L. G., Hatano, S., Ishikawa, H. and Mooseker, M. S. (1973). *J. Cell Biol.* **59**, 109–126.
Tilney, L. G., Kiehart, D. P., Sardet, C. and Tilney, M. (1978). *J. Cell Biol.* **77**, 536–564.
Tilney, L. G. and Mooseker, M. S. (1971). *Proc. Natn. Acad. Sci. USA* **68**, 2611–2615.
Tsao, T. C., Kung, T. H., Peng, C. M., Chang, Y. S. and Tsou, Y. S. (1965). *Scientia Sinica* **14**, 91–105.
Ts'o, P. O. P., Eggman, L. and Vinograd, J. (1957). *Biochim. Biophys. Acta* **25**, 532–548.
Tyler, J. M., Anderson, J. M. and Branton, D. (1980). *J. Cell Biol.* **85**, 489–495.
Ullrick, W. C., Toselli, P. A., Saide, J. D. and Phear, W. P. C. (1977). *J. Mol. Biol.* **115**, 61–74.
Ungewickell, E., Bennett, P. M., Calvert, R., Ohanian, V. and Gratzer, W. B. (1979). *Nature* **280**, 811–814.
Vibert, P. and Craig, R. (1982) *J. Mol. Biol.* **157**, 299–319.
Vibert, P. J., Haselgrove, J. C., Lowy, J. and Poulsen, F. R. (1972a). *Nature New Biol.* **236**, 182–183.
Vibert, P. J., Haselgrove, J. C., Lowy, J. and Poulsen, F. R. (1972b). *J. Mol. Biol.* **71**, 757–767.
Wakabayashi, T., Huxley, H. E., Amos, L. A. and Klug, A. (1975). *J. Mol. Biol.* **93**, 477–497.
Weber, A. and Murray, J. M. (1973). *Physiol. Rev.* **53**, 612–673.
Weeds, A. (1982). *Nature* **296**, 811–816.
Wilkinson, J. M. and Grand, R. J. A. (1975). *Biochem. J.* **149**, 493–496.
Wohlfarth-Botterman, K. E. (1962). *Protoplasma* **54**, 514–539.
Wohlfarth-Botterman, K. E. (1964a). *Int. Rev. Cytol.* **16**, 61–131.
Wohlfarth-Botterman, K. E. (1964b). *In* "Primitive Motile Systems in Cell Biology" (Allan and Kamiya, eds), Academic Press, New York and London.
Wolpert, L., Thompson, C.M. and O'Neill, C. H. (1964). *In* "Primitive Motile Systems in Cell Biology" (Allan and Kamiya, eds), Academic Press, New York and London.
Woods, E. F. (1967). *J. Biol. Chem.* **242**, 2859–2871.
Yamaguchi, M. and Greaser, M. L. (1979). *J. Mol. Biol.* **131**, 663–667.
Yamaguchi, M., Greaser, M. L. and Cassens, R. G. (1974). *J. Ultrastr. Res.* **48**, 33–58.
Yamamoto, K., Yanagida, M., Kawamura, M., Maruyama, K. and Noda, H. (1975). *J. Mol. Biol.* **91**, 463–469.

2. Myosin Molecules, Thick Filaments and the Actin-myosin Complex

ROGER CRAIG AND PETER KNIGHT

Department of Anatomy, University of Massachusetts Medical School, 55 Lake Avenue North, Worcester, USA and Muscle Biology Division, ARC Meat Research Institute, Langford, Bristol, England

I. Introduction 98
II. Myosin Molecule and Filament 103
III. Structure of the Myosin Molecule 107
   A. Structure of Vertebrate Striated Muscle Myosin 107
   B. Structure and Function of the Myosin Head 112
   C. Diversity of Myosin Structure. 115
IV. Structure of Vertebrate Striated Muscle Thick Filaments . . . 118
   A. Native Thick Filaments 118
     1. Organization of myosin molecules 118
     2. Non-myosin components of the thick filament 124
     3. Arrangement of thick filaments in the myofibril lattice . . 131
     4. Summary 133
   B. Synthetic Assemblies of Myosin and its Fragments 134
     1. Assemblies of intact myosin 134
     2. Assemblies of myosin fragments 136
   C. Models of Thick Filament Structure 142
V. Structure of Thick Filaments from Sources Other Than Vertebrate Skeletal Muscle 145
   A. Vertebrate Smooth Muscle and Non-muscle Thick Filaments . . 145
   B. Invertebrate Striated Muscle Thick Filaments 151
   C. Invertebrate Catch Muscle Thick Filaments 157
VI. Actin-myosin Interactions 161
   A. *In Vitro* Studies of Actin-myosin Interaction 165
     1. Decorated actin: the arrowhead structure 165
     2. The arrowhead structure—a structural assay for F-actin . . 171
     3. Partial decoration of actin 173
     4. Non-rigor modes of cross-bridge attachment 174
     5. Regulation: actin-linked 175

6. Regulation: myosin-linked 178
B. Actin-myosin Interaction in Intact Muscle 181
 1. Evidence for actin-myosin interaction in the filament lattice . 181
 2. Geometric constraints in actin-myosin interaction in the filament lattice. 183
 3. Actin-myosin interaction in contracting muscle 192
VII. Acknowledgements 194
VIII. References 195

I. INTRODUCTION

One of the most beautiful images in the electron microscope is that of the ultrastructure of striated muscle (Fig. 1). It is this aesthetic appeal, as well as more scientific reasons, that has led to such an intensive electron microscope study of muscle in all its forms. It is gratifying to the microscopist that frequently the prettiest micrographs are also the most informative.

Muscles are the essence of animal motion. They are fascinating to study because they undergo large and dramatic structural changes that can be studied and interpreted in molecular terms (see Offer, 1974, for review). Actin and myosin are the principal proteins of muscle, and two fundamental experiments on them provided the basis for our understanding of contractility. Engelhardt and Ljubimowa (1939) showed that actomyosin had ATPase activity, whilst Szent-Györgyi (see Szent-Györgyi, 1951) found that an actomyosin solution extracted from muscle, when extruded into water, formed insoluble threads which contracted on addition of ATP. This established the link between chemical energy consumption and mechanical work.

The interaction of myosin, actin and ATP is the basis not only of muscular contraction but also of many microscopic forms of cell motion, such as amoeboid movement, cell division and even cytoplasmic streaming in plants (see Huxley, 1973; Pollard and Weihing, 1974). The major goal of research into these processes is therefore an understanding of the structure and function of these molecules and their interaction, and it is the purpose of this chapter to show what major contributions electron microscopy has made towards this end.

Our greatest insights into biological motion have come from studies of

Fig. 1. (a) Longitudinal section of fixed and embedded frog sartorius muscle. Scale marker indicates 0·5 μm. (b) Schematic interpretation of (a) in terms of discontinuous, interdigitating thick and thin filaments. (c) Transverse section of frog sartorius muscle showing double hexagonal array of thick and thin filaments. Scale marker indicates 200 nm (reprinted from Huxley (1965) courtesy of H. E. Huxley and Scientific American Inc. Copyright © 1965 by Scientific American Inc. All rights reserved).

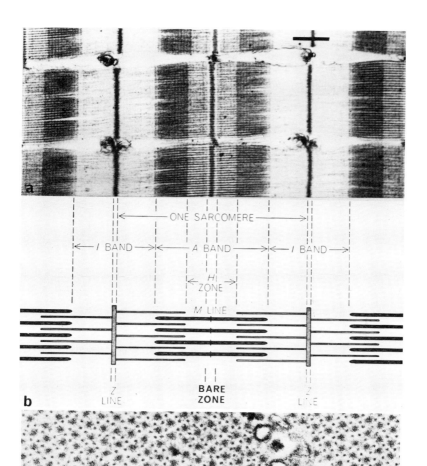

striated muscle. Polarized light microscope observations in the nineteenth Century had indicated that the myofibrils of muscle consisted of rodlets arranged parallel to the fibril axis (see Huxley, 1980 for review). The earliest electron micrographs (of fragmented muscle) confirmed this view: myofibrils appeared to consist of a single set of longitudinal filaments extending continuously through each sarcomere (e.g. Hall et al., 1946; Draper and Hodge, 1949; Rozsa et al., 1950). Contraction of muscle was thought at this time to be a result of the large scale shortening of these filaments, brought about by some internal folding mechanism (see Needham, 1971; Huxley, 1980).

Further experiments soon suggested a rather different picture, however. From X-ray diffraction patterns of muscle, Huxley (1953a) predicted that the filaments were in fact present in a *double* hexagonal array, and electron microscopy of transverse sections of intact muscle confirmed that there were indeed *two* sets of filaments (Huxley, 1953b; Fig. 1c). Light and electron micrographs of myofibrils from which actin or myosin had been extracted showed that the thicker, myosin filaments were confined to the A-bands while the thinner, actin filaments extended through the I-band and into the A-band where they interdigitated with the myosin filaments (Hanson and Huxley, 1953; Fig. 1b). Light microscope observations of the change in striation pattern during muscle shortening were then interpreted as showing an increase in overlap of the two sets of filaments without shortening of the filaments themselves (Huxley and Niedergerke, 1954; Huxley and Hanson, 1954). These findings laid the foundation of the sliding filament hypothesis of muscle contraction, which has formed the basis of all our thinking since, and has been amply supported by subsequent, more refined experiments.

The earlier picture of one array of continuous filaments was most convincingly shown to be wrong by H. E. Huxley (1957). His beautiful electron micrographs (of sections only 15 nm thick) demonstrated that the filaments were indeed discontinuous (Fig. 2) and thereby successfully defended the sliding model against some of its early detractors. These micrographs also showed increased filament overlap with increased shortening of the sarcomere (confirming directly the earlier inferences from light microscopy), and better controlled and more accurate measurements later supported the same notion and revealed no change in filament length on contraction (Page and Huxley, 1963). The strongest confirmatory evidence for the sliding model has come from X-ray diffraction of living muscle, which has

---

Fig. 2. Very-thin longitudinal section of vertebrate striated muscle showing longitudinally discontinuous thick and thin filaments and cross-bridges joining them in the zone of filament overlap. The axial dimension is considerably foreshortened owing to compression during sectioning. Scale marker indicates 0·2 μm (reprinted from H. E. Huxley (1957), courtesy of H. E. Huxley and The Rockefeller University Press).

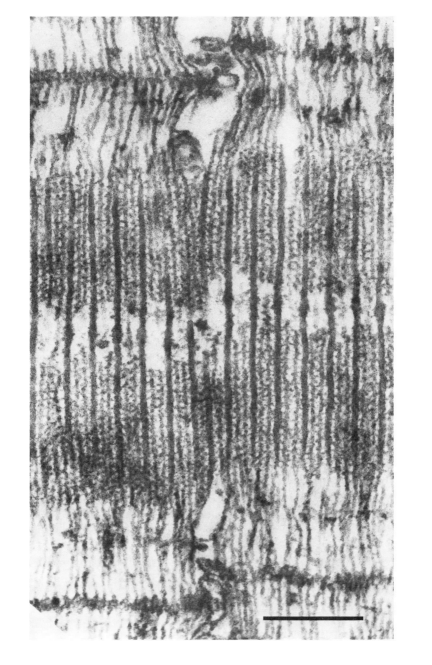

revealed essentially no change in the periodicities characteristic of the thick or the thin filaments when a muscle contracts (Huxley and Brown, 1967; Elliott *et al.*, 1967).

A crucial feature of Huxley's micrographs was the presence of cross-bridges extending between the surfaces of the thick myosin filaments and the thin actin filaments (Fig. 2). These bridges formed part of the thick filaments, since they were seen as projections even when no thin filaments were present. It was suggested that the bridges generated the sliding of the thin past the thick filaments. Since the distance that such bridges could move appeared to be small in relation to how far the filaments could slide, it was suggested that some sort of cyclic attachment, translation and detachment might occur to "row" the filaments past each other (see Hanson and Huxley, 1955; A. F. Huxley, 1957, 1980; H. E. Huxley, 1969)—still the current view of how muscles contract (see Section VI).

The early success of electron microscopy in revealing the secrets of striated muscle contraction has continued, and the technique is now applied to all types of muscle and to non-muscle motile systems. Striated muscle, nevertheless, remains the best understood contractile system because it is the best ordered, and therefore amenable to other structural techniques such as X-ray diffraction, spectroscopy and physiology, and because it is available in ample quantities and thus open to biochemical investigation. It is this accessibility to techniques that complement electron microscopy that has made the study of straited muscle such a success. Although electron microscopy is a powerful and direct visual technique, it is limited in that only static, dried and denatured states have generally been observed: other approaches allow one to build up an overall picture of muscle function which would be incomplete if any one of them were omitted.

In this chapter we show how electron microscopy has played a pivotal role in elucidating the structure of myosin molecules and filaments and their interaction with actin (actin and thin filament structure are discussed, in this Volume, by O'Brien and Dickens; see also Squire, 1981). We also point to current problems where it may be hoped that the technique will make further contributions. We discuss first the structure of the myosin molecule (Section III; see also Lowey, 1971, for an excellent general review). Then we consider how the molecules are assembled in the thick filaments of different muscles and of non-muscle cells (Sections IV and V). Lastly the structure of the actin-myosin complex is reviewed (Section VI), including the contribution that electron microscopy has made to our understanding of the nature of the force-generating event in actin-myosin interaction.

II. MYOSIN MOLECULE AND FILAMENT

Myosin in vertebrate striated muscle exists as a polymer—the thick filament. The filaments are spindle shaped objects 1·6 μm long and 10–12 nm in diameter. In the centre they show a smooth "bare zone" about 150 nm long, but along the rest of their length they have a rough surface (Huxley, 1963; Fig. 3a–e). This unusual form is reflected in the equally unusual structure of the myosin molecule: when thick filaments are depolymerized by treatment with high ionic strength (e.g. 0·5 M KCl) the myosin monomers are found to be elongated molecules, having a globular region, consisting of two heads, at one end of a long, thin tail (see Fig. 7).

Fig 3. Negatively-stained thick filaments from vertebrate striated muscle. (a–e) Native thick filaments. (f) Synthetic filament made from a myosin solution. Scale marker indicates 0·3 μm (reprinted from Huxley (1963), courtesy of H. E. Huxley and Academic Press Inc.).

When the ionic strength of a myosin solution is dropped to physiological levels (0·15 M), filaments similar in appearance to native filaments, but variable in length, are once again formed (Fig. 3f), suggesting that the myosin molecule itself contains most of the information necessary to specify thick filament structure. Intermediate stages of assembly—filaments shorter in length, but with bare zones of constant length—are also observed (Fig. 4). On the basis of these observations and of the solubility properties of myosin (see Section III.A), Huxley (1963) suggested a simple scheme for myosin filament assembly (Fig. 5). Myosin molecules first aggregate tail-to-tail, in bipolar fashion, and the filament then elongates by the polar addition of myosin molecules. The insoluble myosin tails form the backbone of the filament while

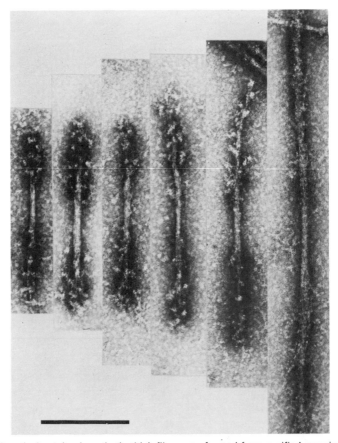

Fig. 4. Negatively-stained synthetic thick filaments, formed from purified myosin by rapid dilution, showing variable lengths. Scale marker indicates 0·3 µm (reprinted from Huxley (1963), by courtesy of H. E. Huxley and Academic Press Inc.).

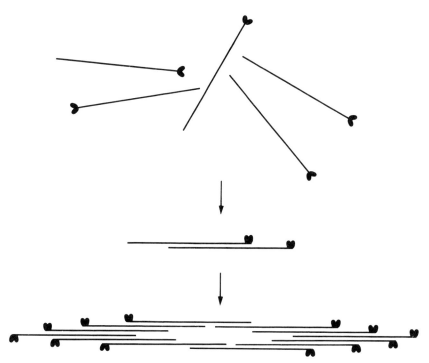

Fig. 5. Scheme for self-assembly of myosin to form bipolar filaments (after Huxley, 1963). The dimer was not in Huxley's original scheme, but has more recently been proposed as an intermediate in assembly (see Section IV.B.1) (reprinted from Offer (1974), by courtesy of G. Offer and Longman Group Ltd).

the globular portions (the cross-bridges in Fig. 2) are on the surface, giving the filament its rough appearance. The central region of antiparallel overlap contains no globular portions and is therefore smooth or "bare". With recent improvements in technique, which have enabled the myosin heads to be unambiguously identified, it has been confirmed that the heads do form the surface projections and that the polarity of the structure reverses across the bare zone (Fig. 6; Trinick and Elliott, 1979; Knight and Trinick, in preparation). As Huxley pointed out, this reversal of polarity is of crucial importance: if it is assumed that during contraction myosin heads exert a directed force on the thin filaments, it is essential that this force change sense at the filament centre in order that the thin filaments in each half of the sarcomere be drawn together.

To understand the structure and function of myosin, a study of the molecule is first essential. This is discussed in detail in the following section.

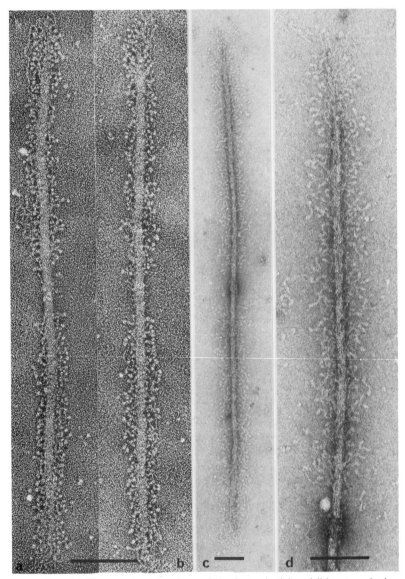

Fig. 6. Native thick filaments from vertebrate skeletal muscle. (a) and (b): rotary shadowed; (c) and (d): negatively stained. The backbone diameter in (a) and (b) is twice that in (c), possibly owing to washing the filaments with distilled water before shadowing. In addition to showing the bipolar packing of myosin molecules these recent images show previously undescribed fine filamentous material associated mainly with the filament tips. Scale markers indicate: (a) and (b), 0·2 μm; (c) and (d), 0·1 μm ((a), (b) reprinted from Trinick and Elliott (1979), by courtesy of J. Trinick and Academic Press Inc.; (c), (d) Knight and Trinick).

III. STRUCTURE OF THE MYOSIN MOLECULE

A. Structure of Vertebrate Striated Muscle Myosin

Myosin purified from high ionic strength extracts of muscle shows the characteristics of a highly asymmetric molecule when examined by viscosity, velocity sedimentation and light scattering; early physico-chemical studies further led to the suggestion that the intact molecule was thicker at one end (see Lowey, 1971). This picture of myosin was confirmed when the molecule was first seen in the electron microscope by platinum shadowing (Rice, 1961; Huxley, 1963; Zobel and Carlson, 1963). It appeared to be an approximately 110–150 nm by 2 nm rod with a roughly 20–40 nm by 3·5 nm globular region at one end. Some substructure could be seen within the globular region, but there was not sufficient detail to settle the prevailing controversy concerning the number of polypeptide chains in the molecule. Physico-chemical studies had suggested between one and three subunits in the globular region (Small et al., 1961; Lowey and Cohen, 1962; Woods et al., 1963; Young et al., 1965; Mueller, 1965), but this uncertainty was directly resolved by electron microscopy, when the shadowing technique was improved. The globular region was then seen to consist of two similar looking "heads" attached to a long flexible tail (Slayter and Lowey, 1967; Lowey et al., 1969). This structure was readily consistent with the postulate, from spectroscopic and X-ray diffraction data (Cohen and Szent-Györgyi, 1957; Lowey and Cohen, 1962), of a two-chain coiled-coil of α-helices forming the tail with each chain folding separately to form a head, and inconsistent with the alternative one- and three-chain structures that had been proposed. The two-headed structure for myosin has been confirmed in the more recent studies of Elliott et al. (1976) and Elliott and Offer (1978) using shadowing methods (Fig. 7) and by Takahashi (1978) using negative staining. The best estimate of the overall structure of the myosin molecule suggests two rather elongated, pear-shaped heads (19 nm long and 4·5 nm wide at their widest) attached to a flexible tail that is 156 nm long and 2 nm in diameter (Elliott and Offer, 1978; Figs 7 and 8a).

The two myosin heads are very flexibly attached to the tail (Slayter and Lowey, 1967). Elliott and Offer (1978) have found that they behave independently, and both can adopt a wide range of angles of attachment to the tail (Fig. 7a–d). Using fluorescence depolarization, Mendelson and Cheung (1976) come to similar conclusions and in addition, using saturation transfer electron paramagnetic resonance (EPR), Thomas et al. (1980) found that the heads can still move freely in synthetic thick filaments and in resting muscle. A flexible attachment between the myosin head and tail is an essential feature of current models of cross-bridge activity (Huxley, 1969).

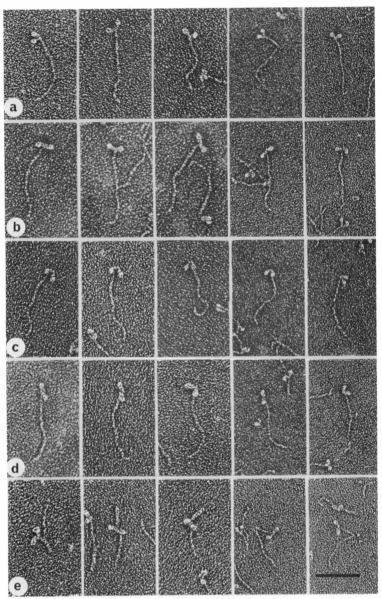

Fig. 7. Rotary shadowed myosin molecules from rabbit skeletal muscle. Rows (a–d) show four different dispositions of heads relative to the tail. Row (e) shows molecules bent back on themselves at a point in the tail ∼43 nm from its junction with the heads. Scale marker indicates 100 nm (reprinted from Elliott and Offer (1978), courtesy of G. Offer and Academic Press Inc.).

A flexible hinge also appears to be present in the myosin tail. Although the tail shows gentle curvature throughout its length, both shadowed and negative-stained molecules often show tails bent back on themselves at a point about 43 nm from the head-tail junction (Elliott and Offer, 1978; Walker, Knight and Trinick, unpublished observation). The flexibility of this hinge is not due to helix-disrupting proline residues, as these are absent from the myosin tail (Lowey *et al.*, 1969). Instead it appears to originate from a local loss of the hydrophobic interactions that stabilize the coiled-coil structure (McLachlan and Karn, personal communication). This hinge may function to enable the heads to move away from the thick filament backbone as proposed in the models of Pepe (1967b) and Huxley (1969) though direct evidence for this is lacking.

The tail may not always exist in a purely α-helical form. Thermal melting studies have shown that, in solution at $3°C$, 98% of the tail is in the α-helical form, but at $38°C$ (physiological temperature for rabbit muscle) this value has dropped to 85%, corresponding to the melting to random coil of a 20 nm section of the tail (Burke *et al.*, 1973). Cyclic melting and recrystallization of a discrete section of the tail has been proposed by Harrington (1971, 1979) as the origin of the force generated by a contracting muscle (see also Section VI.A.4).

Biochemical studies of myosin have established that it is a hexamer comprising two "heavy" chains of about 200 000 daltons each and four "light" chains of about 20 000 daltons each (Weeds and Lowey, 1971). The two heavy chains wind around each other in a coiled-coil of α-helices to form the tail of the molecule (Lowey and Cohen, 1962) and then separately fold to produce the two heads. The light chains are of two chemically distinct classes, and there is one of each class associated with each head.

Myosin can be enzymatically digested to produce well-defined fragments, which can be isolated and studied individually. This approach has greatly simplified the task of understanding this large and complex molecule (see Lowey, 1971). Two main sites are sensitive to proteolysis: digestion may occur 2/5 of the way along the tail from the heads, producing light meromyosin and heavy meromyosin (LMM and HMM) or it may occur at the head-tail junction producing two heavy meromyosin subfragment-1 (S-1) molecules and myosin rod (Figs 8a and 9). HMM can be further digested to produce HMM subfragment-2 (S-2) and S-1 (Figs 8a and 9). The site of enzymatic attack depends on the specificity of the enzyme, the presence or absence of divalent cations, and the state of aggregation of the myosin. The three most frequently used proteases are trypsin, α-chymotrypsin and papain. Trypsin is used principally to produce HMM and LMM (Mihalyi and Szent-Györgyi, 1953). α-chymotrypsin will digest the molecule at both sites unless divalent cations are present, in which case the head-tail junction is protected (Weeds and Pope, 1977). Papain also digests at both sites unless the myosin is

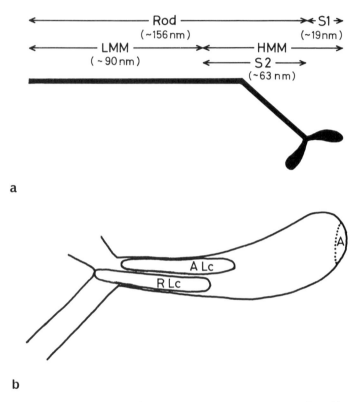

Fig. 8. (a) Schematic representation of myosin molecule, showing the location of fragments produced by brief enzymatic digestion. The bend in the tail is shown to indicate the position of the hinge observed in shadowed myosin molecules. (b) Schematic representation of myosin head showing approximate position of actin binding site (A) and likely positions along the length of the head of alkali light chain (ALc) and regulatory light chain (RLc).

polymerized, in which case the HMM-LMM junction is protected (Lowey *et al.*, 1969). Prolonged digestion by any of these enzymes degrades the initial products to smaller species. In the case of LMM and S-2 this digestion chiefly removes material that is to either side of the site of initial cleavage of the tail converting "long" LMM and S-2 to "short" species that have different aggregation properties (Sutoh *et al.*, 1978; Lu, 1980; Yagi and Offer, 1981).

Of the fragments produced by enzymatic digestion, only LMM and rod aggregate stongly in physiological salt conditions and only HMM and S-1 possess ATPase and actin-binding properties. S-2 neither hydrolyses ATP nor binds to actin, and forms aggregates only near its isoelectric point. It was these observations, as well as those on the morphology of thick filaments, that led to the current model of thick filament structure, where LMM forms the

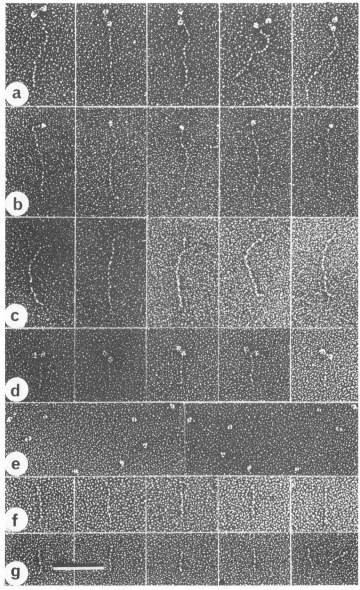

Fig. 9. Vertebrate muscle myosin molecules and their proteolytic fragments observed by rotary shadowing. (a) Intact myosin; (b) single headed myosin (one head removed by enzymatic digestion); (c) myosin rod; (d) HMM; (e) S-1; (f) LMM; (g) S-2. Scale marker indicates 100 nm (reprinted from Lowey et al. (1969), courtesy of S. Lowey and Academic Press Inc.).

backbone and the myosin heads are at the surface where they can interact with actin. The S-2 portion of the tail is the connecting link between the heads and the LMM. It may be only weakly associated with the underlying LMM, so that in the myofibrillar lattice it may tilt away from the surface of the filament to allow the heads to attach to actin over a range of interfilament distances (Huxley, 1969). Crosslinking experiments have indicated, however, that, at least in myofibrils in rigor (see Section VI.A.1) at rest length, such tilting is not required (Sutoh and Harrington, 1977).

B. Structure and Function of the Myosin Head

The structure of the myosin head is of the greatest interest, since the energy-transducing properties of muscle are centred here. Much of our understanding of the myosin head has come from study of myosin subfragment-1. S-1 containing a full complement of light chains has a molecular weight (130 000 dalton; Margossian *et al.*, 1981) close to that expected for an intact myosin head (based on the molecular weights of myosin and rod). It possesses one high affinity nucleotide binding site and one actin binding site, and its ATPase activity is activated by actin (Margossian and Lowey, 1973; Eisenberg *et al.*, 1968). HMM and myosin have two nucleotide binding sites and two actin binding sites. The actin-activated ATPase rate per head of HMM is similar to that of S-1 (Moos, 1972; Margossian and Lowey, 1973). Thus the total ATPase and actin binding capacities of the myosin molecule reside equally in the two myosin heads and the heads appear to act independently of one another.

The light chains together constitute about one third of the mass of the myosin head, but their functions in vertebrate skeletal myosin are unknown. S-1 heavy chain free of all light chains can combine reversibly with actin and retains most of its actin-activated ATPase activity (Wagner and Giniger, 1981; Sivaramakrishnan and Burke, 1982). Therefore, neither chain is essential for these activities. A convenient nomenclature to distinguish the two classes of light chain is derived from the history of their isolation: the DTNB light chain is released by treatment of myosin with dithiobisnitrobenzoate (DTNB) (Gazith *et al.*, 1970) and the alkali light chain (which is not released by DTNB) is released at alkaline pH (Kominz *et al.*, 1959).

On SDS gels of vertebrate fast skeletal myosin the alkali light chains appear as two bands. The two species are closely related. The smaller A2 light chain (mol. wt 17 000 dalton) has a sequence almost identical to the C-terminal part of A1 light chain (mol. wt 21 000 dalton) and differs principally in the lack of a 41-residue N-terminal peptide (Frank and Weeds, 1974). Homodimers of myosin (containing two A1 or two A2 light chains) and heterodimers (containing one of each light chain) all occur *in vivo* (Holt and Lowey, 1977;

Lowey et al., 1979; Hoh and Yeoh, 1979), but there are no obvious functional differences between A1 and A2 containing heads at physiological ionic strength (Wagner and Weeds, 1977; Silberstein and Lowey, 1981). Fluorescent antibody labelling shows that the two light chain species occur in all myofibrils, and electron micrographs of antibody-labelled filaments suggest an even distribution along all filaments (Silberstein and Lowey, 1981). This is to be expected in view of the recent finding (Sivaramakrishnan and Burke, 1981) that the light chains exchange between different S-1 molecules at the physiological temperature (38°C) of a rabbit.

The alkali light chains of slow vertebrate skeletal muscle are chemically distinct from those of fast muscle (Lowey and Risby, 1971) and appear to be partly responsible for the differences observed between the actomyosin ATPase activities of these two types of muscle (Wagner and Weeds, 1977; Marston and Taylor, 1980).

Attempts to crystallize S-1 have not yet succeeded, and we therefore lack the detailed molecular structure of the myosin head that could come from X-ray crystallography. Physico-chemical studies have indicated that S-1 is not spherical (Lowey et al., 1969; Mendelson et al., 1973; Kretzschmar et al., 1978), and this is borne out by the shape of the heads of myosin molecules as seen in the electron microscope. Elliott and Offer (1978) found, by shadowing, an elongated, pear-shaped structure about 19 nm long and 4·5 nm at its widest near the tip (Figs 7 and 8). This degree of asymmetry has been confirmed by both shadowing (Trinick and Elliott, 1979; Flicker et al., 1983) and negative staining (Takahashi, 1978; Craig et al., 1980; Trinick, 1981) and in addition it has been shown that the long axis of the head is often curved (Flicker et al., 1983; Fig. 10a; Knight and Trinick, unpublished observation; Fig. 6d). Reconstructions from micrographs of S-1-decorated actin (see Section VI.A.1) first suggested an elongated and curved structure, and further indicated that S-1 is somewhat flattened (Moore et al., 1970; Toyoshima and Wakabayashi, 1979; Taylor and Amos, 1981), as does analysis of low angle X-ray scattering from S-1 in solution (Mendelson and Kretzschmar, 1980). The length of S-1 in most reconstructions is less than 19 nm however (Moore et al.,1970; Taylor and Amos, 1981; Toyoshima and Wakabayashi, 1979). This is due to loss of part of the light chain complement (when the light chain is present the length is close to 19 nm (Vibert and Craig, 1982; see VI.A.6), and to disorder in the end of S-1 furthest from actin (i.e. the part that adjoins the tail of myosin) which is therefore lost in the helical averaging process of reconstruction (Vibert and Craig, 1982).

To describe in detail the functions of the myosin head it is necessary to know the locations of the actin binding and ATPase sites and the positions of the light chains. The actin binding site for the rigor state has been defined fairly well by reconstructions from micrographs of decorated actin. It appears to be slightly to one side of the tip of the head (Fig. 8b; see Taylor and Amos, 1981 and Section VI.A.1).

The location of the ATPase site within the myosin head is unknown. Since the ATPase is activated by actin it has long been suggested that the site is close to the actin binding site but direct evidence is lacking. Observation by scanning transmission electron microscopy of heavy metal atoms specifically bound in the active site would solve this problem.

The DTNB light chain is thought to be located on the myosin head near its junction with the tail. This is suggested by digestion experiments which have shown that when the light chain has a divalent cation bound to it, it can protect the head-tail junction from proteolysis by α-chymotrypsin (Yagi and Otani, 1974; Weeds and Taylor, 1975). Light chains lacking the metal ion are rapidly digested, following which the head-tail junction can be severed (Bagshaw, 1977). This location is supported by direct observation of the site of antibody labelling of the homologous "regulatory" light chain in scallop (see Section III.C) and by observation of actin decorated by S-1 with and without the light chain present (see Section VI.A.6).

The location of the alkali light chain is less clear. Fluorescence energy transfer measurements indicate that the C-terminal end of the light chain is within 10 nm of the tip of the head, i.e. is in the distal half (Marsh and Lowey, 1980; Takashi, 1979). Cross-linking experiments, on the other hand, show that at least part of the homologous essential light chain of scallop lies adjacent to the regulatory light chain, i.e. in the base of the head (Wallimann et al., 1982). This conclusion is supported by electron microscopy of antibody-labelled scallop myosin, which shows that some antigenic sites on the light chain are located near the head-tail junction (Fig. 10d; Flicker et al., 1983), although the antibody technique is limited in resolution by the bulk of the antibody molecule itself, the Fab fragment being 6 nm long. More precise information might be obtainable by the use of specific heavy atom labelling of the light chains in conjunction with high resolution scanning transmission electron microscopy. On balance, however, it appears that the region of the myosin head near its junction with the tail comprises parts of three polypeptide chains: the heavy chain and the two light chains (Flicker et al., 1983).

It might be anticipated that the energy transducing mechanism in muscle would entail structural changes within the myosin head. Nucleotide binding can enhance the intrinsic fluorescence of the head (see Taylor, 1979), and increase the rate at which bifunctional reagents cross-link two cysteine residues in the head (Reisler et al., 1974; Wells et al., 1980). However, large scale changes have not been detected in the sedimentation coefficient (Gratzer and Lowey, 1969), α-helical content (Chantler and Szent-Györgyi, 1978) or proton nuclear magnetic resonance (proton nmr) spectrum (Highsmith et al., 1979), so the change induced by nucleotide binding may be subtle. Actin binding (in the absence of ATP) can enhance the intrinsic fluorescence of the head (see Taylor, 1979) and suppresses the mobility of amino-acid residues within the head that is observed by proton nmr (Highsmith et al., 1979). A gross change in the shape of the head is not induced by actin, since the shape of the head seen in decorated actin (Section VI.A) and in myosin alone is similar. Highsmith et al. (1979) found that 22% of the amino acid residues in the head were highly mobile, and Wells et al. (1980) have found that two cysteine residues in the head move relative to each other by at least 1·6 nm. Such fluidity in conformation could permit the head to change shape during the ATPase cycle or become elastically distorted during force generation.

C. Diversity of Myosin Structure

The structure of the myosin molecule is remarkably constant in a wide variety of species and tissues. The two headed appearance of the molecule and the length of the tail are the same in myosins from adult and embryonic vertebrate skeletal muscle, vertebrate cardiac muscle, vertebrate smooth muscle, scallop striated muscle, vertebrate brain and blood platelet (Elliott et al., 1976). There is also a very high degree of homology between the amino-acid sequences of myosin tails from such diverse sources as insect, nematode and mammal, with several regions of total identity (McLachlan and Karn, 1982). The one exception seems to be the myosin-like proteins isolated from *Acanthamoeba*. One species—myosin II—is similar to other myosins except that the tail is only 90 nm long (Pollard et al., 1978). Others—myosin IA and IB—appear to be quite different, possessing only a single heavy chain and two light chains per molecule (see Gadasi et al., 1979).

Myosin molecules from diverse sources are also functionally very similar, being capable of forming bipolar filaments, of hydrolysing ATP and of binding to actin. There are some important variations on this scheme, however, reflecting some chemical diversity between different myosins. Smooth and non-muscle myosins have different solubility properties from striated muscle myosin, and possibly assemble into filaments only when the tissue is activated (see Section V.A). Myosins from fast and slow muscles have high and low actin-activated ATPase activities respectively, and this is apparently related to differences in their heavy chain and alkali light chain sequences (Sreter et al., 1973, 1975; Weeds et al., 1974; Weeds and Burridge, 1975; see also Whalen et al., 1981, and references therein).

Most importantly, the DTNB light chain, which in vertebrate striated myosin has no clear function, is in many other myosins replaced by a light chain of very similar structure (Kendrick-Jones and Jakes, 1977) that is involved in regulating the interaction of myosin with actin in response to Ca^{2+} (in vertebrate striated muscle the regulatory switch is located on the thin filaments; see Section VI.A.5). In the case of scallop striated muscle, and a variety of other invertebrate striated muscles, binding of Ca^{2+} directly to the myosin head requires this light chain and produces structural changes that allow the heads to interact with actin (Szent-Györgyi *et al.*, 1973; Chantler and Szent-Györgyi, 1980). With vertebrate smooth muscle and with non-muscle cells, phosphorylation of the homologous myosin light chain is a trigger that switches on actin-myosin interaction. This phosphorylation is in turn triggered by Ca^{2+}, via a calmodulin-dependent specific myosin light chain kinase (see review by Adelstein and Eisenberg, 1980). Regulatory function appears to have been lost from the DTNB subunit in vertebrate striated myosin. Nevertheless, this total class of light chains is often referred to as regulatory light chains, whether or not they actually regulate their own myosin *in vivo*.

There has been considerable effort to understand the structural role of this light chain in regulation, especially in the scallop, since here it can be reversibly removed simply by treating the myosin with 10 mM EDTA. Shadowed, intact myosin molecules show a normal morphology, with long, pear-shaped heads (Fig. 10a). When the regulatory light chains are removed the heads appear to be round and the narrow necks have become invisible (Flicker *et al.*, 1983; Fig. 10b). In the case of S-1, a comma-shaped structure becomes spherical on removal of light chain. These observations, as well as those on actin decorated with S-1 with and without light chain (Section VI.A.5) suggest that the regulatory light chain is located in the neck of the myosin head, probably extending along its length from near the head-tail junction to a point about half-way up to the actin binding site (Fig. 8b; see also Scholey *et al.*, 1981). This is supported by the finding that antibodies to the regulatory light chain bind to the neck region of the myosin head, often near to its junction with the tail (Flicker *et al.*, 1983; Fig. 10c). Removal of the light chain from this region could cause the neck to become too thin to see, explaining why the head now appears round instead of elongated. The similarity of the scallop regulatory light chain to the regulatory light chains in other myosins (including the DTNB light chain in vertebrate skeletal myosin) suggests a similar location in all cases (see Scholey *et al.*, 1981); this is also implied by the appearance of actin decorated with vertebrate skeletal S-1 with and without the DTNB light chain (see Section VI.A.6).

There has been much effort to try to detect structural changes in scallop myosin heads in response to Ca^{2+} binding. So far neither electron microscopy (Craig *et al.*, 1980; Flicker *et al.*, 1983) nor spectroscopic measurements (Chantler and Szent-Györgyi, 1978) have revealed any effects, and it seems that the regulatory mechanism must be a rather subtle one. A response is observed, however, in the case of vertebrate striated muscle. When thick filaments bind Ca^{2+} their sedimentation coefficient increases by 3% and the

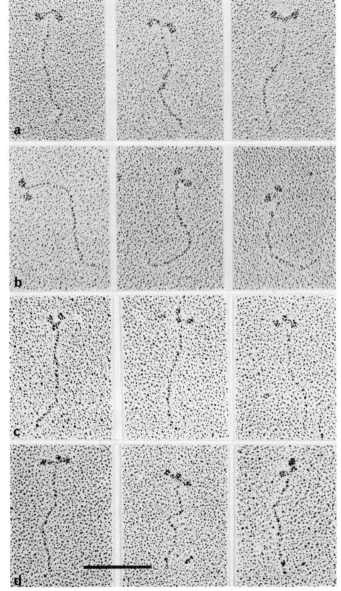

Fig. 10. Rotary shadowed scallop myosin molecules. (a) Intact myosin; (b) myosin with regulatory light chains removed; (c) myosin labelled with antibody to the regulatory light chain; (d) myosin labelled with antibody to the alkali light chain. In (c) and (d) the antibody used is an Fab fragment and appears as the smaller lump generally occurring between the two myosin heads. This figure is printed with the contrast reversed compared to Figs 7 and 9. Scale marker indicates 100 nm (courtesy of P. Flicker).

viscosity of the solution falls (Morimoto and Harrington, 1974a), suggesting a structural change, presumably in the heads, which contain the Ca^{2+} sites. The significance of this result is unclear, because X-ray diffraction has failed to detect any change in the ordering of the myosin heads on stimulation of muscle stretched to no overlap of thick and thin filaments (Yagi and Matsubara, 1980; Huxley et al., 1980).

IV. STRUCTURE OF VERTEBRATE STRIATED MUSCLE THICK FILAMENTS

Since Huxley (1963) deduced the basic organization of myosin molecules in the thick filament, progress towards a more detailed description has been slow. This is due to the labile nature of the array of heads on the filament surface, which has made it extremely difficult to preserve their 3-dimensional organization for examination in the electron microscope. It is also due to the difficulty in resolving the overlapping arrangement of tails, only 2 nm in diameter, in the filament backbone.

A variety of approaches has been used to study the filament in its native state, and these are discussed in the following section (IV.A). Synthetic assemblies of purified myosin or myosin fragments have also been studied and these are discussed in Section IV.B. Models of thick filament structure are discussed in the final section (IV.C).

A. Native Thick Filaments

1. Organization of Myosin Molecules

The technique of X-ray diffraction can provide detailed information on the arrangement of macromolecules in living tissue. X-ray patterns from resting vertebrate striated muscle (fast and slow skeletal, and cardiac) show a series of layer lines that are orders of 43 nm; the third order at 14·3 nm has high intensity on the meridian (Fig. 11a; Huxley and Brown, 1967; Matsubara, 1974; Matsubara and Millman, 1974). These layer lines are thought to arise mainly from the myosin cross-bridges and are consistent with a helical arrangement, having an axial repeat of 43 nm and a subunit repeat of 14·3 nm (Huxley and Brown, 1967; Fig. 11b). The distribution of intensity along the layer lines is consistent with cross-bridges whose centre of mass is at a radius (r) between 9 and 17 nm from the filament axis, i.e. further out than the surface of the backbone (Huxley and Brown, 1967; Squire, 1975; Haselgrove, 1980). The rapid fade-out of intensity at relatively low resolution implies considerable disorder of the cross-bridges about their mean positions (Huxley and

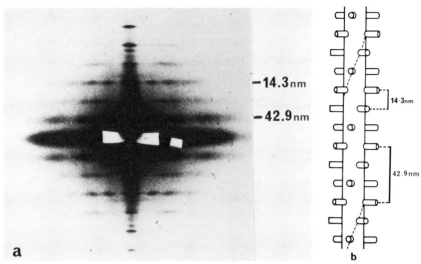

Fig. 11. (a) X-ray diffraction pattern of live, resting frog sartorius muscle. Muscle axis vertical. This pattern shows the series of layer lines at orders of 42·9 nm, but is too long an exposure to show clearly the meridional intensities on the first three layer lines. The periodicities in the structure giving rise to two of the layer lines are indicated. (b) Diagram showing the positions of cross-bridges deduced from X-ray patterns such as (a). The X-ray pattern does not indicate how many cross-bridges are at each 14·3 nm period. The diagram shows three, as indicated by other data (see text), and not two as originally proposed by Huxley and Brown (1967). The dashed line indicates the path of one of the helices on which cross-bridges lie. Each cylinder represents two myosin heads ((a) reprinted from Huxley and Brown (1967), courtesy of H. E. Huxley and Academic Press Inc.; (b) reprinted from Offer (1974), by courtesy of G. Offer and Longman Group Ltd).

Brown, 1967), as is also suggested by EPR measurements of relaxed muscle (Thomas and Cooke, 1980). Little of the diffracted intensity is thought to come from the myosin tails, so little information is obtained on the structure of the filament backbone (but see Section V.B).

The helical arrangement of cross-bridges proposed by Huxley and Brown (1967) should give rise to meridional reflections on only every third layer line, at orders of 14·3 nm. Huxley and Brown also observed meridional intensity on other layer lines, however, and therefore concluded that the helical symmetry was only approximate. Bennett (1977) has shown that a systematic displacement of the cross-bridges to either side of an exact 14·3 nm repeat could account for much of this "forbidden" meridional intensity (see Section IV.B).

Since the X-ray patterns contain no phase information, they do not reveal the hand of the helix. Because of lattice sampling and other problems, they also give no unambiguous information on the number of cross-bridges n (= the rotational symmetry) present at each 14·3 nm level. They would be consistent with $n=2$, $r=9$ nm; $n=3$, $r=13$ nm; or $n=4$, $r=17$ nm (Squire, 1975; Haselgrove, 1980). The arrangement for $n=3$ is shown in Fig. 11b. The disposition of the cross-bridges is also quite ambiguous and can only be deduced in a very general way. They do not appear to project directly out from the backbone (Fig. 11b is schematic and meant to show only the symmetry), but instead are probably wrapped around it, with the two heads pointing on average in opposite directions, up and down the filament axis (Haselgrove, 1980).

The rotational symmetry (n) of the cross-bridge arrangement is an important parameter as it will affect the potential interactions that the thick filaments can make with the thin filaments in the lattice of the myofibril. Because n has not been determined by electron microscopy or X-ray diffraction, there have been several attempts to estimate it by less direct methods. Almost all of these methods involve estimates of the total amount of myosin in a thick filament, which is then divided equally between the total number of 14·3 nm levels of cross-bridges along the filament to obtain n. Estimates of the number of nucleotide binding sites in myofibrils lead, on the assumption of one nucleotide per myosin head, to $n=3·5$ (Maruyama and Weber, 1972; Marston and Tregear, 1972). The myosin content of muscle coupled with the number of thick filaments expected per unit volume (Huxley, 1960) leads to 432 molecules per filament, i.e. $n=4·4$. Estimation of the number of thick filaments per unit volume of a myosin solution of known concentration, using the classical particle counting method of Hall, gives $n=4·3$ (Morimoto and Harrington, 1974b).

Our detailed knowledge of the thin filament structure, of the thick:thin filament ratio and of the molecular weights of actin and myosin permits an estimate of myosin content of the thick filament from measurement of the ratio of masses of myosin to actin. From biochemical analysis and interference microscopy of intact myosin extracted fibres, Huxley and Hanson (1957) and Hanson and Huxley (1957) obtained a myosin:actin ratio which would give $n=3·2$ (see Tregear and Squire, 1973). Using quantitative staining of sodium dodecyl sulphate-polyacrylamide gels of muscle, two papers report actin: myosin ratios giving $n=2·7$ (Tregear and Squire, 1973) and $n=2·5$ (Potter, 1974) and two others giving $n=3·9$ (Morimoto and Harrington, 1974b) and $n=3·8$ (Pepe and Drucker, 1979).

The lack of agreement between these indirect techniques, and between different workers using the same technique, suggests that more direct approaches are needed. Determination of mass per unit length of the filament by quantitative electron scattering, using the scanning transmission electron microscope (STEM) has yielded $n=2·7$ (Lamvik, 1978). Although careful account was taken of mass loss due to radiation damage, some uncertainty may exist in this result because the filaments used were rather fragmented.

Nevertheless, using the same STEM method and a better preparation of filaments, Reedy et al. (1981) have obtained a similar result ($n = 2·9$).

At present, then, we are left with the possibility that n is either 3 or 4. We favour the value of 3 since it is supported by the most direct technique so far used (STEM) and by observations of 3-fold symmetry in the thick filament backbone (see below). The problem will not be solved convincingly, however, until direct visualization of the cross-bridges has been achieved. New methods of filament preparation have recently lead to the preservation of ordered arrays of myosin heads on the surfaces of some invertebrate thick filaments (see Section V.B), and similar techniques applied to vertebrate filaments may reveal n and other parameters directly.*

Although the three-dimensional order of the cross-bridges has not yet been well preserved in vertebrate filaments, some details of the axial arrangement have been clearly observed. Negatively stained A-segments (thick filament arrays held together at the M-line) reveal much more detail than single filaments (Hanson et al., 1971). When stained with ammonium molybdate, axially ordered cross-bridges are sometimes seen (Craig, 1977). The bare zone can then be unambiguously identified and is found to be 149 nm long. In addition, small perturbations in the 14·3 nm periodicity are visible near the centre and ends of the A-band, implying changes in myosin packing in these regions. Similar but clearer detail is obtained with negatively-stained longitudinal cryosections of muscle (Fig. 12a; Sjöström and Squire, 1977a). A 14·3 nm periodicity can be seen extending from the bare zone to the ends of the filament (except for a gap three repeats from the end) and there are slight perturbations in the periodicity near the bare zone and the ends. A-bands labelled with antibodies to S-1 show that cross-bridges occur along the whole length of the thick filament apart from breaks at the centre and near the ends of the array (Craig and Offer, 1976a). The simplest interpretation of these results is that each of the stain-excluding lines on (or near) the 14·3 nm periodicity in cryosections and A-segments is due to a row of cross-bridges, and that there is a missing row of cross-bridges two periods from the end. On this basis one can count that there are 49 rows of bridges in each half filament, and the filament length, assuming that it ends at the last row of bridges, is 1·57 μm (Fig. 15a).

When stained with uranyl acetate, A-segments reveal considerable structure in the thick filament backbone (Hanson et al., 1971; Craig, 1977; Fig. 12b). Extending from each side of the bare zone is a series of eleven polar bands with a repeat of 43 nm—the axial thick filament repeat seen in X-ray

---

* Kensler and Stewart (1983) have recently reported the successful preservation of the array of myosin heads in frog filaments. They find that $n = 3$ and that the heads (which extend out to a radius of about 14·5 nm) lie on three right-handed helices.

patterns of muscle. Each band starts with a prominent stain-excluding stripe and is followed distally by a complex subsidiary banding before the strong stripe of the next band. The first three bands differ from each other and from the next seven. Comparison of this staining pattern with the labelling pattern obtained with antibodies to some non-myosin thick filament components (Section IV.A.2) shows that the prominent stripes are due to these non-myosin components (Craig, 1977). In contrast, the similarity of optical diffraction patterns of A-segments to diffraction patterns from LMM paracrystals suggests that the subsidiary banding comes largely from the LMM backbone of the filament (Hanson et al., 1971; O'Brien et al., 1971). The differences in banding in the first three bands, corresponding to the region of cross-bridge perturbation mentioned earlier, suggest that the myosin packing is continually changing in this region, as the molecules change from anti-parallel to purely parallel interactions (Hanson et al., 1971; Craig and Offer, 1976b; Craig, 1977). The four non-myosin proteins in this region are probably also different from each other. The next eight bands are similar in appearance and the outer seven of them contain the same non-myosin component, C-protein (Section IV.A.2.). These represent a region of approximately constant myosin packing. The cross-bridge pertubations near the ends of the filament again indicate a region of changing myosin interactions as the filament tapers towards its tip. The appearance of a 43 nm repeat in the backbone when viewed in projection (as in A-segments) is inconsistent with a strictly helical arrangement of myosin molecules (in projection one would expect a 14·3 nm repeat—see Fig. 11b). As discussed earlier, some perturbation of the myosin helix must be present, even in the region of "constant" packing (Craig and Offer, 1976b; Craig, 1977).

The cryosections of muscle also reveal a longer axial repeat than the 14·3 nm periodicity of the cross-bridges (Fig. 12a). At the position of every third level of cross-bridges is a stripe that excludes stain more strongly than the others and, as in the case of A-segments, this has been attributed to non-myosin components superimposed on the cross-bridges at these levels (Sjöström and Squire, 1977a).

There have been many indications of backbone substructure when thick filaments are observed in cross-section. Pepe and his associates (Pepe and Drucker, 1972; Pepe and Dowben, 1977; Pepe et al., 1981) observe about 12

Fig. 12. (Left) Longitudinal cryosection of human skeletal muscle, negatively stained. (Right) Negatively-stained A-segment from frog skeletal muscle. The numbering scheme refers to the prominent, stain-excluding stripes. (Left) and (Right) Are the same magnification: (Right) appears shorter owing to reduced uptake of stain near ends of segment. Scale marker indicates 0·3 µm ((Left) reprinted from Sjöström and Squire (1977a), courtesy of J. Squire and Academic Press Inc.; (Right) reprinted from Craig (1977), courtesy of Academic Press Inc.).

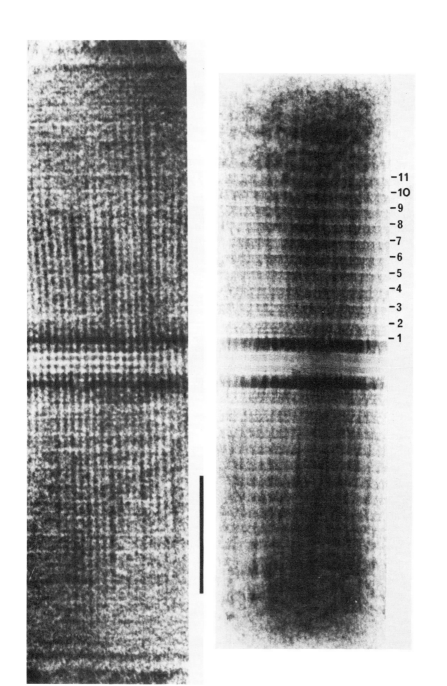

discrete units hexagonally packed with a centre-to-centre spacing of approximately 4 nm, and the change in appearance observed on tilting the specimen suggests that the units represent subfilaments running parallel to the filament axis. Gilëv (1966a, b) has also made observations of subunit structure, in invertebrate thick filaments. However, the variability of the images and the uncertainty in interpreting sectioned material at this resolution suggest caution in accepting the fine details of such images at face value. Nevertheless, the presence of structural components of about this diameter is also suggested by a near-equatorial reflection at a spacing of ~ 3.5 nm in optical diffraction patterns of A-segments (O'Brien et al., 1971) and by the observation of substructure in enzymatically treated filaments (Tsao et al., 1966). In addition, X-ray patterns (of invertebrate muscle) show weak backbone reflections at about this spacing (Wray, 1979a; see Section V.B), and similar reflections have been reported from vertebrate muscle (Millman, 1979). Since myosin tails have a diameter of approximately 2 nm, indications of a 4 nm spacing in the thick filament backbone would suggest discrete groupings of tails forming larger units (see Section IV.C).

An even larger grouping of tails (perhaps groups of the 4 nm units) is suggested by two other observations. Recently, Maw and Rowe (1980) have found that thick filaments treated with distilled water and then negatively stained, reversibly fray into three similar subfilaments, about 6 nm in diameter, on either side of the bare zone (Fig. 13a). This is an unambiguous demonstration of substructure in the organization of the backbone of the filaments. The micrographs suggest that the subfilaments are packed in the intact thick filament in a simple linear array, rather than winding around the filament axis. In thin cross-sections of fish and frog muscles, near the tip of the filament and near the M-line, Luther and Squire (1980) and Luther et al. (1981) (Fig. 13b and c) have also observed a division of the filament into three subunits, of similar size to those seen by Maw and Rowe (1980). Both of these observations would strongly suggest that the rotational symmetry of the filament, n, is 3.

2. Non-myosin Components of the Thick Filament

Indications of non-myosin proteins in vertebrate striated muscle thick filaments first came from X-ray diffraction and electron microscopy. X-ray patterns showed, in addition to the 43 nm myosin layer line system, a strong meridional reflection at 44·2 nm (Huxley and Brown, 1967). This could be correlated with a series of stripes, seen at the time (Huxley, 1967), in the A-band of muscle (Fig. 1a), and it was suggested that they represented a non-myosin thick filament component, since the periodicity appeared to be different from that of the myosin.

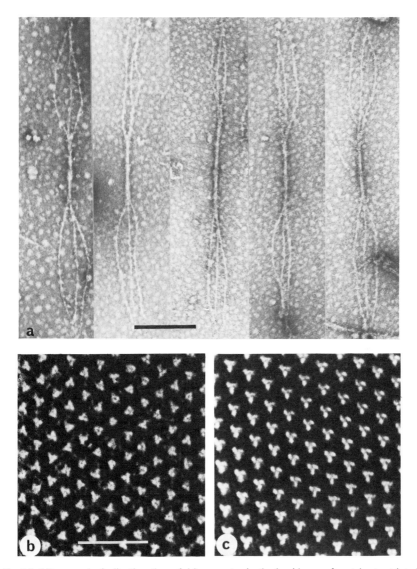

Fig. 13. Micrographs indicating three-fold symmetry in the backbone of vertebrate striated muscle thick filaments. (a) Filaments frayed into 3 subfilaments by rinsing with distilled water before negative staining. Scale marker indicates 0·3 μm. (b) and (c) Transverse sections of (b) frog and (c) fish muscle in the bare zone, showing 3 subunits in the filament backbone. In (b) the filaments lie on a superlattice and have varying orientations. In (c) the orientations are all the same and the image has been enhanced by optical filtering. (b) and (c) Have both been printed in reverse contrast so that the filaments appear white. Scale marker in (b) and (c) indicates 200 nm ((a) reprinted from Maw and Rowe (1980), courtesy of A. Rowe and Macmillan Journals, Ltd; (b) from Luther and Squire (1980), and (c) from Luther *et al.* (1981), courtesy of P. Luther and Academic Press Inc.).

It was later shown by SDS gel electrophoresis that several polypeptides were present as persistent impurities in myosin preparations, and it was suggested that these might be components of the thick filaments. They were named alphabetically according to their mobilities on the gels (Starr and Offer, 1971). C-protein, the most abundant, has been purified and characterized (Offer *et al.*, 1973). It is a monomer with a molecular weight of 140 000; it is elongated but contains no α-helix. It has been shown to bind tightly to myosin at physiological ionic strength, both to the LMM and S-2 regions, but not to the myosin heads (Moos *et al.*, 1975; Starr and Offer, 1978; Safer and Pepe, 1980). It binds to LMM paracrystals (Section IV.B.2) with the same repeat as the LMM and if added in excess disrupts the regularity of myosin filaments formed *in vitro*.

The location of C-protein has been determined by antibody labelling. Immunofluorescence of myofibrils labelled with antibodies to C-protein shows that it occurs only in the middle one-third of each half of the A-band. Electron microscopy of antibody-labelled muscle supports this and shows that it occurs in seven discrete stripes, about 43 nm apart, in each of these regions (Offer, 1972; Pepe and Drucker, 1975; Craig and Offer, 1976b; Fig. 14). When papain fragments of the antibody are used for labelling, the antibody stripes are very narrow, suggesting that C-protein has a very limited axial extent (8–10 nm). The positions and widths of these stripes are the same as those of stripes 5–11 in A-segments, suggesting that these latter stripes are due to C-protein (Craig, 1977; Fig. 15). Estimates of the quantity of C-protein in myofibrils or thick filaments predict that two to four C-protein molecules are bound at each of these axial positions on the filament (Offer *et al.*, 1973; Morimoto and Harrington, 1974b; Craig and Offer, 1976b). The components responsible for stripes 1–4 are currently being investigated (Starr, Bennett, Offer and Craig, unpublished data), and it has recently been shown that H-protein (Starr and Offer, 1971) lies on stripe 3 (see Craig and Megerman, 1979; Fig. 15e).

The function of these non-myosin components is unknown. C-protein in cardiac muscle appears to be partially phosphorylated and this phosphorylation is markedly enhanced on adrenaline stimulation, suggesting a possible regulatory function (Jeacocke and England, 1980). Huxley and Brown (1967) suggested that the component giving rise to the A-band stripes might be responsible for determining the very constant (~ 1.6 μm) length of the thick filament. Since the 44·2 nm meridional reflection suggested that the stripes had a slightly different period from myosin, it was suggested that the two might co-polymerize, forming a vernier system that was particularly stable at a certain length. However, based on X-ray diffraction of muscle labelled with antibodies to C-protein, Rome *et al.* (1973) have suggested that the 44·2 nm reflection might arise by splitting of a 43 nm reflection, owing to interference

Fig. 14. Longitudinal section of rabbit psoas muscle labelled with antibodies to C-protein. Scale marker indicates 0·7 μm (reprinted from Craig and Offer (1976b), courtesy of The Royal Society).

between diffracted rays from the two halves of the A-band. This interpretation is supported by optical diffraction patterns of whole and half A-bands (Craig and Megerman, 1979), which also show that the fine splitting of other meridional reflections in the high resolution X-ray pattern from muscle (Huxley and Brown, 1967; Haselgrove, 1975) may arise from similar interference effects (Rome, 1972; Craig, 1975).

The interference concept shows that, in principle, the C-protein repeat could be determined by the 43 nm repeat of the underlying myosin but still give rise to a 44·2 nm reflection. Whether or not the myosin and C-protein repeats are *actually* both 43 nm, however, is not yet decided. Optical diffraction patterns of half A-segments and of half A-bands labelled with antibodies to C-protein, where both myosin and C-protein repeats are seen but interference effects are avoided, suggest that the repeats are the same (Craig, 1977). Similar patterns from cryosections led Squire *et al.* (1976) to conclude that C-protein has a 44 nm repeat and more recently, Squire *et al.*

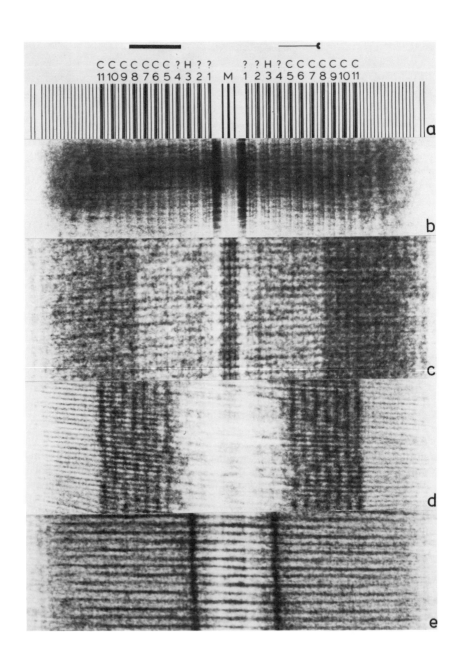

(1982) have refined this value to 43·4 nm. If the C-protein and myosin repeats are indeed unrelated, then C-protein would have to extend along the thick filament from one axial site to the next in order to generate its own periodicity. The narrow stripes seen in A-segments and A-bands labelled with antibodies to C-protein show no evidence of this. Moreover, C-protein molecules, seen by shadowing, are less than 43 nm long (Offer and Elliott, personal communication).

The non-myosin proteins could also serve a purely structural role in stabilizing the organization of the myosin molecules, for example by wrapping around the backbone of the filament (Offer, 1972). In support of such a stabilizing role, Trinick and Cooper (1980), in studies of the sequential disassembly of native thick filaments at elevated salt concentrations, found a highly co-operative disassembly of a section of the filament corresponding approximately in length to the C-protein region. It may be that the transition coincides with the dissociation of C-protein from the backbone.

In addition to non-myosin proteins in the cross-bridge region, almost all vertebrate striated muscles have another specialized structure in the bare zone. This is the M-line, which appears at low resolution in longitudinal sections as a general increase in electron density of the filaments, superimposed on which is a series of dense transverse stripes (Fig. 1a). The number of stripes is generally three, spaced about 20 nm apart, though in some fibres two weaker stripes are seen in the series, one on each side of the main three. This may reflect different fibre types (Sjöström and Squire, 1977a, b). In negatively-stained A-segments and cryosections, additional finer stripes are seen (Fig. 12) which Sjöström and Squire suggest arise from staining of the anti-parallel packed myosin tails.

In transverse section the M-line is seen to consist of a hexagonal array of bridges joining each thick filament to its six neighbours (Knappeis and Carlsen, 1968; Pepe, 1971; Luther and Squire, 1978). Fine "M-filaments" (5 nm in diameter) attached to these bridges run parallel to the thick filaments and are joined to each other by finer bridges (Fig. 16; Knappeis and Carlsen, 1968; Luther and Squire, 1978).

The proteins responsible for each of these structural components have not yet been identified but it has been established by antibody labelling at the light and electron microscope level that one isoenzyme of creatine kinase and a

Fig. 15. Myosin and non-myosin proteins in vertebrate skeletal muscle thick filaments. (a) Schematic diagram of non-myosin proteins superimposed on array of cross-bridges with 14·3 nm periodicity; (b) negatively-stained frog A-segment with non-myosin proteins showing as prominent white stripes; (c) positively-stained section of frog muscle with non-myosin proteins showing as narrow black lines; (d) rabbit muscle with antibodies to C-protein labelling stripes 5–11; (e) rabbit muscle with antibodies to H-protein labelling stripe 3. All micrographs have the same magnification: the A-segment appears shorter owing to poor contrast at the ends. Scale marker indicates 0·2 μm (reprinted from Craig and Megerman (1979), courtesy of Academic Press Inc.).

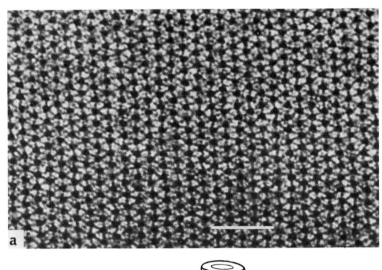

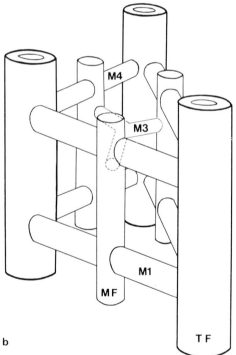

Fig. 16. (a) Transverse section of frog sartorius muscle in the M-line region. Scale marker indicates 200 nm. (b) Perspective drawing of the M-line region. TF are the thick filaments; MF, the M-filaments; M1 and M4 the main M-bridges and M3 the finer M bridges (reprinted from Luther and Squire (1978), by courtesy of P. Luther and Academic Press Inc.).

165 000 dalton protein are present in the M-line (Morimoto and Harrington, 1972; Turner et al., 1973; Wallimann et al., 1977, 1978; Masaki and Takaiti, 1974; Trinick and Lowey, 1977).

The M-line presumably stabilizes the transverse and longitudinal order of the thick filament lattice, though it has been shown (Knappeis and Carlsen, 1968) that high degrees of stretch reversibly disrupt the register of the filaments and the M-line structure. The thick filaments of muscle without an M-line (Page, 1965) are generally in poorer register than those with an M-line. When a muscle shortens, the interfilament distance increases by as much as 7 nm (Elliott et al., 1963). The M-line bridge structure may therefore be capable of stretching, but nothing is yet known about this process.

An additional component of the thick filament, termed the end filament, has recently been described (Trinick, 1981; Fig. 17a and b). When a rabbit thick filament is frayed at low ionic strength, the subfilaments frequently coalesce near the filament tips, and the end filament, one per tip, is seen in this region (Fig. 17a). It is about 85 nm long and 5 nm wide, with a pronounced axial periodicity of 4.2 ± 0.3 nm (Fig. 17b), and frequently it terminates in an amorphous almost globular structure about 10×15 nm (Fig. 17b). The precise position of the end filament in the intact filament, its composition and function are all unknown. A similar structure has also been seen emerging to a variable extent from the tips of intact thick filaments from frog (Craig, unpublished observation; Fig. 17c).

3. Arrangement of Thick Filaments in the Myofibril Lattice

The pattern of sampling along the myosin layer lines in X-ray patterns of resting frog muscle led Huxley and Brown (1967) to suggest that nearest neighbour thick filaments are not identically arranged, but that next nearest neighbours are, i.e. that the thick filaments lie on a superlattice. The alignment of thick filament staining patterns in A-segments and sections indicates that the non-equivalence of neighbours is not due to axial displacement of filaments and therefore must be due to rotation around the filament axis (Craig, 1977).

Luther and Squire (1980) have studied the thick filament arrangement of frog muscle in detail by observing transverse sections of the bare zone, where the triangular profiles of the thick filaments reveal the relative rotations directly. They find that the superlattice is not constant over a whole myofibril but exists in small domains separated by less crystalline packing (Fig. 13b). From their observations they suggest a simple mechanism for the assembly of the lattice of thick filaments, in which a thick filament adding to the lattice does so in accordance with two simple rules which determine its orientation with respect to its nearest neighbours. Their modelling studies show that

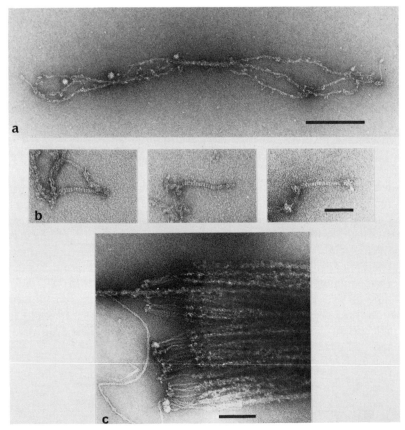

Fig. 17. End filaments of vertebrate skeletal muscle thick filaments seen by negative staining. (a) Frayed rat psoas muscle thick filament showing an end filament at each tip. Scale marker indicates 0·2 μm. (b) Montage of end filaments at higher magnification. Scale marker indicates 50 nm. (c) A-segment from frog sartorius muscle showing end filaments. Scale marker indicates 100 nm ((a) reprinted from Trinick (1981), by courtesy of J. Trinick and Academic Press Inc.; (b) courtesy of J. Trinick; (c) Craig).

independent addition of filaments to adjacent parts of the lattice easily leads to forced violations of the rules where these zones of addition join, with a consequent breakdown of crystalline packing, as is observed in the sections. Not all muscles show a superlattice: some fish have a single orientation of thick filaments (Franzini-Armstrong and Porter, 1964; Pepe, 1971; Luther et al., 1981; Fig. 13c). Whether or not a superlattice is formed is presumably determined by the preferred pattern of interactions between the M-line proteins of adjacent filaments (Luther and Squire, 1980). Because of its simple lattice, fish muscle has the advantage over other muscles that it allows the use of image averaging for determining the detailed structure of the filament backbone (Luther *et al.*, 1981; see, for example, Fig. 13c), and makes simpler the analysis of the pattern of actomyosin interactions in the myofibril lattice.

4. Summary

From the preceding discussion we can make the following main conclusions about vertebrate striated muscle thick filaments:

(i) The myosin molecules are packed with their tails in the backbone and heads at the surface; the packing is anti-parallel near the middle of the filament and parallel for the rest of the length. The myosin tails are grouped into three "subfilaments".

(ii) The first heads occur 75 nm from the centre of the M-line and emerge with an approximately 14·3 nm periodicity all the way to the end of the filament except for one missing level two periods from the tip.

(iii) The arrangement of cross-bridges is probably *approximately* helical, but the presence of a 43 nm backbone repeat, which is emphasized by non-myosin proteins, and of forbidden meridional X-ray reflections means that adjacent 14·3 nm levels of cross-bridges are not equivalent, indicating departure from exact helical symmetry.

(iv) The number of myosin molecules per 14·3 nm cross-bridge repeat is three (or possibly four), and there are therefore 294 (or 392) myosin molecules per filament, assuming 98 levels of cross-bridges in each filament.

(v) Various non-myosin components of unknown function, for instance C-protein, are arranged on the filament backbone in part of the cross-bridge region. They are about 43 nm apart and the first lies just proximal to the first row of cross-bridges.

(vi) The myosin packing changes continuously during the first three 43 nm periods as the molecular interactions change from anti-parallel to parallel. Changes in packing also occur near the ends of the filament. In between there is a region of approximately constant myosin packing where C-protein is bound.

(vii) When present, M-line bridges maintain the register and rotational relationship of the thick filaments.

(viii) The tip of the filament includes an end filament of unknown function.

We are still ignorant of many major features of thick filament structure, for instance:

(i) The precise symmetry of the cross-bridge arrangement and disposition of bridges are unknown.

(ii) We do not know at the molecular level how the tails pack nor how they interact with non-myosin components.

(iii) We have no idea how thick filaments assemble *in vivo* and how their length and width are so precisely determined.

(iv) While there is evidence for large changes in thick filament structure during muscular activity or on passing into rigor (see Section VI.B.3), these changes are not understood at the molecular level.

B. Synthetic Assemblies of Myosin and its Fragments

Although much has been learned about thick filaments from observations of the native structure, the complexity of the problem has led to the study of myosin interactions in simpler systems where assembly of purified myosin and of myosin fragments are studied *in vitro*. While many of these studies are of interest for the insights they provide on protein-protein interactions, we concentrate on those results that bear on our understanding of the native thick filament structure.

1. Assemblies of Intact Myosin

The physical chemistry of the myosin monomer-polymer transformation has provided a basis for understanding the observed synthetic structures. Hydrodynamic studies have suggested that at high salt (where thick filaments are depolymerized), myosin molecules are in a rapid monomer-dimer equilibrium (Godfrey and Harrington, 1970a, b; Herbert and Carlson, 1971; Harrington and Burke, 1972). Szuchet (1977) and Emes and Rowe (1978), however, found no dimers under these conditions and this discrepancy is not resolved. Nevertheless, evidence for a dimer intermediate in the monomer-polymer system has come from pressure jump studies of the kinetics of polymer formation (Davis, 1981). Chemical crosslinking, followed by shadowing and electron microscopy, has indicated that the dimer comprises two myosin molecules staggered by about 44 nm (Davis *et al.*, 1982). In the following discussion we do not distinguish between monomer and dimer and use the term "monomer" to imply either.

As the ionic strength of a myosin solution is lowered, filaments are formed (Fig. 4), their structure depending critically on the method of formation and on the exact ionic environment. At physiological ionic strength and pH, polymerization can be essentially complete within 20 ms and only about 10 μg/ml myosin monomer remains in equilibrium with polymer regardless of the polymer concentration (Katsura and Noda, 1971, 1973a). The filaments formed by rapid dilution are about 0·5 μm long; as dilution time increases so does both filament length and length dispersion: slow dialysis can yield filaments between 1 and 12 μm long (Josephs and Harrington, 1966). Although the monomer concentration reaches equilibrium with polymer quickly, the lengths of the filaments in these conditions never move towards an equilibrium value, presumably because of the very low concentration of monomer. By contrast, at higher pH or ionic strength where the monomer concentration at equilibrium is higher, the filaments are more dynamic and do attain an equilibrium length regardless of the method of preparation (Kaminer and Bell, 1966; Josephs and Harrington, 1966, 1968; Katsura and Noda, 1971). These polymers are always short (about 0·5 μm or less) with a narrow size range. Thus long polymers formed at pH 7 become indistinguishable, when brought to pH 8, from polymers formed at pH 8, whereas polymers formed at pH 8 and then brought to pH 7 grow in length only by the amount corresponding to the amount of monomer taken into polymer form (Kaminer and Bell, 1966).

All these properties of the myosin monomer-polymer system show it to be of the nucleation and growth type. The monomer-polymer equilibrium is insensitive to temperature, the reaction being driven in the direction of polymer by an increase in entropy accompanying release of structured water. About 25 bonds (ionic and probably also hydrophobic) are made per monomer incorporated (Harrington and Josephs, 1968); such a small number could make the filament structure sensitive to its local environment in the cell, a point especially relevant to the very labile smooth and non-muscle myosin filaments (see Section V.A). For instance Harrington and Himmelfarb (1972) have shown that the binding of as few as one or two molecules of ATP, ADP or pyrophosphate to the rod portion of rabbit skeletal myosin at pH 8·3 destabilizes the polymer, and Pinset-Härström and Truffy (1979) have shown that the width of filaments generated by dilution at pH 7 is markedly reduced (to that of the native filament) by the inclusion of MgATP. While the influence of a range of cations and anions on filament formation has been reported (Brahms and Brezner, 1961), we are unaware of any study of the structure of filaments formed in a solvent of physiological composition.

The appearance of synthetic thick filaments in the electron microscope is quite variable. Huxley (1963) showed that many filaments prepared by dilution at pH 7 were bipolar and had a central bare region (Fig. 4) and this appearance is commonly seen in preparations where filaments are short, for instance at high pH (Josephs and Harrington, 1966) or in hypertonic salt (Kaminer and Bell, 1966). On the other hand, filaments prepared slowly at pH 7 typically show no bare zone at any position (Moos et al., 1975; Hinssen et al., 1978). Such filaments are bipolar in the sense that the two ends of each filament look alike and are also, at least locally, packed with a polar arrangement of molecules (Moos et al., 1975), but their structure is otherwise obscure. The short length and bipolar appearance usually characteristic of filaments at high pH or ionic strength suggest that anti-parallel rather than parallel interactions between tails are predominant under these conditions.

When observed by negative staining, synthetic thick filaments may display thin "whiskers" projecting at an angle from the filament backbone (Pollard,

1975; Pinset-Härström and Truffy, 1979). The shadowed native filaments of Trinick and Elliott (1979; Fig. 6a and b) show the tips of the myosin heads up to 50 nm from the filament backbone. These observations are simply explained if one assumes, as suggested earlier, that the whisker-like S-2 is only loosely associated with the tightly-packed LMM part of the backbone, thus allowing the heads to extend a considerable distance from the filament axis (Pollard, 1975). This interpretation has been directly confirmed recently using improved negative staining of both native and synthetic thick filaments. Pairs of myosin heads are seen lying at some distance from the filament backbone and connected to it by the S-2 part of the tail (Knight and Trinick, unpublished observations; Fig. 6c and d). The whiskers frequently appear straight and sharply angled to the filament surface. The length of the whiskers is variable. Therefore the bend that allows this angling does not necessarily occur at the hinge observed in the tails of shadowed myosin molecules (Elliott and Offer, 1978; Section III.A).

Synthetic myosin filaments may show some degree of axial order. Both 14·3 nm repeats (Moos et al., 1975; Hinssen et al., 1978) and 43 nm repeats (Eaton and Pepe, 1974) have been seen, the same as those in native filaments. The 43 nm repeat of Eaton and Pepe occurs in filaments formed at 0·3 M KCl, pH 7, a condition where myosin only just starts to form filaments. These structures may be related to the subfilaments of the thick filament, which have myosin molecules staggered by 43 nm (Squire, 1975; Wray, 1979a) since one would expect subfilaments to be more stable than the filament as a whole.

Although these *in vitro* studies have led to general concepts of filament formation and stability, we should stress that the assembly process that occurs *in vivo* remains entirely unknown.

2. Assemblies of Myosin Fragments

The arrangement of the myosin tails in the backbone of the native thick filament is unknown. This is because of the presence of the myosin heads, which obscure the underlying backbone structure, because of the confused appearance produced in the electron microscope by superimposed tails only 2 nm in diameter, and because of the distortion of the structure that occurs during preparation for electron microscopy. It is also due to the limited extent of the array of tails in both the transverse and axial dimensions, which limits the usefulness of image averaging methods. Attention has therefore focussed on the more extended arrays that can be formed by the head-less parts of the myosin molecule, namely rod, LMM and S-2 (see Section III.A). Of these, LMM, which is the core of the native filament, forms the most extended structures and has been the most widely studied. Although these structures were first seen over 25 years ago (Philpott and Szent-Györgyi, 1954), the packing of LMM molecules within them is only now being described in detail. In this section we discuss the contribution these studies have made to our understanding of the organization of myosin tails in the native filament.

LMM is highly polymorphic. Spindle-shaped structures (called tactoids), sheets, ribbons, tubes and open net structures are all formed (see Bennett, 1981). The types of structure depend on the method of preparation of LMM from myosin (because this produces molecules of different length) and also on the pH and the ionic composition of the solution used for polymerization (e.g. Katsura and Noda, 1973b; Chowrashi and Pepe, 1977). Like synthetic thick filaments, the largest structures are formed by slow polymerization: tactoids up to 500 μm long have been produced by dialysis to low salt over several days (Yagi and Offer, 1981). The tactoids and sheets often show pronounced axial periodicities that are usually integral multiples of 14·3 nm, and the most common values are 43 nm and 14·3 nm (Fig. 18a and b). The occurrence of these periodicities, characteristic of the packing of myosin in the thick filament, encourages the belief that some of the interactions between molecules in the LMM aggregates are the same as those between myosin tails in the thick filament.

Within the larger periodicities of the tactoids and sheets, a complex sub-banding pattern is seen by uranyl-acetate staining that probably reflects the distribution of charged amino acids along the molecules. This pattern varies between LMM tactoids prepared from red and cardiac muscles on the one hand and white muscle on the other suggesting that different myosin heavy chains are present in these muscle types (Nakamura *et al.*, 1971), as has since been confirmed by other techniques (see, for instance, Whalen *et al.*, 1981).

Models of LMM packing have generally attempted to explain the 43 nm and 14·3 nm axial periodicities in only the axial dimension and in terms of longitudinally-oriented overlapping LMM molecules, since there has been little evidence of organization of molecules in the transverse dimensions (Huxley, 1963; Katsura and Noda, 1973b; Safer and Pepe, 1980; Bennett, 1981). Analysis of the appearance of tactoids with a 43 nm axial repeat (see Fig. 18a) indicates that an important interaction is between parallel molecules staggered by this amount. This interaction is likely to be involved in formation of the cross-bridge array in the thick filament, but in order to form a 14·3 nm periodicity, it would be necessary to use an additional parallel interaction that is a different order of 14·3 nm, e.g. 57 nm ($= 4 \times 14\cdot3$ nm). Another important interaction in the tactoids is between anti-parallel molecules overlapping by 84 nm. This may be related to a similar overlap observed in rod segments (see below).

Two recent studies of three-dimensionally ordered LMM aggregates have led to three-dimensional packing models that may have important implica-

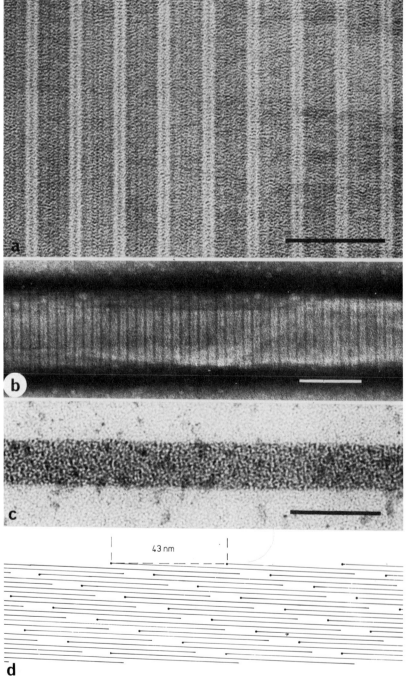

tions for the native thick filament structure. Bennett (1976, 1977) has shown that a sheet aggregate that shows no obvious periodicity by negative staining comprises of layers, in each of which molecules are staggered axially by 43 nm and are close-packed transversely. Neighbouring layers are displaced axially by 0.375×43 nm (Fig. 18c and d). If cross-bridges in the 43 nm repeat of the native filament were staggered axially by 0·375 and 0·625 and of 43 nm (instead of the normally accepted 0·333 and 0.667×43 nm) the perturbation of the helix thus produced could account for the unexplained forbidden meridional reflections seen in the X-ray pattern of relaxed muscle (see Section IV.A.1; Bennett, 1977).

Yagi and Offer (1981) have shown by X-ray diffraction and electron microscopy that tactoids, if grown slowly, also have crystalline order in three dimensions. They find an axial repeat of 42·9 nm, and in the plane perpendicular to the axis a rectangular unit cell which, at pH 6·6, has dimensions 6.5×3.9 nm. Their three dimensional model is again a layered structure. In each layer, parallel molecules are staggered by 43 and 86 nm and anti-parallel molecules overlap by 41 and 84 nm. Both Pepe's (1971) and Squire's (1973) models of thick filament backbone structure assume the packing of the myosin tails to be hexagonal or nearly so, but neither study of LMM packing shows evidence for this, favouring instead a rectangular or tetragonal arrangement. The recently proposed structure for the paramyosin

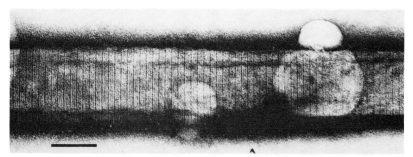

Fig. 19. S-2 paracrystals formed at pH 4.5. Scale marker indicates 100 nm (reprinted from Lowey et al. (1967), by courtesy of S. Lowey and Academic Press Inc.).

Fig. 18. Aggregates of LMM. (a) Negatively-stained sheet showing a 43 nm periodicity. Tactoids with a 43 nm repeat look similar but, being thicker, are normally much more heavily stained. (b) Negatively-stained tactoid, showing a 14·3 nm periodicity. (c) Section of an epoxy-embedded sheet (a different type from (a)) showing an approximately longitudinal, edge-on view. Oblique striations can be seen that indicate displacements between layers of the sheet. Scale markers indicate 100 nm. (d) Model interpreting the appearance of (c), indicating the lattice on which the molecules lie ((a) reprinted from Yagi and Offer (1981), by courtesy of G. Offer and Academic Press Inc.; (b) from Bennett (1981), by courtesy of P. Bennett and Academic Press Inc.; (c) and (d) from Bennett (1977), by courtesy of P. Bennett).

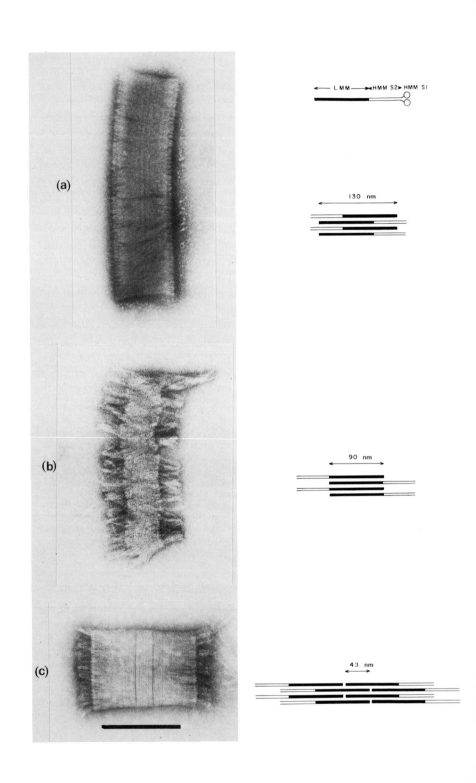

filament (see Section V.C), though not at molecular resolution, does not suggest hexagonal packing either. It is also of interest that side polar filaments formed by smooth muscle myosin (Section V.B) are approximately square in cross-section (Craig and Megerman, 1977). Here too the tails may have a tetragonal arrangement.

Myosin S-2 is soluble under most conditions, but can precipitate at its isoelectric point. Under these conditions it forms tactoids with a 14 nm repeat (Lowey et al., 1967; Fig. 19). Thus although S-2 interactions are only weak, they do exist and may be significant in the native filament where the S-2 portions of the myosin molecules are brought into close contact with one another by the arrangement of the underlying LMM (see Section III.A).

Myosin rod has not been shown to produce tactoids. At 0·1 M KCl, pH 7, sheets are formed which show a 14 nm periodicity (Moos et al., 1975), while at 0·17 M KCl, pH 8·3, filaments 490 nm long are formed (Harrington and Himmelfarb, 1972). By analogy with the synthetic myosin filaments formed in similar conditions the rod filaments would be bipolar structures, but the type of packing has not been determined in either sheets or filaments. Divalent cations cause rods to aggregate into bipolar segments which can grow indefinitely laterally to form ribbons in which the molecules run transversely (Fig. 20). Similar structures are formed by intact myosin and by LMM. The existence of three distinct forms of rod segment that show anti-parallel overlaps of about 43, 86 and 130 nm suggests that these interactions may be involved in the anti-parallel packing of myosin tails in the bare zone of the thick filament (Cohen et al., 1970; Kendrick-Jones et al., 1971; Harrison et al., 1971; Fig. 21), but other anti-parallel interactions may also be required (Bennett, 1981).

From these studies it can be seen that all parts of the myosin molecule influence the structure of the thick filament. LMM is capable of adopting many modes of assembly, some containing the elements of the periodicities of the native thick filament. The whole rod is less polymorphic but will form filaments. The additional bulk of the heads of the intact myosin molecule appears to limit the lateral extent of aggregation, precluding the formation of sheets and tactoids and favouring instead cylindrical filaments in which the heads can avoid contact with one another by being distributed around the circumference.

Fig. 20. Negatively-stained segments formed from myosin rod (by precipitation with divalent cations) and their interpretation in terms of antiparallel molecular overlaps. (a) 130 nm overlap; (b) 90 nm overlap; (c) 43 nm overlap. Scale marker indicates 200 nm (reprinted from Cohen and Szent-Györgyi (1971), courtesy of C. Cohen and Prentice-Hall Inc.).

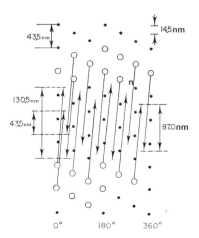

Fig. 21. Schematic surface lattice of myosin filament. Circles represent myosin cross-bridges having a repeat of 43·5 nm and an axial separation of 14·5 nm, similar to the values derived from X-ray diffraction (Section IV.A.1). The diagram shows how myosin molecules with a tail length of 145 nm lying on this lattice could have antiparallel overlaps of 43·5, 87 and 130 nm as observed in rod segments. If the bare zone or rod length were different from those shown, the overlaps might also be altered, but the differences between them would still be ~43 nm, as observed (reprinted from Harrison *et al.* (1971), courtesy of C. Cohen and Academic Press Inc.).

C. Models of Thick Filament Structure

Because the structure of the thick filament has not been solved experimentally, there have been a number of attempts to build models of the molecular packing in the filament backbone that can plausibly account for the data available. We shall consider the two best known here. A third, outlined in Section V.B, is also mentioned.

Pepe's model (Pepe, 1967a, 1971; Pepe and Drucker, 1972, 1979; Pepe and Dowben, 1977; Pepe *et al.*, 1981) attempts to explain the structure of vertebrate skeletal thick filaments only. It is constructed from 12 subfilaments, with centre-to-centre spacing of 4 nm, containing two myosin tails (without any axial stagger) at any cross-sectional level. The 4 nm subfilaments are hexagonally close packed and run parallel to the filament axis but the tails within a subfilament twist gradually around each other. The subfilaments comprise the LMM portions of the myosin molecules, and S-2 does not form part of the backbone. The major interactions between subfilaments are staggers of 14·3, 28·6 and 43 nm and each subfilament gives off a cross-bridge every 86 nm. Three subfilaments are buried in the core of the filament and the myosin molecules involved must protrude between the surface subfilaments to give off their cross-bridges. The backbone has a *three*-fold screw axis (equivalent environments being related by a translation of 28·6 nm and a rotation of 120°), but no rotation axis; the axial repeat is 86 nm. There are four myosin molecules per 14·3 nm, but the cross-bridge distribution has an approximate *two*-fold rotation axis; there are thus four myosin heads per cross-bridge. The backbone and cross-bridge symmetries are thus quite different from each other. The myosin molecules are not equivalent to each other: if the 4 nm subfilaments were straight there would be eight quite different environments for myosin molecules; since they twist, but only very slowly, every myosin molecule lies in unique surroundings.

Pepe's model is not consistent with the 43 nm axial repeat observed in projection (in A-segments) in the backbone of vertebrate skeletal thick filaments (Hanson *et al.*, 1971; Craig, 1977) since it would predict a projected repeat of 28·6 nm. It may be wrong in assuming that the number of myosin molecules per 14·3 nm is four and not three, since this point is not yet resolved. The model is inconsistent with three (Pepe and Drucker, 1979) which seems, at present, to be the more likely value; it would be untenable if three were shown to be correct. The model is not based on the use of a small number of specific interactions placing the molecules in equivalent or quasi-equivalent environments as in the case with other biological polymers (Caspar and Klug, 1962). The most important feature of the model is the presence of 4 nm diameter subfilaments, for which considerable evidence is accumulating.

Squire (1971, 1973) suggested a *general* model in which the myosin molecules of filaments from diverse muscles have a single basic mode of packing. The rotational symmetry n can vary from one particular case of the model to another without significant changes in packing, and n is proportional to filament diameter. The myosin molecules are not associated into subfilaments but occur singly, all being in equivalent environments. They are not parallel to the filament axis but run at a small angle to it, following a helical path, starting near the filament axis and progressing outwards as the helix turns, placing the heads at the surface (Fig. 22a); the whole myosin tail forms part of the backbone. In the case of vertebrate skeletal muscle, uncertainty in the filament diameter permits models with rotational symmetries of either three or four, but based on other experimental observations, three is strongly favoured (Squire, 1975). The vertebrate model has a nine-fold screw axis (with $n=3$), the translation between equivalent myosin molecules being 14·3 nm, with a rotation of 40° (see Fig. 11b). In later papers Squire (1975, 1979) retains the principle of equivalence of interactions but suggests alternative structures built of subfilaments of varying size (quite different from those of Pepe) containing several molecules in cross-section. In some structures several subfilaments aggregate to form an intermediate filament and the intermediate filaments themselves aggregate to form the filament (Fig. 22b–d). To maintain equivalence, the molecules in a subfilament twist around each other, as do the

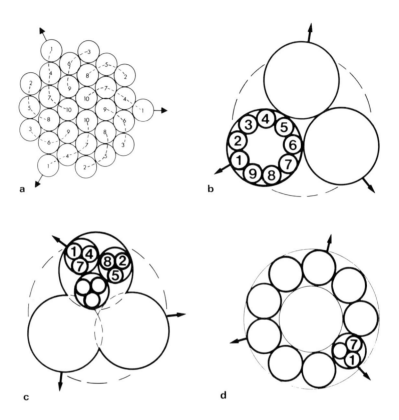

Fig. 22. Models proposed by Squire (1973, 1979) for the packing of myosin tails in the vertebrate skeletal thick filament, seen in transverse section. (a) Packing without subfilaments. The diagram can be viewed in two ways. In one, the relative staggers of individual molecules at a single cross-sectional level are indicated as follows: a molecule numbered n is laid down $(14\cdot3 \times p)$ nm closer to the M-line than molecule $(n+p)$. In the other way the dashed lines can be considered as a projection of the helical path of the molecules. The numbers in this case yield the axial level at which the molecules reach the radial and azimuthal positions shown. (b) A three subfilament model in which nine myosin tails seen in cross-section form a hollow cylinder in each subfilament. Each tail winds round the circumference of the subfilament, and the subfilaments wind round the filament axis. (c) A nine subfilament model, in which hollow subfilaments are absent. Three tails form a subfilament, and three of these small subfilaments are grouped to form an intermediate filament. Three intermediate filaments form a filament. (d) A different organization of the subfilaments of (c) which produces a hollow core. In (b) to (d) the numbers have the same meaning as in (a). The arrows represent cross-bridges in every case (reprinted from Squire (1973, 1979) by courtesy of J. Squire and Academic Press Inc.).

subfilaments in a filament or intermediate filament and the intermediate filaments in a filament.

Squire's modelling is a theoretical attempt to enumerate all conceivable modes of myosin packing in which a small number of specific interactions are used and the molecules are in equivalent environments. He also makes suggestions as to how slight perturbations in packing may produce the non-equivalence sometimes observed, e.g. in vertebrate filaments (Squire, 1979). At present, insufficient experimental evidence is available to decide whether any of these hypothetical models can explain the native structure (although Maw and Rowe's (1980) demonstration of three large subfilaments in vertebrate striated muscle would tend to exclude Squire's (1973) early model). Squire's work has, however, been most important in providing a unifying concept which relates a range of apparently diverse structures.

Wray (1979a) has obtained X-ray evidence from invertebrates suggesting the presence of 4 nm diameter subfilaments as common building blocks in a variety of muscles with filaments of differing symmetries (Section V.B.). His general packing scheme, when applied to muscles with the vertebrate symmetry, gives a structure with straight subfilaments, parallel to the filament axis, similar in cross-section to Fig. 22d. This leaves a large core, however, which is not observed in electron micrographs. Wray therefore suggests that there may be an orderly collapse of the tubular structure, to fill in the core. Because the more central subfilaments would necessarily adopt slightly different resting configurations, this arrangement might explain the 43 nm repeat seen in projection in the filament backbone, and also the imperfect helical symmetry apparent from the X-ray diffraction data.

Which of these models most closely resembles the native structure is uncertain, and the true structure may well combine features of several of them. Only more experimental data can help to distinguish these possibilities.

V. STRUCTURE OF THICK FILAMENTS FROM SOURCES OTHER THAN VERTEBRATE SKELETAL MUSCLE

A. Vertebrate Smooth Muscle and Non-muscle Thick Filaments

The contraction of vertebrate smooth muscle is based on an actin-myosin-ATP interaction similar to that in striated muscle. For many years, however, it was not known whether myosin in vertebrate smooth muscle occurred as filaments or not, since thick filaments are sometimes absent from sectioned smooth muscle and from fresh homogenates (see Shoenberg and Needham, 1976, for review).

The state of myosin in vertebrate smooth muscle has been clarified by electron microscopic and X-ray studies. First, electron microscopy has shown that smooth muscle myosin *is* capable of forming filaments *in vitro* (Hanson and Lowy, 1964; see Shoenberg and Needham, 1976, for later references). Secondly, X-ray diffraction patterns of living smooth muscle (Lowy *et al.*, 1970; Shoenberg and Haselgrove, 1974) can show a 14·3 nm meridional reflection—characteristic of the cross-bridge spacing in myosin filaments of striated muscle (Section IV.A). Although this reflection is very weak at 37°C, it does suggest that some of the myosin in vertebrate smooth muscle must be in filament form under physiological conditions (see Shoenberg and Needham, 1976).

When "filaments" were first observed in sectioned muscle, they were of two distinct classes. One appeared as broad, ribbon-shaped structures (Lowy and Small, 1970) which had cross-bridges regularly arrayed on the two faces of the ribbon (Small and Squire, 1972). The cross-bridges on opposite faces had opposite polarities; thus, the bipolar filament of striated muscle was replaced by a "face-polar" ribbon in smooth muscle. Thin filaments of opposite polarity could still, in principle, be pulled past each other in such a system, so face-polar ribbons were, in a functional sense, still bipolar. Somlyo and co-workers, on the other hand, found filaments that were cylindrical in cross-section, simiar to striated muscle filaments (see Shoenberg and Needham, 1976). There was considerable controversy over which of these structures represented the *in vivo* state: were filaments produced by the break-up of ribbons (Small and Squire, 1972) or were ribbons artifactual aggregates of filaments (Somlyo *et al.*, 1971)? There is now agreement that filaments, not ribbons, are closer to the *in vivo* structure (see Somlyo *et al.*, 1977; Small, 1977; Fig. 44d), but uncertainty remains about the type of polarity possessed by these filaments. They may be simple bipolar structures similar to those of striated muscle, although the cross-bridge detail in the micrographs is not sufficient to confirm this point (Ashton *et al.*, 1975; Somlyo *et al.*, 1977). An alternative and intriguing possibility is that the unusual face-polarity of the ribbons is reflecting some equally unusual polarity in their constituent filaments (Squire, 1975), since it is not easy to see how bipolar filaments could aggregate to form face-polar ribbons.

In vitro studies indicate that filaments with a different type of polarity from that of vertebrate skeletal muscle thick filaments can indeed be formed from smooth muscle myosin. Under certain ionic conditions smooth muscle myosin can assemble into long "side-polar" filaments, having cross-bridges of the same polarity (and with a 14 nm spacing) along the whole length of one side of the filament and the opposite polarity along the other side (Craig and Megerman, 1977, 1979; Fig. 23a). There is no central bare zone, but bare regions occur at the ends. These observations suggest a simple arrangement of myosin molecules with specific anti-parallel interactions occurring along the

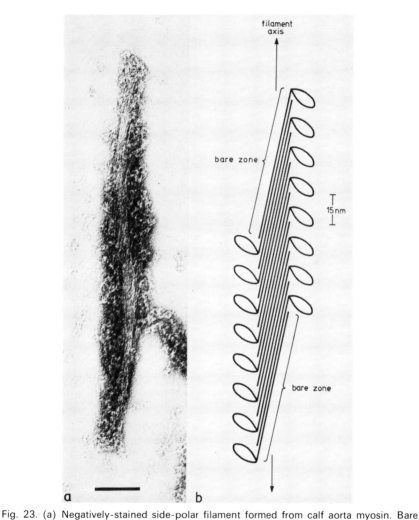

Fig. 23. (a) Negatively-stained side-polar filament formed from calf aorta myosin. Bare regions occur at the top left and bottom right of the filament. Scale marker indicates 50 nm. (b) Schematic diagram of side-polar filament to illustrate construction from myosin molecules with antiparallel overlaps along the whole length of the filament. The three-dimensional structure and the precise values of the overlaps are unknown: the diagram is intended to convey only the general mode of packing ((a) and (b) reprinted from Craig and Megerman (1977), courtesy of The Rockefeller University Press).

whole length of the filament, not just at the centre (Fig. 23b). The filaments do not have cylindrical symmetry but in cross-section are approximately square with bridges on only two opposite sides of the square. Such filaments would be functionally bipolar since the bridges on opposite sides are of opposite polarity.

Under different conditions, smooth muscle myosin can form short (0·5 μm) bipolar filaments with a central bare zone, similar to those formed by striated muscle myosin (Hanson and Lowy, 1964; Kaminer, 1969; Sobieszek, 1972; Wachsberger and Pepe, 1974; Craig and Megerman, 1977; Hinssen et al., 1978), or it can form short, bipolar segments (Kendrick-Jones et al., 1971). The rarity of *longer* bipolar filaments, where pure polar interactions would occur along most of the length, emphasizes the importance of *anti*-parallel interactions between molecules of smooth muscle myosin that is already implied by the side-polar structure. Long filaments, with cross-bridges apparently distributed around the whole circumference at each 14 nm level, *have* been formed with several types of smooth muscle myosin (Sobieszek, 1972; Sobieszek and Small, 1972; Small, 1977; Hinssen et al., 1978; Shoenberg and Stewart, 1980). However, the presence of terminal bare regions and the absence of a central bare zone of polarity reversal (Fig. 24a) indicate that these filaments are not bipolar. Instead, anti-parallel interactions similar to those in side-polar filaments seem to be implied, although the circumferential distribution of cross-bridges shows that the packing has been modified. Hinssen et al. (1978) suggest that the anti-parallel dimers lie on a helix, that at each 14 nm level cross-bridges of opposite polarity could alternate around the circumference, and at any one azimuth going along the filament the bridges could all be of one polarity. This unusual but hypothetical structure could thus, in principle, function to slide thin filaments in opposite directions.

These polymorphic aggregates indicate the sensitivity of smooth muscle myosin to its environment, only small changes in ionic conditions sometimes being needed to produce the different types. Such *in vitro* studies are important in that they illustrate the potential interactions of the myosin molecule and thus provide a guide to what may occur *in vivo*.

Thick filaments having an appearance similar to the synthetic modified side-polar form described above (Fig. 24a) have been observed in smooth muscle homogenates (Fig. 24b; Small, 1977; Hinssen et al., 1978). Structures based on the side-polar mode of assembly are attractive candidates for the *in vivo* state of myosin in smooth muscle, since they could help to account in a simple way for the high degrees of shortening of which these muscles are capable (Small and Squire, 1972; Craig and Megerman, 1977, 1979). The key question, however, is whether such structures do in fact reflect the *in vivo* form. Shoenberg (1969) found thick filaments in smooth muscle homogenates in the presence of ATP, Mg^{2+} and free Ca^{2+}, but not when the Ca^{2+} was chelated. It

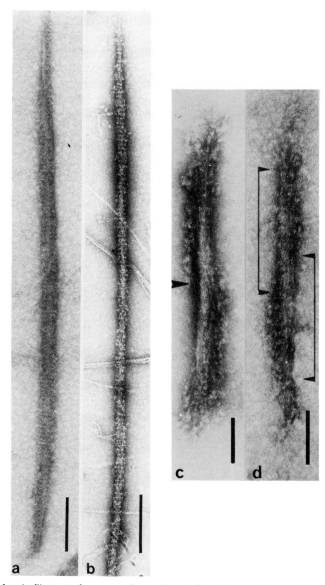

Fig. 24. Myosin filaments from smooth muscle and *Amoeba*, seen by negative staining. (a) Filament formed from purified chicken gizzard myosin. (b) Filament from a homogenate of smooth muscle cells from the taenia coli of the guinea pig. (c) Synthetic filament from a crude myosin preparation from *Amoeba proteus*. The arrow points to cross-bridges flared out from the filament backbone. (d) Filament from a homogenate of *Amoeba*. The relative shift of the two edges of the bare zone, marked by the arrows, indicates a similarity to smooth muscle filaments. Scale markers indicate 0·2 μm (reprinted from Hinssen *et al.* (1978) by courtesy of J. V. Small and Academic Press Inc.).

was therefore suggested that thick filaments might be absent *in vivo* from resting muscle and form only at the onset of contraction, when Ca^{2+} was released. As emphasized by Shoenberg and Needham (1976), simple preparative procedures such as glutaraldehyde fixation or cooling can cause rapid changes in the ion permeability of the cell membrane and may also cause Ca^{2+} release. Either factor might cause the very labile smooth muscle myosin to form thick filaments during fixation or isolation, possibly in some unphysiological way (see below).

Suzuki *et al.* (1978) and Scholey *et al.* (1980) have recently made the very striking observation that unphosphorylated smooth muscle myosin *in vitro*, at physiological ionic strength and ATP concentrations, is in the depolymerized state (see also Shoenberg and Stewart, 1980), whilst when the myosin regulatory light chain is phosphorylated (by Ca^{2+} activation of a calmodulin-dependent light chain kinase), the myosin assembles into filaments. The filaments so formed typically have features in common with those of the side-polar structure (Scholey, Kendrick-Jones and Craig, unpublished work). This effect of phosphorylation strengthens the view above that activation might cause formation or augmentation of smooth muscle thick filaments, thus providing a simple contractile regulatory effect. However, the relevance of these *in vitro* obervations to living muscle remains to be established. Electron microscopy of properly relaxed or contracting muscle, rapidly frozen without chemical fixation or antifreeze treatment, would be the most direct approach. Bond *et al.* (1981) and Somlyo *et al.* (1981) have recently reported results obtained using this approach and conclude that in resting smooth muscle, thick filaments actually are present and that they are not phosphorylated. Further work from more laboratories will probably be needed before a consensus is reached on the relationship between the *in vitro* and *in vivo* findings.

As with smooth muscle, many of the motile activities of non-muscle cells are based on an actin-myosin-ATP interaction (see Huxley (1973) and Pollard and Weihing (1974) for reviews). Despite this fact it has been very difficult to demonstrate the presence of thick filaments in sectioned non-muscle cells. There are two possible reasons for this. First, in non-muscle cells the concentration of myosin can be very low, so the chance of including a thick filament in a section is correspondingly small. Secondly, the myosin in non-muscle cells is even more labile than smooth muscle myosin. Furthermore, *in vitro* experiments similar to those performed on smooth muscle myosin show that non-muscle myosin in the presence of ATP forms filaments only when the regulatory light chain is phosphorylated (Scholey *et al.*, 1980). Thus, some functionally bipolar myosin filaments may exist, but only during periods of contractile activity.

Considerable insight into the mechanism of the reversible polymerization of both smooth muscle and non-muscle myosins *in vitro* has been obtained recently by electron microscopy. Because of its high sedimentation coefficient (10–12S) compared to myosin monomers in high salt (6S), it was initially proposed that the depolymerized myosin was dimeric (Suzuki *et al.*, 1978). The molecular weight of the material was, however, found to be that of monomeric myosin (Trybus *et al.*, 1982; Suzuki *et al.*, 1982). This discrepancy was resolved when the depolymerized myosin was viewed by heavy-metal shadowing in the electron microscope. A new structure of the myosin monomer was seen, in which the tail was bent back on itself twice to form three segments of roughly equal length (Trybus *et al.*, 1982; Onishi and Wakabayashi, 1982; Craig *et al.*, 1983). The segments were arranged such that the more distal hinge region appeared to be bound to the base of the myosin heads, where the regulatory light chain is thought to be (see Section III.C). This folding produces a more compact structure than myosin has in high salt, which accounts for the higher sedimentation coefficient that was observed. Furthermore, it has been shown that when the regulatory light chain becomes phosphorylated under physiological conditions of ionic strength and pH, the folded myosin unfolds and assembles into filaments (Craig *et al.*, 1983).

As with smooth muscle, the native structure of non-muscle thick filaments is unknown. Non-muscle myosin *in vitro* tends to form assemblies similar to those obtained with smooth muscle myosin. Short bipolar filaments are observed (Pollard and Weihing, 1974; Hinssen *et al.*, 1978) as well as filaments based on the side-polar mode of assembly (Fig. 24c; Hinssen *et al.*, 1978). "Native" filaments, with side-polar characteristics, have been observed in *Amoeba proteus* homogenates (Fig. 24d; Hinssen *et al.*, 1978), though whether these represent the *in vivo* state or were induced to form during preparation (as was suggested above for smooth muscle myosin) is unknown. Nevertheless, it seems likely that side-polar like or short bipolar filaments will be involved in non-muscle motility.

B. Invertebrate Striated Muscle Thick Filaments

The diverse and specialized movements encountered among the invertebrates are served by an equally diverse and specialized array of striated muscles. These range from the indirect flight muscles of insects which shorten $\sim 2\%$ of their length in the working cycle (see Pringle, 1957, 1965) to the supercontracting scutal depressor muscles of barnacles which shorten by up to 70% (Hoyle *et al.*, 1965). The near-crystalline array of thick and thin filaments in the insect flight muscles is apparently an adaptation permitting the high frequency (up to 1000 Hz) of the wing-beat, while the perforated Z-disc of the barnacle permits the thick filaments to enter adjacent sarcomeres during supercontraction. Comparative studies of such muscles have been important in revealing general principles of molecular organization.

The thick filaments of the invertebrates exhibit considerable variation in structure depending on the species and the anatomical location of the muscle. X-ray diffraction patterns of a range of muscles have led to the determination of the approximate rotational symmetry (n) and general conclusions concerning the disposition of the cross-bridges of their thick filaments (Wray et al., 1975; Millman and Bennett, 1976). This has been possible because the patterns

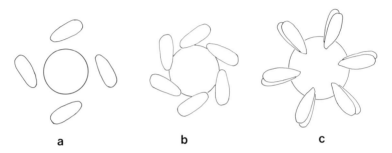

Fig. 25. Scheme illustrating the diversity of bridge location and configuration in invertebrate myosin filaments (deduced from X-ray diffraction). (a) *Limulus*; (b) *Homarus* and (c) *Placopecten*. The bridges at one level in each filament are shown, and the circles represent the backbone diameters as estimated from electron microscopy. The bridges represent pairs of heads, shown separated in (c). The rotational symmetries shown for *Limulus* and *Placopecten* are 4 and 6, but symmetries of 3 and 7 respectively are equally compatible with the X-ray data; the symmetry for *Homarus* is hypothetical (reprinted from Wray et al. (1975), courtesy of J. Wray and Macmillan Journals Ltd).

Fig. 26. Electron micrographs and three-dimensional reconstructions of invertebrate thick filaments. (a) Negatively-stained *Limulus* filament. The helical tracks of cross-bridges can be seen by viewing along the thick filament at a glancing angle. Scale marker indicates 100 nm. (b) Computer filtered image of one surface of *Limulus* thick filament. The helical tracks of cross-bridges (the solid density contours) are clearly seen, the long axis of each bridge making an angle with the helix path. The bare zone would be at the bottom of the picture. (c) Cross-sectional view of *Limulus* reconstruction, showing four-fold symmetry and azimuthal twisting of bridges. The filament backbone is omitted from the reconstruction. (d) Negatively-stained *Placopecten* filament with helical tracks of cross-bridges visible. Scale marker indicates 200 nm. (e) Longitudinal view of three-dimensional reconstruction of *Placopecten* filament showing cross-bridges but not the backbone. Both lower and upper surfaces of the filament are shown, as if one were seeing through the filament: where lower bridges are hidden by those on the upper surface, they are shown with dotted lines. For ease of illustration, the reconstruction is shown with a helical repeat of 43 nm. In reality the helix is slightly less tight, with a 48 nm repeat. In this reconstruction, the bare zone would be at the bottom. (f) Cross-sectional view of one level of cross-bridges in *Placopecten* showing seven-fold symmetry and absence of any strong azimuthal twisting of bridges. In (b) and (e) each cross-bridge is thought to include two myosin heads ((a) from Kensler and Levine (1982), courtesy of R. Kensler and The Rockefeller University Press; (b) and (c) from Stewart et al. (1981), courtesy of M. Stewart and Academic Press Inc.; (d) and (f) from Vibert and Craig (1983) by courtesy of Academic Press Inc.).

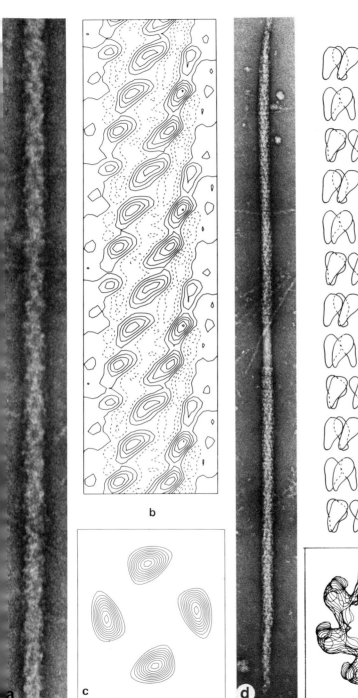

b

c

d

e

| 14.5mm |

f

are not sampled by the filament lattice, so that the continuous distribution of intensity along each layer line can be easily measured (cf. Section IV.A.1). The rotational symmetry is found to vary from three or four in *Limulus* to six or seven in *Placopecten* (Fig. 25) roughly in proportion to filament backbone diameter, as predicted by Squire (1971). The disposition of the cross-bridges is also variable. In most cases, as with the vertebrate (Section IV.A.1), the bridges are tilted and have considerable azimuthal twist. In the case of the scallop, however, the twist is minimal.

With recent improvements in preparative technique, ordered arrays of cross-bridges on invertebrate thick filaments have now been observed directly, by electron microscopy of negatively stained specimens (Kensler and Levine, 1982; Vibert and Craig, 1983; Fig. 26a and d). The order approaches the best that can be anticipated on the basis of the X-ray diffraction patterns, and the three-dimensional reconstructions that have been calculated confirm directly the predictions based on this earlier work. Furthermore, they remove ambiguities still present in the X-ray interpretations, such as the precise rotational symmetry and disposition of the cross-bridges. In *Limulus* there is a four-fold rotation axis (Fig. 26c; Stewart *et al.*, 1981), while the rotational symmetry for *Placopecten* is apparently seven (Fig. 26f; Vibert and Craig, 1983). In *Limulus* the cross-bridges have considerable slew, meaning that as one views down the filament axis, the cross-bridge tends to wrap around the filament circumference; in *Placopecten* the slew is much less; in both cases there is considerable axial tilt of the bridges (Fig. 26b and e), and the direction of tilt is found to be away from the bare zone. Heavy-metal shadowing of filaments shows the hand of the helices along which the cross-bridges lie. In both *Limulus* (Levine and Kensler, 1982) and *Placopecten* (Vibert and Craig, 1983) the helices are right-handed.

The X-ray patterns also contain considerable detail about the backbone of invertebrate thick filaments. By comparing patterns from several crustacean muscles, Wray (1979a) has suggested a general subfilament structure for the backbone, which resembles a multi-stranded cable (Fig. 27). The filament is hollow, in agreement with electron micrographs (Jahromi and Atwood, 1969). Each 4 nm diameter subfilament (strand) contains in cross section a cylindrical group of three myosin tails twisting around each other and gives off a cross-bridge at intervals of 43·5 nm along its length. Neighbouring subfilaments are staggered by 14·5 nm, leading to a regular helical array of cross-bridges, the subunit repeat of which is constant at 14·5 nm, but the helical repeat of which varies between muscles. The helical arrangements are all closely related, the differences being due simply to changes in the number of subfilaments (related to the different rotational symmetries), and to small changes in twist of the subfilaments around the filament axis (compare Fig. 27a and b). Wray's model thus shares the feature of Squire's that the myosin

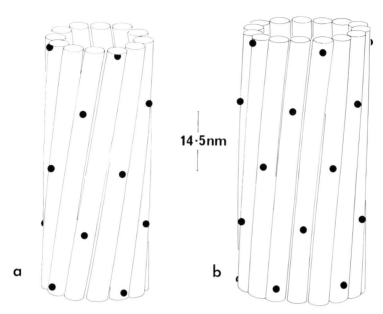

Fig. 27. Models for the organization of subfilaments in the backbone of invertebrate muscle thick filaments, deduced by Wray (1979a) from X-ray diffraction data. (a) Lobster fast muscle. (b) Lobster slow muscle. The rotational symmetries shown are four and five respectively based on the diameters of the filaments observed in electron micrographs. Note that the filaments differ in the number of subfilaments ($=3n$) and in the amount that they twist around the filament axis. The positions of cross-bridges are indicated by dots (reprinted from Wray (1979a), by courtesy of J. Wray and Macmillan Journals Ltd).

molecules in all these muscles pack in a very similar way, only small variations being necessary to produce the different filament structures observed. Although Wray's model is based on a comparison of invertebrate muscles, the myosin filament symmetry in the case of vertebrate muscle suggests a further closely related backbone structure, the subfilaments now running parallel to the filament axis (Wray, 1979a; Section IV.C).

The diversity of invertebrate thick filament structure is also apparent from electron microscope sectioning studies. In transverse section, as mentioned above, filaments from many sources often have an apparently hollow core (Jahromi and Atwood, 1969; Pringle, 1972), in contrast to the compact structure of vertebrate filaments. Evidence for a hollow core has also recently come from equatorial X-ray diffraction data (Yagi and Matsubara, 1977). There is considerable variation in the diameter (12–25 nm) and length of filaments from one muscle to another (Pringle, 1972). Lengths vary between species, e.g. from 1·8 μm in scallop striated muscle (Millman and Bennett, 1976) to 7·5 μm in crab eyestalk muscle (Hoyle and McNeill, 1968), and between muscles within a species, e.g. the thick filaments in slow crustacean muscles were found to be twice as long (6 μm) as those in fast muscles (Jahromi and Atwood, 1969). Even within a single fibre the length may be variable. The most striking finding is that the two halves of a single A-band may differ in length, indicated by an asymmetrically placed bare zone. The thin filaments in the two halves of the sarcomere are of correspondingly different lengths (Franzini-Armstrong, 1970). Thick filaments are often in only approximate register and the M-line in many muscles is correspondingly weak or absent (Jahromi and Atwood, 1969; Millman and Bennett, 1976).

A most surprising variability in thick filament length has been observed in *Limulus* striated muscle, where it has been suggested that the filaments themselves may shorten during contraction (Dewey *et al.*, 1973, 1977). Both the A-band and isolated thick filaments appear to be shorter in contracted than in relaxed muscles. Brann *et al.* (1979) find that filaments in ATP and 10^{-8} M Ca^{2+} (relaxing conditions) shorten when exposed to high Ca^{2+} (10^{-6} M) and that ATP is necessary for this shortening. Experiments with phosphatase are consistent with a dephosphorylation mechanism for the lengthening of shortened thick filaments. It must be noted, however, that Wray *et al.* (1974) find no change in the 14·5 nm meridional X-ray reflection in *Limulus* muscles at different sarcomere lengths, and that phosphorylation in other systems has been found to alter the thick filaments by changing the solubility of the myosin (Scholey *et al.*, 1980; Kuczmarski and Spudich, 1980). Changes in myosin packing in the thick filament, if they occur, may be best investigated by electron microscopy, now that preservation of structure is sufficiently good (Kensler and Levine, 1982).

There are, of course, exceptions to our generalizations concerning filament structure and organization in invertebrates. For example, the thick filaments of scallop striated muscle are approximately constant in length (Millman and Bennett, 1976). Insect flight muscle thick filaments are constant at 2·4 μm (in *Lethocerus*), are organized in a highly regular array and are connected together by an M-line, although this structure differs from the vertebrate M-line. In general such a high degree of filament organization, in both vertebrate and invertebrate muscles, seems to reflect a rapid and efficient contractile ability.

In insect fibrillar flight muscle the I-bands are only about 50 nm long and the thick filaments are apparently connected to the Z-line across this distance by some non-myosin "connecting protein". This may be responsible for the high resting stiffness of insect flight muscle and may play a role in its stretch-activation properties (Auber and Couteaux, 1963; Pringle, 1978; Bullard *et al.*, 1977; Saide, 1981).

Counterparts of the non-myosin components (e.g. C- and H-protein) found in vertebrates (see Section IV.A.2) have not been demonstrated in invertebrate muscles. On the other hand, one component not present in vertebrates is found very commonly in invertebrate muscles. This is the protein paramyosin. In invertebrate striated muscles it is present in variable quantities (Winkelman, 1976; Levine et al., 1976) and is thought to occur in the core of these filaments along part or all of their length (Elfvin et al., 1976; Millman and Bennett, 1976; Bullard et al., 1977). In invertebrate smooth muscles paramyosin is the most abundant protein and fulfils a major structural role. Its structure and function are discussed in the following section.

C. Invertebrate Catch Muscle Thick Filaments

Molluscs possess smooth muscles that can maintain tension with minimal energy expenditure. These are called the catch muscles. For instance in littoral bivalves it is a catch muscle that keeps the two halves of the shell tightly closed during exposure at low tide. Electron microscopy of this muscle shows that the thick filaments are very variable in size, but always large: up to 150 nm in diameter and 30 μm in length. The filaments taper at their ends, are very roughly circular in transverse section (Lowy and Hanson, 1962; Elliott, 1964), and are bipolar (Szent-Györgyi et al., 1971). The greater part of the filaments (at least 70% by mass) is not myosin but paramyosin, the myosin lying on the surface of a paramyosin core (Szent-Györgyi, et al., 1971; Hardwicke and Hanson, 1971). The paramyosin molecule is a two chain, coiled-coil of α-helices (Cohen and Szent-Györgyi, 1957; Cohen and Holmes, 1963). In this sense it resembles the tail of the myosin molecule, although it is about 130 nm long compared with 155 nm for the myosin tail (Lowey et al., 1963; Cohen et al., 1971; Elliott and Offer, 1978). Synthetic assemblies of paramyosin are usually long filaments or ribbons with a variety of polar and non-polar patterns of transverse striations. Cohen et al. (1971) have interpreted these patterns in two dimensions in terms of arrays of longitudinally oriented molecules, varying in the stagger and polarity of nearest neighbour molecules.

The core of the native filament is very highly ordered and provides a detailed X-ray diffraction pattern but variable electron microscope images. The wide angle pattern indicates that the molecules are oriented along the long axis of the filament (Cohen and Holmes, 1963). The small angle X-ray data from dry specimens have been analysed in terms of a net structure (Bear and Selby, 1956; Fig. 28). The longer axis of the unit cell of the net coincides with the filament axis and the dimensions of the unit cell ($72 \times \sim 30$ nm) can be correlated with one of the observed structures of stained filaments (compare Fig. 28 with Fig. 29, $-5°$, $0°$ and $5°$ tilts; Hall et al., 1945; see also Hanson and Lowy, 1964).

Ambiguity in the interpretation of the X-ray pattern permitted the proposal of two models for the molecular packing: Elliott (1964) proposed a crystalline

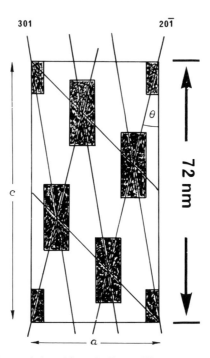

Fig. 28. The Bear-Selby net deduced from the X-ray diffraction pattern of molluscan smooth muscle. The rectangles show how one of the appearances of a negatively stained paramyosin filament (e.g. Fig. 29, 0°) correlates with the net. The *b* axis is normal to the page. Note that the structure has a longitudinal 14·4 nm periodicity when viewed along the *a* axis. The transverse striations seen in tilted cross-sections (Fig. 31) correspond to the 301 and $20\bar{1}$ planes marked (reprinted from Bennett and Elliott (1981), by courtesy of P. Bennett and Chapman and Hall, Ltd).

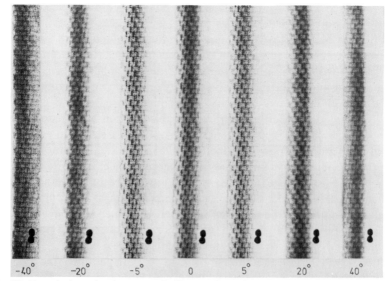

Fig. 29. Negatively-stained paramyosin filament from *Ostrea edulis*, showing the different appearances observed upon rotation about the filament axis by the angles shown. The filament happened to lie with the b axis approximately normal to the grid plane, so the Bear Selby net is seen clearly at low angles of tilt. At higher tilt angles the 14·4 nm periodicity is dominant. Scale marker indicates 200 nm (reprinted from Elliott (1979), by courtesy of A. Elliott and Academic Press Inc.).

model and Elliott and Lowry (1970) proposed a helicoidal model in which the net structure was rolled up like a carpet to form a cylindrical filament. By observing isolated filaments at a series of tilt angles (Fig. 29) Elliott (1979) has now demonstrated that the various appearances of the filament have their origin in a three-dimensional crystalline ordering of the molecules, which form a layered structure (Fig. 30), though the arrangement is different from that proposed by Elliott (1964). The Bear-Selby net structure is seen most distinctly when the filament lies such that the crystal axis normal to the net plane is parallel to the electron beam. The more that axis deviates from the beam direction, the more diffuse the net appears, until only the longitudinal subunit repeat of the pattern (a 14·4 nm periodicity) is seen. The dark-staining rectangles visible in Fig. 29, which indicate gaps between successive paramyosin molecules along the filament axis, run through the filament from one side to the other (Dover and Elliott, 1979; Fig. 30).

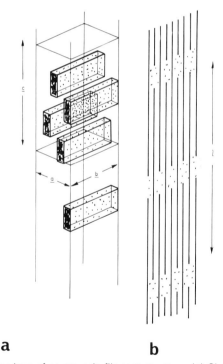

a   b

Fig. 30. Perspective drawings of paramyosin filament structure. (a) Shows regions of stain extending through the filament. (b) Diagram indicating organization of molecules in a single layer of the structure in (a). Each paramyosin molecule (length, l) is 129·1 nm long and is displaced 72 nm in the c-direction relative to its neighbours. Stain lodges in the stippled zones (reprinted from Elliott (1979), by courtesy of A. Elliott and Academic Press Inc.).

This interpretation is supported by observations of tilted transverse sections of smooth muscles from a variety of molluscs (Bennett and Elliott, 1981). Roughly transverse striations are seen in essentially all filaments when they are viewed obliquely (Fig. 31) corresponding to views down particular lattice planes of the proposed crystal structure (Fig. 28). Since the striations are frequently curved, the crystallinity is not perfect, as is also suggested by layer-line streaks associated with the net reflections in the X-ray pattern of intact muscle (Elliott and Lowy, 1970).

Myosin purified from catch muscle (for instance by its selective removal from the surface of the paramyosin core) will readily form synthetic filaments very similar in structure and dimensions to those of rabbit striated muscle myosin. If the paramyosin core is present during polymerization, however, much of the myosin binds directly to the core. Rabbit striated muscle myosin and LMM behave similarly to molluscan myosin (Szent-Györgyi et al., 1971).

There is disagreement on the relative amounts of myosin and paramyosin in catch muscles. Szent-Györgyi et al. (1971) found enough myosin to cover the entire surface of the filament whereas Elliott (1974) found only about one sixth as much. In the former case myosin-myosin interactions could dominate the ordering of the myosin molecules on the surface while if the latter estimate were correct or if myosin-paramyosin interactions were dominant, the distribution of myosin should reflect the underlying paramyosin core structure (see Cohen, 1982). The structure at the surface of the core cannot be uniform: on two sides the Bear-Selby net will be the dominant surface lattice, while on two sides, at roughly 90° to these, the simpler long repeat of the net will be dominant (see Fig. 30). Studies with antibodies to S-1 or with well-preserved filaments (cf. Section IV.B) should establish whether the surface myosin does indeed have the non-uniform distribution thus predicted. Elliott (1974) has observed that round objects on the filament surface, interpreted as groups of myosin heads, are indeed arranged on a Bear-Selby

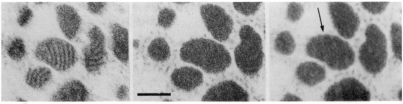

Fig. 31. A transverse section of the smooth adductor muscle of *Ostrea edulis* that shows how tilting the section in the electron microscope produces views along lattice planes of the crystalline structure (cf Fig. 28). Left, grid tilted +34°, showing broad striations corresponding to 20$\bar{1}$ planes; centre, tilted 10°; right, tilted −6°: arrow indicates fine striations corresponding to 301 planes. Scale marker indicates 100 nm (reprinted from Bennett and Elliott (1981), by courtesy of P. Bennett and Academic Press Inc.).

net, but it is now in question how far around the surface of the filament this arrangement extends.

VI. ACTIN-MYOSIN INTERACTION

The interaction of actin, myosin and ATP is believed to be at the centre of a great many forms of cell motility, including muscle contraction. The structure and biochemistry of this interaction is therefore of the greatest interest.

The structural changes during muscle shortening are thought to be brought about by an interaction between actin and myosin. The filaments themselves do not shorten and the movable elements, the cross-bridges, have a range of movement of the order of only 10 nm. Since sliding can involve a relative movement of hundreds of nanometers, it is accepted that some sort of cyclic attachment and detachment of cross-bridges must occur. This is generally pictured as attachment of a myosin head to actin, some type of conformational change of this actomyosin complex (such as a tilting of the myosin head) generating the sliding force, detachment of the head, followed by a change back to its initial structure before reattachment to actin at a point further along the filament (Fig. 32 bottom; see e.g. Huxley, 1969). The total process may be simply pictured as a kind of rowing of the filaments past each other. A detailed understanding of the structural basis of this process requires knowledge of the structures of both thick and thin filaments. Thick filaments have already been discussed in detail (Sections IV and V). Thin filaments are fully described in this volume by O'Brien and Dickens (see also Squire (1981) for a recent review). For easy reference for the rest of this section, the essential features of the structure of the thin filament have been summarized in Fig. 33 and its legend.

The hydrolysis of ATP by actomyosin is also a cyclic process. Each head of myosin binds and hydrolyses ATP, and each can bind to actin. The biochemistry of these interactions has been much illuminated by kinetic studies (Fig. 32 top). Early, it was shown that the steady-state Mg-ATPase rate of myosin, S-1 and HMM is very slow (~ 0.03 s^{-1}) but is activated markedly by F-actin (e.g. Eisenberg and Moos, 1967; Eisenberg et al., 1968). Transient kinetic studies (see Taylor, 1979, for review) have shown that the hydrolysis step of ATP by myosin is actually fast, and that it is the release of the products of hydrolysis that sets the overall low rate for myosin ATPase (Lymn and Taylor, 1970; Fig. 32 top). F-actin accelerates the ATPase rate by enhancing the rate of the step associated with dissociation of ADP and Pi from myosin (Fig. 32 top). The rate and equilibrium constants of the different steps mean that the cycle of ATP hydrolysis by myosin in the presence of actin (as occurs in the filament lattice of the myofibril), will tend to follow the heavy line

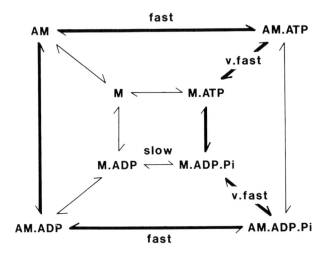
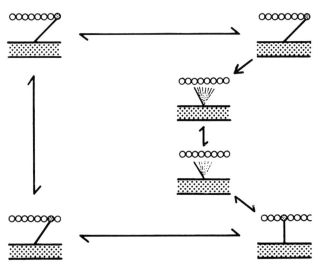

Fig. 32. Correlation between Mg-ATPase cycle and cross-bridge cycle. (Top) Simplified scheme of the myosin and actomyosin Mg-ATPases, shown in the form of two concentric cycles. A is F-actin; M is myosin, HMM or S-1. For clarity, indication of the uptake or release of actin, ATP, ADP or inorganic phosphate (Pi) has been omitted from the scheme. Heavy arrows indicate the predominant pathway *in vitro*. (Bottom) Hypothetical structural stages in the cross-bridge cycle. The scheme would be appropriate for a muscle shortening under no load. The situation in an isometric muscle (where no filament translation occurs) may be more complex. The Z-line would be to the left of the diagram.

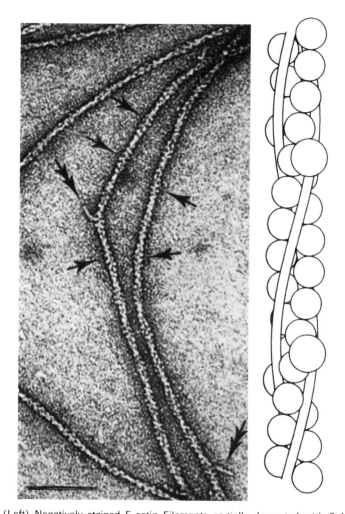

Fig. 33. (Left) Negatively-stained F-actin Filaments partially decorated with S-1. Scale marker indicates 50 nm. (Right) Model of thin filament structure. Scale marker indicates 10 nm. The thin filament contains two chains of globular actin subunits, that are staggered by half the subunit periodicity (5·5 nm) and that wind around one another in two right-handed helices. In projection these helices cross over one another at roughly 38 nm intervals. The actin subunits are somewhat elongated and tilted with respect to the filament axis. (Left) Groups of actin subunits form chevrons (arrows) which have opposite polarity to the S-1 attachment (double arrows). In the grooves between the chains, elongated tropomyosin molecules, about 40 nm long and 2 nm wide, lie end to end with a small overlap between adjacent molecules. The more globular troponin molecule binds at an identical site on each tropomyosin molecule ((Left) from Vibert and Craig (1982), by courtesy of Academic Press Inc.; (Right) reprinted from O'Brien et al. (1975), by courtesy of J. Couch and Academic Press Inc.).

shown in Fig. 32 top. The crucial point is the weakening of the actomyosin interaction by ATP and the strengthening of it following ATP hydrolysis, since this corresponds simply with the detachment/reattachment steps of the structural cycle, which are mandatory if the filaments are to slide past each other by any large distance.

The major goal of muscle research is to unite the biochemical and structural aspects of the cross-bridge cycle into an overall picture of cross-bridge action. A simplified and speculative way of doing this has been shown in Fig. 32. Over the last few years there has been much effort, including electron microscope approaches, to obtain evidence for such a scheme, and to describe it in detail (see Taylor, 1979; Eisenberg and Hill, 1978; Huxley and Simmons, 1971). Much is known about certain *stable* cross-bridge states (rigor and relaxed) and these will be discussed in this section. However, the transitions from one state to the next in a cycling cross-bridge are extremely rapid and it is not a simple process to stop the system in a known state and then to examine its structure. More indirect methods have generally been used. These include:

(i) the use of analogues of ATP which may produce long-lived structural states analogous to intermediate states of the actomyosin ATPase cycle;

(ii) rapid stretch or release of whole muscle fibres: X-ray diffraction and tension and stiffness measurements that give information on the structural state of the molecules of the myofilaments at very short intervals after perturbation of a steadily contracting muscle.

Very recently rapid-freezing techniques have been developed to a point where direct methods may be feasible, by arresting cross-bridges instantaneously for direct examination in the electron microscope (Heuser *et al.*, 1979; Heuser, 1981).

The structural mechanism by which actin-myosin interaction is regulated is also very important, for a muscle must be able to relax as well as contract. In a relaxed muscle the rate of ATP hydrolysis is very low, the tension and stiffness are very small and X-ray studies indicate that there is little actin-myosin interaction. When the muscle is stimulated to contract all of these factors increase greatly. The immediate trigger for contraction is the release of Ca^{2+} ions (to a level of $\sim 10^{-5}$ M) from the sarcoplasmic reticulum into the sarcoplasm surrounding the filaments. Withdrawal of Ca^{2+} ($< 10^{-8}$ M) by the sarcoplasmic reticulum brings about relaxation (see Ebashi and Endo, 1968; Weber and Murray, 1973). It is believed that some step in the cross-bridge cycle is inhibited in the absence of Ca^{2+}, and that this inhibition is relieved when Ca^{2+} is present (see Taylor, 1979; Adelstein and Eisenberg, 1980, for reviews).

There are three main mechanisms of regulation of actin-myosin interaction. In one, which occurs in vertebrate striated muscle and many invertebrate striated muscles (Lehman and Szent-Györgyi, 1975), the regulatory switch (the protein complex of troponin and tropomyosin) is located on the actin filaments (actin-linked regulation), and the binding of Ca^{2+} to troponin is thought to bring about structural changes in the complex that control the interaction of myosin with actin (see O'Brien and Dickens, this volume). In the other two mechanisms, regulation occurs on the myosin filament (myosin-linked regulation) and involves a light chain of myosin, located on the myosin head (see Section III.C).

In all three types of regulation it is of great interest to know what are the molecular changes in structure that *control* the actin-myosin interaction. Both electron microscopy and X-ray diffraction have been applied to this problem and considerable information has been obtained concerning the location, disposition and alterations in structure of the regulatory components. These will be considered in Sections VI.A.5 and 6.

The interaction between actin and myosin, and its regulation, can be studied in the electron microscope either *in vitro*, by negative staining of the complex of purified proteins (a situation analagous to that studied by the kineticists), or in whole muscle (by sectioning) corresponding more closely to the *in vivo* state. The latter method can be correlated more directly with X-ray diffraction and physiological studies and may reveal effects, such as those which the filament lattice may impose on possible actin-myosin interaction, which do not occur in the solution studies. We discuss the whole muscle results in Section VI.B but we start with the *in vitro* methods.

A. *In Vitro* Studies of Actin-myosin Interaction

1. Decorated Actin: The Arrowhead Structure

The most stable actin-myosin interaction occurs in the absence of ATP. This is analogous to the state of *rigor mortis*, where muscle becomes stiff owing to depletion of ATP and consequent strong binding of the actin to the myosin filaments. By analogy with the cross-bridge cycle (Fig. 32) this "rigor" state is thought to be related to the end of the "drive stroke" of the myosin cross-bridge (when actin and myosin are attached but hydrolysis products are displaced). It is therefore a significant state for our understanding of the cross-bridge mechanism and, owing to its stability, more is known of its structure than of any other attached state.

Huxley (1963) obtained the first detailed electron microscope images of rigor actin-myosin interaction. He observed that actin filaments treated with myosin or heavy meromyosin in the absence of ATP displayed a very clear polar structure consisting of a series of "arrowheads". A similar appearance was also generated when S-1 was used (Moore *et al.*, 1970; cf Fig. 34). The repeat of the arrowheads was 37 nm, very close to the crossover repeat of the

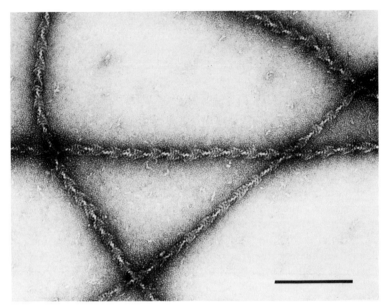

Fig. 34. Negatively-stained arrowheads formed by decorating rabbit F-actin with rabbit S-1 in the absence of nucleotide. The S-1 was prepared by chymotryptic digestion and contains no DTNB light chain. Scale marker indicates 100 nm (reprinted from Craig et al. (1980), courtesy of Academic Press Inc.).

double helix of undecorated F-actin. The arrowheads also showed a clear shorter axial period of about 6 nm similar to the period of the subunits along the F-actin helical strands (Fig. 33; O'Brien and Dickens, this Volume). It was concluded that the S-1, HMM or myosin molecules had simply combined with the F-actin, with little change in its structure, giving rise to a double helix of myosin heads wound around the outside of the original actin filament. The arrowhead was thus simply the envelope of ∼ 14 myosin heads attached to the ∼ 14 actin subunits in one crossover repeat of the F-actin. The clear polarity of the complex revealed by the arrowheads indicated that the underlying actin was also polarized, though this had not been apparent from micrographs of actin alone. By decoration of I-segments (isolated Z-discs with their array of thin filaments projecting from either side) Huxley (1963) found that the polarity of the thin filaments reversed at the Z-line: the arrowheads always pointed away from the Z-line. Such reversal of polarity is, of course, an essential feature of the sliding filament mechanism, where direction of force development is determined by specific interaction of actin and myosin molecules (a similar and equally essential reversal occurs at the centre of the thick filament—see Section II).

Because each arrowhead consists of a complicated superposition of myosin heads and actin subunits, it was not possible to deduce by simple visual inspection the structure of a single actin-myosin head complex. An understanding of the structure had to await the development of three-dimensional reconstruction techniques, using Fourier methods (DeRosier and Klug, 1968). Mathematical procedures allow the calculation of the three-dimensional structure from the observed two-dimensional projection on the micrograph, on the assumption of a known helical symmetry for the particle in question. By these means, Moore *et al.* (1970) calculated the first three-dimensional model of the arrowhead complex. In this model the S-1

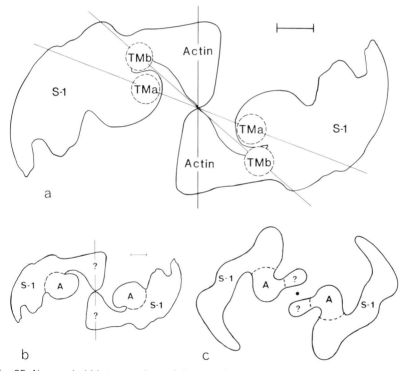

Fig. 35. New and old interpretations of the arrowhead structure. (a) Projection down the filament axis, towards the Z-line, of two acto-S-1 units in the three-dimensional reconstruction of Taylor and Amos (1981). TMa is thought to be the position of tropomyosin in active muscle and TMb its position in relaxed muscle where it could sterically block attachment of myosin heads. (b) An interpretation of the Taylor and Amos reconstruction along the lines of the original Moore *et al.* (1970) interpretation. A is actin. (c) A diagram redrawn from Fig. 5 of Huxley (1972), which was based on the helical projection of the reconstructed image obtained by Moore *et al.* (1970). Scale markers indicate 2 nm (reprinted from Taylor and Amos (1981), courtesy of L. Amos and Academic Press Inc.).

Fig. 36. Solid model of a decorated thin filament based on the three-dimensional reconstruction calculated by Taylor and Amos (1981). This model is based on the average of several images obtained under conditions of minimal electron exposure. One complete arrowhead and parts of two others are shown, pointing towards the top. S-1 is the long, curved, tilted structure on the outside of the filament while actin is near the filament axis and not so readily discerned. Scale marker indicates 10 nm (courtesy of L. Amos).

molecule appeared elongated (15 × 4 nm) and curved, and was tilted at about 50° to the filament axis, such that the subfragment pointed in the direction of the arrowheads. The S-1 also appeared to be slewed (Fig. 35c). It was suggested that it was the shape of S-1 and this specific geometry of attachment that gave rise to the characteristic arrowheads of decorated actin.

In recent work with more refined electron microscope techniques (e.g. minimal dose exposure and image averaging), it has been suggested that some details of this interpretation are incorrect. Toyoshima and Wakabayashi (1979) and Wakabayashi (1980) conclude from their reconstructions that S-1 is tilted at an angle of 75–90° to the filament axis (cf 50° of Moore et al., 1970). Thus they believe that the arrowhead appearance is generated principally by the slewing and curvature of S-1s that are attached approximately perpendicularly to the filament.

From their three-dimensional reconstruction of decorated thin filaments (Fig. 36), Taylor and Amos (1981) conclude that the identification of S-1 and actin in the model of Moore et al. is partially incorrect. The component interpreted by Moore et al. as actin is here thought to be part of S-1 (compare Fig. 35a with 35b and c), and the gap in the centre of the Moore et al. structure (Fig. 35c) is here filled in (apparently due to the greater radial resolution obtained by Taylor and Amos) and interpreted as actin (Fig. 35a). Thus the actin–S-1 link is quite different from that proposed by Moore et al.

The shape of an actin subunit in these reconstructions, and the contacts it makes with other subunits (principally with subunits in the other long-pitched strand) are in good agreement with reconstructions of actin in paracrystals (Wakabayashi et al., 1975). The S-1 binding site is on the other side of the actin subunit from that in the Moore et al. model (compare Fig. 35a with b and c). S-1 is about 13 nm long (shorter than the true head length, partly owing to the absence of the DTNB light chain (see Section VI.A.6)), and 5–6 nm thick near its attachment to actin. The S-1 is curved and slewed as in the Moore et al. model, and there is a bend about 6 nm from its tip, beyond which it is narrower (Figs 35a, 36) as found in shadowed myosin molecules (see Section III.B). The actin-binding site is just to the outside of the tip of the myosin head (not on the side as found by Moore et al.) so that a portion of S-1 extends into the long pitch helical groove of actin (Fig. 35a).

The angle of tilt is about 60° (similar to Moore et al.), although, as pointed out by Taylor and Amos, the definition of this angle is rather a semantic problem and depends on the choice of the main axis of S-1. The axis chosen by Taylor and Amos (from the centre of the actin–S-1 binding site, along the main body of the molecule to the distal portion at highest radius from the particle axis) means that S-1 *as a whole* is considerably tilted. The *main* region of density in their model (and chosen by Wakabayashi as the axis) is indeed closer to 90°. The axis most appropriate to describing any changes in angle of attachment during contraction, however, would seem to be that which passes through the head-tail junction. Direct observation of single attached heads (both S-1 and HMM) supports the view that the axis of the head, as a whole, *is* considerably tilted (Fig. 39; Section VI.A.3).

Vibert and Craig (1982) have similarly calculated three-dimensional reconstructions from improved images of decorated filaments, in this case using scallop proteins (see Section VI.A.6 and Craig *et al.*, 1980), and come to the same conclusions as those of Taylor and Amos. Such agreement between two independent groups working on different systems and with a large number of excellent images would suggest that this model, despite its considerable changes from the long-accepted model of Moore *et al.*, should be considered very seriously. It should be pointed out, however, that Wakabayashi (1980) has not committed himself to either interpretation, so that agreement is not yet unanimous.

A more complex geometry for the actin-S-1 interaction has been proposed following a refinement of the Taylor and Amos reconstruction by Amos *et al.* (1982). An additional criterion imposed during selection of filaments for reconstruction was that their optical diffraction patterns should be consistent with the X-ray diffraction patterns from hydrated insect flight muscle fully decorated with S-1 (Holmes *et al.*, 1982). The most significant new feature of the revised reconstruction is that each S-1 makes contact with two actin subunits. The major contact demonstrated by Taylor and Amos is still present, but an additional contact is made with an actin subunit lying adjacent on the other long-pitched helical strand. This result is consistent with chemical cross-linking data that have shown that S-1 binds to two actin subunits (Mornet *et al.*, 1981).

Decorated actin has recently been observed by the rapid freezing technique of Heuser *et al.* (1979). Following freezing, the specimen is etched and rotary shadowed producing the appearance shown in Fig. 37 (Heuser, 1981; Heuser and Cooke, 1983). The decorated actin filament appears very similar to a two-stranded rope, corresponding to the two long-pitch acto-S-1 helices. There is generally little directionality to the rope when compared to the arrows seen by negative staining. We believe this is explained by the fact that this technique gives a view of one surface of the structure. The marked polarity of the arrowheads arises substantially from the fact that the structure, seen by negative staining, is a projection of the upper and lower sides, which are equally contrasted and viewed superimposed on each other. The three-dimensional solid model calculated from the arrowheads (Fig. 36) does not itself have a strong arrowhead appearance, and filaments with this structure, contrasted mainly on one surface (as with shadowing) would indeed be expected to have an appearance resembling a fairly non-polar two-stranded rope, as observed. This is supported by the filament on the right of Fig. 37b

2. Myosin Filaments and Actin-myosin Complex

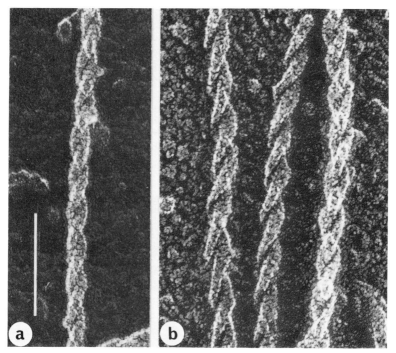

Fig. 37. S-1-decorated actin observed by the procedure of rapid freezing followed by freeze-etching and rotary shadowing. (a) Unfixed, (b) fixed and treated with uranyl acetate before freezing. Scale marker indicates 100 nm (from Heuser and Cooke (1983), courtesy of J. Heuser and Academic Press Inc.).

which has accumulated metal on both upper and lower sides and clearly has downward-pointing arrowheads. The assumption that the arrowheads seen by negative staining do approximate the native structure of decorated actin is supported by X-ray diffraction observations. Optical diffraction patterns of arrowheads closely resemble X-ray diffraction patterns of native insect flight muscle that has been treated with S-1 to decorate fully all the actin filaments in the filament lattice (Amos *et al.*, 1982).

2. *The Arrowhead Structure—A Structural Assay for F-actin*

The characteristic shape and easy identification of the arrowhead structure have proved immensely useful as a means of identifying F-actin filaments in non-muscle cells. The cytoplasmic matrix of most cells abounds in filaments of various shapes and sizes (see Goldman *et al.*, 1976) and their identification by simple visual inspection of electron micrographs is not straightforward.

Fig. 38. Thin section through basal end of a stereocilium from the cochlea of the alligator lizard (*Gerrhonotus multicarinatus*). The preparation was demembranated and decorated with S-1, producing arrowheads on the actin filaments. PM is a remaining fragment of the membrane limiting the basal end of the stereocilium. Scale marker indicates 200 nm (reprinted from Tilney *et al.* (1980), courtesy of L. Tilney and The Rockefeller University Press).

However, when S-1 or HMM is allowed to diffuse into a cell, whose membrane has been disrupted, it binds to those filaments composed of F-actin to form arrowheads, which can then be recognized in sections or by negative staining (Ishikawa et al., 1969; and e.g. Clarke et al., 1975). An example of the arrowheads seen by this method is shown in Fig. 38 (Tilney et al., 1980). The decoration method reveals not only the presence of actin filaments but their polarity as well. This allows comparison with the structural arrangement in muscle (where the arrowheads point away from the Z-line) and therefore may suggest possible models for interaction with myosin, when myosin is present, to produce the motile activity in question.

3. Partial Decoration of Actin

Owing to superposition effects, the only way in which the appearance of fully decorated filaments can be interpreted is by three-dimensional reconstruction (Section VI.A.1). However, if the filaments are only partially decorated, these effects are greatly reduced and certain features of the binding of myosin heads to actin can be observed and interpreted directly. Partial decoration is achieved simply by using lower concentrations of S-1 or HMM.

When S-1 is used for decoration one can observe the attachment of single heads to actin (Fig. 39f and g; Craig et al., 1980). S-1 containing a full complement of light chains is curved, about 15 nm long in projection, and its width near its point of attachment to actin is about 4 nm (Vibert and Craig, 1982). This tapers somewhat in the distal third of the molecule. S-1 lacking its regulatory or DTNB light chain is ~ 10 nm long and less obviously curved (Fig. 39g; Vibert and Craig, 1982). These findings agree with three-dimensional reconstructions (Taylor and Amos, 1981; Vibert and Craig, 1982) and with observations of single myosin molecules (Elliott and Offer, 1978; Takahashi, 1978; Flicker et al., 1981).

When actin filaments are partially decorated with HMM, the two heads can be both clearly seen attached to actin (at about the same angle), generally to neighbouring subunits of the same long-pitch helical strand (Fig. 39a–e; Craig et al., 1980). This shows directly that the second head of HMM can attach to actin. Solution studies had suggested that this was so (Margossian and Lowey, 1973), but it had not been demonstrated directly nor had the precise nature of binding of the two heads been observed. Since both heads attach at the same angle, there must be some distortion distally to allow them to come to their common junction with the tail, assuming the heavy chains are in register (see Offer and Elliott, 1978). This is indeed apparent in the micrographs, the distal ends of the two heads having different curvature (Fig. 39a–e). Sometimes the tail can be seen emerging from the junction of the heads (Fig. 39e).

Trinick and Offer (1979) have shown that under certain *in vitro* conditions it

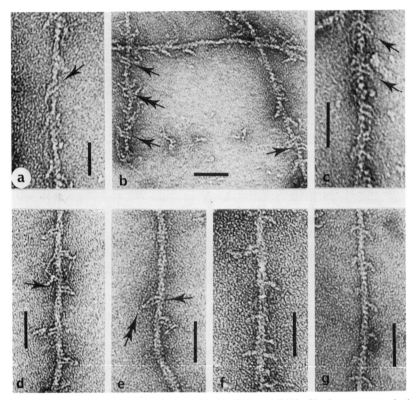

Fig. 39. Thin filaments partially decorated with S-1 or HMM. Single arrows mark the attachment of pairs of HMM heads to neighbouring actin subunits. (a–c) scallop HMM; (d) and (e) rabbit HMM; (f) scallop S-1; (g) rabbit S-1 with no DTNB light chain. Double arrow in (b) points to heads of HMM apparently binding to next nearest neighbour actin subunits. Double arrow in (e) points to part of myosin tail bent back at an acute angle to the two heads. This is best seen by viewing obliquely along the direction of the arrow. Scale markers indicate 30 nm (reprinted from Craig et al. (1980), courtesy of Academic Press Inc.).

is possible for the two heads of HMM to attach to different actin filaments. The filaments thus become crosslinked and often form ordered arrays, usually with adjacent filaments anti-parallel and with links at regular intervals of 38 nm formed by the two HMM heads (Fig. 40). The one-filament or two-filament modes of interaction of HMM with actin assume special importance when the interactions that occur in the filament lattice of muscle are considered (Section VI.B.2).

4. Non-rigor Modes of Cross-bridge Attachment

It is widely accepted that the structure of the actomyosin-head complex changes during the cross-bridge cycle (Fig. 32). It has therefore long been a

2. Myosin Filaments and Actin-myosin Complex

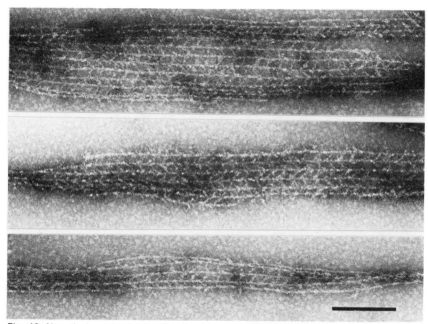

Fig. 40. Negatively-stained rafts showing the cross-linking of F-actin filaments by HMM. Scale marker indicates 200 nm (reprinted from Trinick and Offer (1979), courtesy of J. Trinick and Academic Press Inc.).

goal of muscle research to detect states of cross-bridge attachment other than the rigor state, i.e. stages of the cross-bridge cycle other than the end of the drive stroke. Changes in the angle of attachment to actin of a head of fixed shape have generally been sought. It should be noted, however, that in principle, changes may equally occur in the shapes of the actin subunits or of the heads themselves with no alteration of their interface, and in one model of contraction (Harrington, 1971, 1979) the conformation of the actomyosin-head complex does not actively change at all, shortening occurring in the S-2 portion of the molecule.

Analogues of ATP (principally AMP. PNP) have been used to perturb the rigor acto-S-1 complex in attempts to demonstrate the existence of other attached states. The most direct approach has been to use negative staining to study the decoration of actin by S-1 or HMM in the presence of analogues. However, when clear decoration has been obtained under conditions where the complex of actin, S-1 and analogue might be expected to exist, the structure has not appeared obviously different from rigor arrowheads (Craig and White, unpublished results). The future may lie in the use of chemically crosslinked actomyosin (Mornet *et al.*, 1981) or in rapid freezing techniques where the cross-bridges are stopped at different stages of their interaction with actin and ATP. This is clearly an area deserving of intensive study since only one attached state has ever been observed by negative staining, and the tilting-bridge model implies that there should be at least one other (Huxley, 1969; Huxley and Simmons, 1971). Comparable studies on intact muscle are reviewed in Section VI.B.3.

5. Regulation: Actin-linked

Actin-linked regulation of muscle contraction involves the control proteins troponin and tropomyosin: tropomyosin, an α-helical coiled-coil molecule, ∼40 nm long and 2 nm in diameter, follows the long-pitched grooves of the actin double helix, and troponin, a more globular protein, is attached to tropomyosin at intervals of 38·5 nm (every 7th actin; Fig. 33; see Spudich *et al.*, 1972; Cohen, 1975; O'Brien and Dickens, this volume). Based on changes in the X-ray pattern of contracting muscle (see Squire, 1975, for review) and on electron microscopy of thin filaments in the active and relaxed states (Wakabayashi *et al.*, 1975; Gillis and O'Brien, 1975) it has been suggested that tropomyosin rolls in the actin groove, in response to Ca^{2+} binding to troponin, from a position where it blocks S-1 attachment (the relaxed state) to a position where attachment is not blocked (the active state) (Fig. 41a). This "steric-blocking" mechanism for maintaining the relaxed state has been widely accepted and accounts for much of the biochemical data but, as pointed out by the original authors (see Seymour and O'Brien, 1980), is based on an unproved assumption—that tropomyosin does indeed lie on the side of the groove that coincides with the S-1 binding site on actin (Fig. 41a). If it lay on the other side (Fig. 41b) it is not so clear how steric blocking would operate. This ambiguity is equivalent to an uncertainty in the orientation of the filaments used to produce the actin-tropomyosin model. Seymour and O'Brien (1980) reasoned that if this uncertainty could be removed, for example by reconstructing filaments still attached to a Z-line, which defines their orientation, the ambiguity would no longer exist. One could then superimpose, with the correct relative orientation, reconstructions of undecorated thin filaments and decorated thin filaments (whose orientation is defined by the direction of the arrowheads) to determine the side of the groove on which tropomyosin lies. Using this approach, they concluded that tropomyosin is, in fact, on the opposite side of the groove from that previously postulated for steric blocking (Fig. 41b).

Since troponin-tropomyosin does not prevent the binding of S-1 to actin in the absence of ATP, it should be equally possible to settle this matter directly, by observing the position of tropomyosin in a *decorated* filament and seeing whether it indeed lies on the same side of the groove as the attached S-1. The

2. Myosin Filaments and Actin-myosin Complex

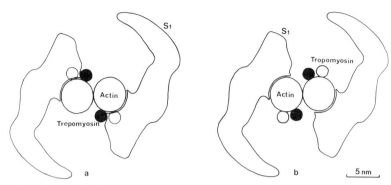

Fig. 41. (a) Steric blocking mechanism for the prevention of head attachment to actin by tropomyosin. This view is comparable to the views down the helix axis of decorated actin shown in Fig. 35 and is based on the Moore *et al.* (1970) interpretation of the arrowhead structure (Fig. 35c). Two positions for tropomyosin are shown: the open circle where tropomyosin coincides with part of the S-1-binding site on actin, and so could physically block attachment; and the filled circle where tropomyosin has moved towards the centre of the actin groove thus unblocking the binding site. (b) Similar view to (a) illustrating the alternative locations of tropomyosin, on the opposite side of the actin groove to (a). Here S-1 attachment would not be blocked. (reprinted from Seymour and O'Brien (1980), courtesy of J. Seymour and Macmillan Journals Ltd).

difficulty with this method lies in identifying the different components in the reconstructed image, and in seeing tropomyosin (2 nm in diameter and apparently tending to be positively stained) in a decorated filament 20 nm in diameter. Nevertheless, Taylor and Amos (1981) and Vibert and Craig (1982) obtained three-dimensional reconstructions where tropomyosin was thought to be visible. It appeared in decorated native thin filaments as a line of continuous density running near the centre of the actin groove, in close contact with S-1. In decorated F-actin, this feature was weaker and less continuous. With this identification Taylor and Amos, using rabbit skeletal proteins, concluded that tropomyosin was indeed on the "wrong" side of the groove for steric blocking if one accepts the location of the S-1 binding site on actin suggested by Moore *et al.* (1970; compare Figs 35a and 41a). But, as discussed earlier (Section VI.A.1), they believed that this S-1 binding site was mechanism was again possible, with the components simply moved around from their old positions (Fig. 35a). The experiments of Vibert and Craig (1982) using scallop proteins (see next section) supported this view.

Recently, however, the situation has changed again. O'Brien *et al.* (1983), have reinterpreted the data of Seymour and O'Brien (1980) as showing only the position of the actin, with the tropomyosin not visible. Their new reconstruction of actin plus tropomyosin alone, places tropomyosin back on the side of the groove originally assumed by Moore *et al.* (1970). Thus, given the new Taylor and Amos identification of the S-1 binding site, steric blocking as originally envisaged is not possible. However, the tropomyosin could interfere with the second contact between S-1 and actin that was demonstrated by Amos *et al.* (1982) (O'Brien *et al.*, 1983; see chapter by O'Brien and Dickens, this Volume).

While the location of tropomyosin must for now be regarded as an open question, current results would point to a model where tropomyosin can only partially block the attachment of myosin heads to actin. The situation may be even more complex however. The S-1 binding sites observed are for the rigor state, which is probably similar to that at the end of the cross-bridge drive stroke. But the binding site at the start of the drive stroke may be quite different, and tropomyosin may be in a position to block attachment at this point completely. This is important since steric blocking has generally been viewed as a blocking of *attachment* of S-1 to actin. On the other hand, recent experiments suggest that S-1 can bind to regulated thin filaments even at low Ca^{2+} levels where no activation of S-1 ATPase occurs (Chalovich *et al.*, 1981; Chalovich and Eisenberg, 1982). This implies that the attachment of S-1 to actin is not greatly inhibited, but possibly the transition to the end of the drive stroke is. In this case the coincidence of the position of tropomyosin at low Ca^{2+} with a rigor S-1 binding site on actin might suggest a steric blocking of this transition.

6. Regulation: Myosin-linked

Structurally, little is known about the mechanism of myosin-linked regulation. X-ray diffraction of contracting muscle would be unlikely to help as it has with actin-linked regulation, since the regulated components, the myosin heads, presumably cycle asynchronously in a contracting muscle, giving rise only to diffuse X-ray scattering. Biochemical studies, however, demonstrate that the regulatory light chains (Section III.C) are essential for Ca^{2+} control (Szent-Györgyi *et al.*, 1973).

To understand the structural mechanism of myosin-linked regulation one would like to know the location of the light chains to determine, for example, whether they might interact directly with the actin-binding site on S-1. One would also like to know what structural changes occur in the heads when they are switched on—by Ca^{2+} binding (in the case of scallops and other invertebrates) or when the light chain is phosphorylated (in vertebrate smooth muscle and non-muscle cells).

Some answers to these questions are now starting to appear from study of changes in the actin-myosin complex when the regulatory light chains are removed. In the case of the scallop this is achieved reversibly by very mild treatment (using 10 mM EDTA) (see Chantler and Szent-Györgyi, 1980). X-ray diffraction patterns of rigor scallop fibres depleted of their regulatory

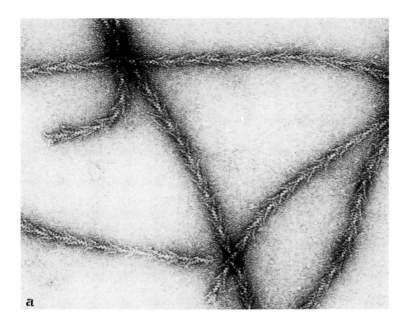

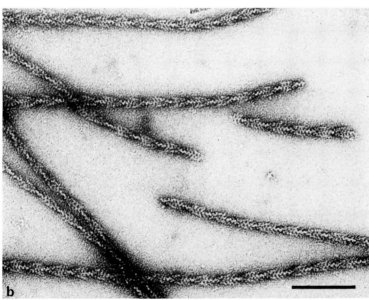

Fig. 42. Decoration of scallop thin filaments with scallop S-1. (a) S-1 with regulatory light chain present. Arrowheads barbed and slightly concave. (b) S-1 from which light chain has been removed. Arrowheads blunt and slightly convex. Scale marker indicates 100 nm (reprinted from Craig *et al.* (1980), by courtesy of Academic Press Inc.).

light chains show changes in diffracted intensity coming from cross-bridges attached to actin, implying a change in the distribution of cross-bridge density along the actin filament (Vibert et al., 1978). Electron micrographs of thin filaments decorated with scallop S-1 with and without the regulatory light chains reveal clear differences in the arrowhead structures (Craig et al., 1980). With the light chains present the arrowheads have a "barbed" appearance (Fig. 42a) and appear slightly concave in outline. They are 26 nm wide at their base. When the light chains are removed the arrowheads lose their barbs and become more "blunt" with a convex outline (Fig. 42b). They are now only 19 nm wide. Similar changes in arrowhead appearance occur when HMM is used instead of S-1. With rabbit skeletal proteins (where there is no myosin-linked regulation) a barbed/blunt transition can also be observed, in this case when the DTNB light chain is removed. The arrowheads in Fig. 34 were obtained using rabbit S-1 depleted of DTNB light chain and appear similar to the "blunt" scallop arrowheads of Fig. 42b.

Three-dimensional reconstructions of the barbed and blunt structures reveal a substantial reduction in mass of the distal end of the head on removal of the scallop regulatory light chain (Fig. 43; Vibert and Craig, 1982). With the

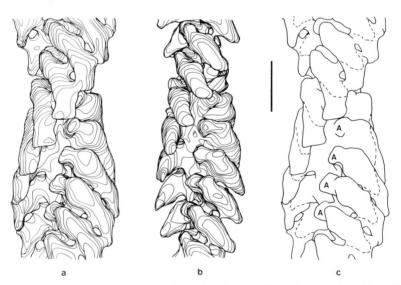

Fig. 43. Three-dimensional reconstructions of scallop thin filaments decorated with scallop S-1. (a) S-1 with regulatory light chain present; (b) S-1 lacking light chain; (c) shows the difference between (a) and (b) (A is actin; actin-S-1 complexes shown in outline only): when the light chain is removed, the S-1 is shorter, as indicated by the dashed lines. Note that in contrast to Fig. 42(a), the arrowhead in (a) does not show an obvious barbed structure. The barbed appearance is the result of superposition of several long S-1 molecules which are seen in projection in the negatively-stained image. Scale marker indicates 10 nm (from Vibert and Craig, 1982, by courtesy of Academic Press Inc.).

light chain present the apparent length of S-1 is ~17 nm whilst when it is removed, the length is only ~13 nm. There is no obvious change in the proximal 9 nm of S-1 nor in its mode of attachment to actin. When the shape of S-1 is observed directly in partially decorated filaments, this length difference is also obvious (Vibert and Craig, 1982). These results suggest that the ~10 nm long regulatory light chain (Stafford and Szent-Györgyi, 1978) is located in the distal half of the 20 nm long head (Section III.B), probably extending to the head-tail junction (Fig. 8). The results of shadowing myosin molecules support this interpretation (Section III.C). A simple view of these results, then, would suggest that the light chain could not sterically block the actin-binding site on S-1, which is near the tip of the head (Fig. 8b), and that some more indirect mechanism must operate.

Similar arrowheads are produced at low and high Ca^{2+} levels, so no clue to the mechanism of myosin-linked regulation has emerged in this way. This may be due to the difficulty of controlling Ca^{2+} levels when staining with uranyl acetate, or to the fact that observations are made in the absence of ATP, whereas regulatory structural changes may occur only in its presence.

Barbed arrowheads are obtained when actin is decorated with gizzard S-1 or HMM containing intact regulatory light chain (Kendrick-Jones et al., 1982), suggesting that this light chain is also located in the neck in vertebrate smooth muscle myosin. No change in arrowhead structure has yet been detected when the S-1 or HMM is phosphorylated. Nevertheless, the similarity in structure and location of the gizzard and scallop regulatory light chains suggests that, although the regulatory signals differ, the structural mechanisms of regulation may be very similar. The fact that the regulatory light chains are present in the neck region of the myosin head suggests, as one possibility, that regulation may involve alterations in the flexibility of this part of the head, possibly at the head-tail junction.

B. Actin-myosin Interaction in Intact Muscle

So far we have considered electron microscopy of actin-myosin interaction in solution, a situation not directly comparable with the *in vivo* state. Here we discuss thin sections of epoxy-embedded whole muscle; whilst generally of lower resolution than negatively-stained arrowheads, these do provide a closer approach to the *in vivo* situation and are also more directly comparable with X-ray diffraction studies of intact muscle.

1. Evidence for Actin-myosin Interaction in the Filament Lattice

As was shown in the Introduction, the filaments of vertebrate striated muscle are parallel to each other and organized into a regular two-dimensional

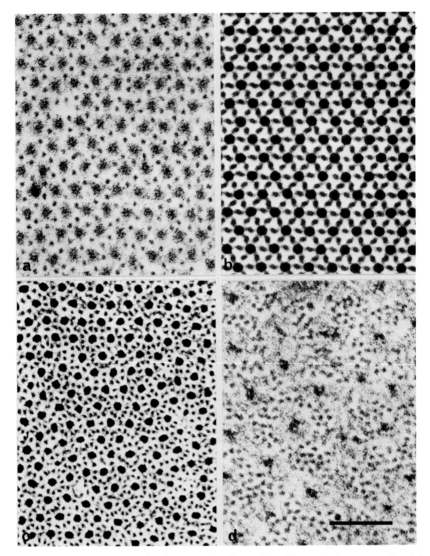

Fig. 44. Transverse sections of various muscles illustrating different lattice symmetries. (a) Vertebrate skeletal muscle with six thin filaments, at triad positions, around each thick filament; (b) insect flight muscle, with six thin filaments at dyad positions; (c) scallop striated muscle, with about eleven thin filaments, not at precise lattice positions, around each thick filament; (d) vertebrate smooth muscle, with many thin filaments around each thick filament. Scale marker is only approximate and indicates about 100 nm ((a) reprinted from Huxley (1965), courtesy of H. E. Huxley and Scientific American Inc.; (d) from Somlyo et al. (1973), courtesy of A. Somlyo and The Royal Society; (b) and (c) Craig).

hexagonal lattice (Fig. 1c). This was deduced from the equatorial X-ray diffraction pattern of muscle and from observation of transverse sections of muscle in the electron microscope (Huxley, 1953a, b). In other muscles the same basic arrangement exists although the details may vary, such as the number and location of the thin filaments around each thick filament. The lattices are sometimes highly ordered (e.g. vertebrate striated, insect flight; Fig. 44a, b) and sometimes less so (e.g. scallop striated, vertebrate smooth; Fig. 44c, d). Under *in vitro* conditions, early stages of lattice formation may occur spontaneously (Hayashi *et al.*, 1977). The arrangement of filaments into a regular lattice makes possible a rapid, efficient and directional development of force on stimulation.

What interactions occur between thick and thin filaments in the lattice of the myofibril? In the relaxed state, muscle has very low stiffness and the array of thick filaments and the array of thin filaments diffract X-rays independently of one another (Huxley and Brown, 1967). In solution under relaxing conditions the association constant of S-1 with thin filaments and the activation of S-1 ATPase are both low (Chalovich *et al.*, 1981). These observations imply that the interactions of thick and thin filaments in a relaxed muscle, if they occur, are short-lived.

When a muscle goes into rigor, its stiffness increases greatly and there are large changes in the intensities of the equatorial X-ray reflections, accompanied by strengthening of the actin layer lines (and an increase in effective radius of the scattering units giving rise to them) and weakening of the myosin layer lines (Huxley and Brown, 1967). The association constant of S-1 with actin measured in solution in the absence of ATP is high (e.g. Greene and Eisenberg, 1980). In transverse section in the electron microscope, the thick filaments in rigor muscle are smaller (in relation to the thin filaments) than they are in relaxed muscle (Huxley, 1968). Together, these observations suggest that when a muscle passes into rigor the myosin cross-bridges move from the surface of the thick filaments to the thin filaments where they form long-lived and extensive interactions with actin (Huxley, 1968).

During contraction the stiffness of a muscle is greater than in the relaxed state and the intensities of the equatorial X-ray reflections are similar to those of rigor muscle (see Huxley, 1979). In solution under contracting conditions S-1 ATPase is strongly activated by thin filaments (see Weber and Murray, 1973). These results suggest a physical contact between thick and thin filaments in a contracting muscle.

2. *Geometric Constraints on Actin-myosin Interaction in the Filament Lattice*

In all striated muscles there are fewer myosin heads than actin subunits in the overlap zone. Thus even if all heads were to attach to actin there would still be

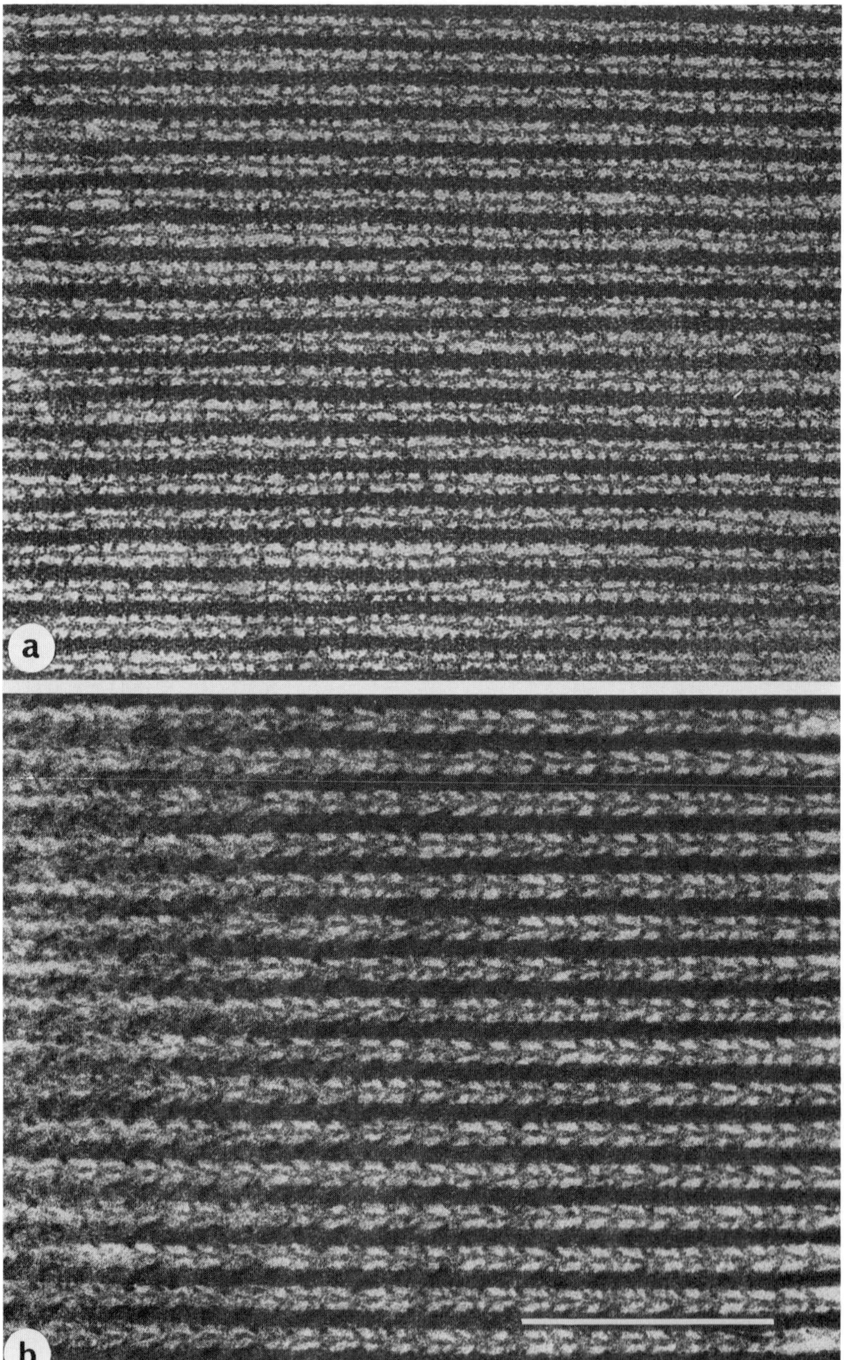
a
b

some subunits unoccupied. On the other hand, although more than sufficient actins are present to satisfy all myosin heads, geometric constraints may prevent a fraction of the heads from binding. The most dramatic example of this is Wray's (1979b) suggestion that at certain longitudinal displacements of thick relative to thin filaments in insect flight muscle there may be virtually no interaction of myosin heads and actin, while if the filaments slide past each other only 20 nm from these minima, maximal interaction is possible. This he suggests is a major factor in the stretch-activation of insect flight muscle (Pringle, 1978).

Most of our knowledge of the detailed way in which thick and thin filaments can interact with each other in the lattice of the myofibril has come from studies of rigor muscle, where the actin-myosin interactions are relatively stable and fixed. Insect flight muscle has provided our greatest insights into the

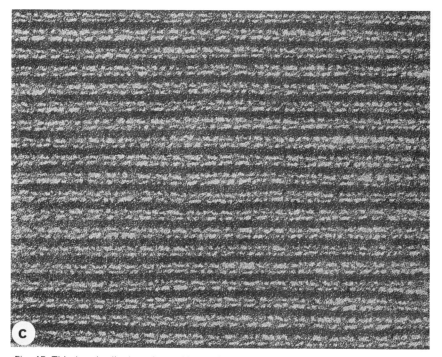

Fig. 45. Thin longitudinal sections of insect flight muscle. (a) Relaxed, with cross-bridges, approximately perpendicular to the filament axis, at 14·5 nm intervals; (b) rigor, with bridges at about 45° to the filament axis and forming double chevrons with a repeat of 38 nm; (c) in AMP.PNP with cross-bridges at varying angles and showing signs of both 38 and 14·5 nm periodicities. The bare zone is to the right of the pictures. Scale marker indicates 300 nm (courtesy of M. Reedy and M. Reedy, unpublished micrographs; see also Reedy et al., 1983).

structure of rigor muscle owing to its highly ordered structure. Longitudinal sections of insect flight muscle show that in rigor the cross-bridges are attached to actin in an angled configuration; the angle is approximately 45° (Reedy et al., 1965; Reedy, 1968; Fig. 45b). A tilted configuration is also supported by arrowhead images (Section VI.A) and by X-ray diffraction of whole muscle (Reedy et al., 1965; Miller and Tregear, 1972). The tilted cross-bridges point away from the Z-line (as in decorated actin) and occur in pairs, forming chevrons. The chevrons themselves are also usually paired longitudinally (with about 15 nm between neighbours), producing double chevrons. The repeat of the double chevrons is 38 nm (the same as the thin filament crossover repeat). In thin transverse sections one observes an array of "flared X"s: each thick filament is surrounded by six thin filaments, four of which receive bridges (the four arms of the X centred on the thick filament), and two of which receive none (Fig. 46). The two bridges on one side of an X frequently originate from the same root.

Reedy's major conclusion from the interpretation of these images (Reedy, 1968) was that the very specific distribution of cross-bridges between thick and thin filaments, with regular alternation of regions of high (the double chevrons) with regions of low (the gaps between the double chevrons) numbers of bridges, implied specific geometric constraints on the interaction of actin and myosin in the filament lattice. The regions of the thin filaments to which heads were bound were termed by Reedy "target areas", and it was suggested that these were determined by optimum azimuthal orientation of actin subunits relative to neighbouring thick filament heads (cf. Fig. 47). The fact that the cross-bridges are arranged with the periodicity of the actin

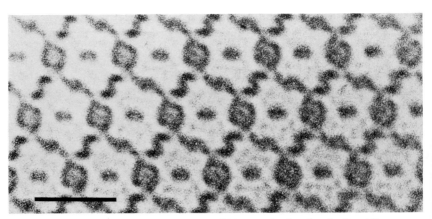

Fig. 46. Thin transverse section of rigor insect flight muscle showing the flared X structure: optically filtered image. The view is from the Z-line towards the M-line. Scale marker indicates 50 nm (courtesy of M. Reedy and M. Reedy, unpublished data).

2. Myosin Filaments and Actin-myosin Complex

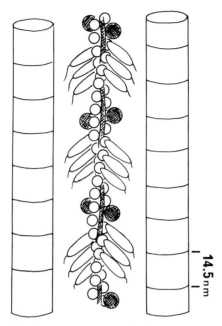

Fig. 47. Diagram illustrating the general concept of a "target area", where the potential for myosin head attachment to thin filaments is confined to a limited group of actin subunits. The diagram shows one thin filament in the dyad position between two thick filaments, as in insect and fast crustacean muscles. The heads originate from the 14·5 nm repeat shown on the thick filaments, but the exact way in which they are connected to the backbone is not known. For them to attach to the unoccupied actins with the same geometry as is shown for the attached heads, large distortions would be necessary. Cross-hatched molecules are troponin and tropomyosin (reprinted from Wray *et al.* (1978), courtesy of P. Vibert and Academic Press Inc.).

filament, rather than that of the myosin filament from which they originate, suggests that the actin filaments are the major constraining factor and that some flexible portion of the myosin molecule enables the heads to "search" in three dimensions for accessible actins, thus enabling them to adopt the thin filament symmetry while still attached to the thick filament (see Wray *et al.*, 1978; Section III.A). The hinge in the myosin tail (Huxley, 1969; Elliott and Offer, 1978) might fulfil this role.

Squire (1972) extended Reedy's interpretation using further geometric arguments. He pointed out that the approximate two-fold axis of the flared-X did not necessarily mean that the myosin filament was two-stranded, as assumed by Reedy (1968). He suggested that there might be six cross-bridges at each level in the thick filament (in accordance with his general model of thick filament structure, Section IV.C) but that the arrangement of thin filaments might be such that only four were in sterically favourable orientations to receive cross-bridges, producing the four arms of the flared-X. He suggested that those heads not stabilized by attachment to actin might not be seen in the electron micrographs (evidence for this has been obtained by Reedy (1971)). Thus the flared-X would not give information on the symmetry of the thick filament but on the arrangement of the surrounding thin filaments.

Recent interpretation of X-ray diffraction patterns from insect flight muscle has supported Reedy's concept of "target areas" on the thin filaments (Holmes *et al.*, 1980). The X-ray patterns are not compatible with an even distribution of myosin heads along the thin filament but imply a periodic alternation of regions of high and low occupancy by myosin heads (Fig. 47). X-ray patterns show that such geometric constraints apply to other muscles as well, e.g. scallop, lobster and crayfish (Wray *et al.*, 1978; Vibert *et al.*, 1978). Electron microscope observations of other rigor muscles support this view and further show that rigor bridges are angled, as in the insect. Chevrons with a 38 nm repeat may be seen in crayfish, lobster (fast and slow muscles), and in scallop striated muscle (Craig, unpublished results). One of the most surprising aspects of the target area concept is that there appear to be some actin subunits (possibly greater than a half in insect) that are not used at all during contraction, even when the filaments slide past one another, since their myosin-binding sites are in positions and orientations that cannot be reached by myosin cross-bridges. The geometric constraints on the interaction of actin and myosin in whole muscle can thus be quite extreme.

Using geometric arguments, Offer and Elliott (1978) have suggested that, in rigor insect flight muscle, the two myosin heads from a single myosin molecule may bind to different thin filaments. They show that, with certain assumptions about the symmetry of the filaments and their arrangement, the two heads may bind with less strain to different filaments than to actin subunits in the same filament (Fig. 48). This would account for Reedy's electron micrographs which often show the two bridges on one side of a flared-X arising from a common origin on the thick filament. A flared-X would thus contain heads from only two myosin molecules, not four. Subsequent calculations (Offer *et al.*, 1981) show that the X-ray diffraction pattern of rigor insect flight muscle is compatible with the view that all actin-myosin interactions in this muscle occur by such two-filament interactions. In support of this idea, Trinick and Offer (1979) have demonstrated that in solution two thin filaments can be crosslinked by a myosin molecule (Section VI.A.3; Fig. 40). It must be emphasized that such a view is not incompatible with observations of both heads bound to a single filament (Craig *et al.*, 1980; Section VI.A.3; Fig. 39a–e). The preferred configuration will depend on the geometry of the system in question (in the work of Craig *et al.* thin filaments were not free to move, so crosslinking could not occur). For the same reason, the arguments of Offer

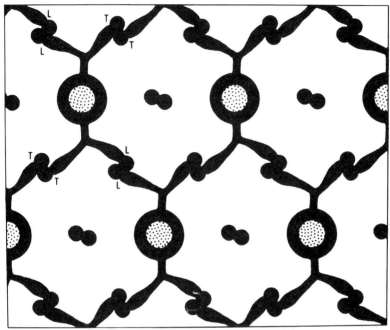

Fig. 48. Two filament interaction model for the appearance of thin transverse sections of rigor insect flight muscle (cf. Fig. 46). The view is from the M-line towards the Z-line. L and T refer to cross-bridges contributing to leading and trailing halves respectively of a double chevron. Note that the four bridges per thick filament are made by two myosin molecules. The heads of the remaining myosin molecules, which are unable to bind to actin, are not shown (reprinted from Offer *et al.* (1981), by courtesy of G. Offer and Academic Press Inc.).

and Elliott for insect flight muscle cannot be extended to other muscles until the structure and arrangement of the filaments are known in these cases.

Although the two filament interaction model is a simple and attractive way of explaining many data, it is not in full agreement with tests applied to it. The model predicts that only 56% of the heads are attached in rigor (Offer *et al.*, 1981), whilst Lovell *et al.* (1981) conclude from digestion experiments that at least 70% are attached. EPR measurements on rigor insect flight muscle should give another estimate of the fraction of heads attached, but have not yet been reported. EPR measurements on vertebrate muscle suggest that in this case all the heads are attached in rigor (Thomas *et al.*, 1980; Thomas and Cooke, 1980), in agreement with digestion experiments on the same system (Lovell and Harrington, 1981). This difference between insect and vertebrate muscles implies differences in interfilament interaction between the two species, as would be expected from their different lattice arrangements (Fig. 44a and b).

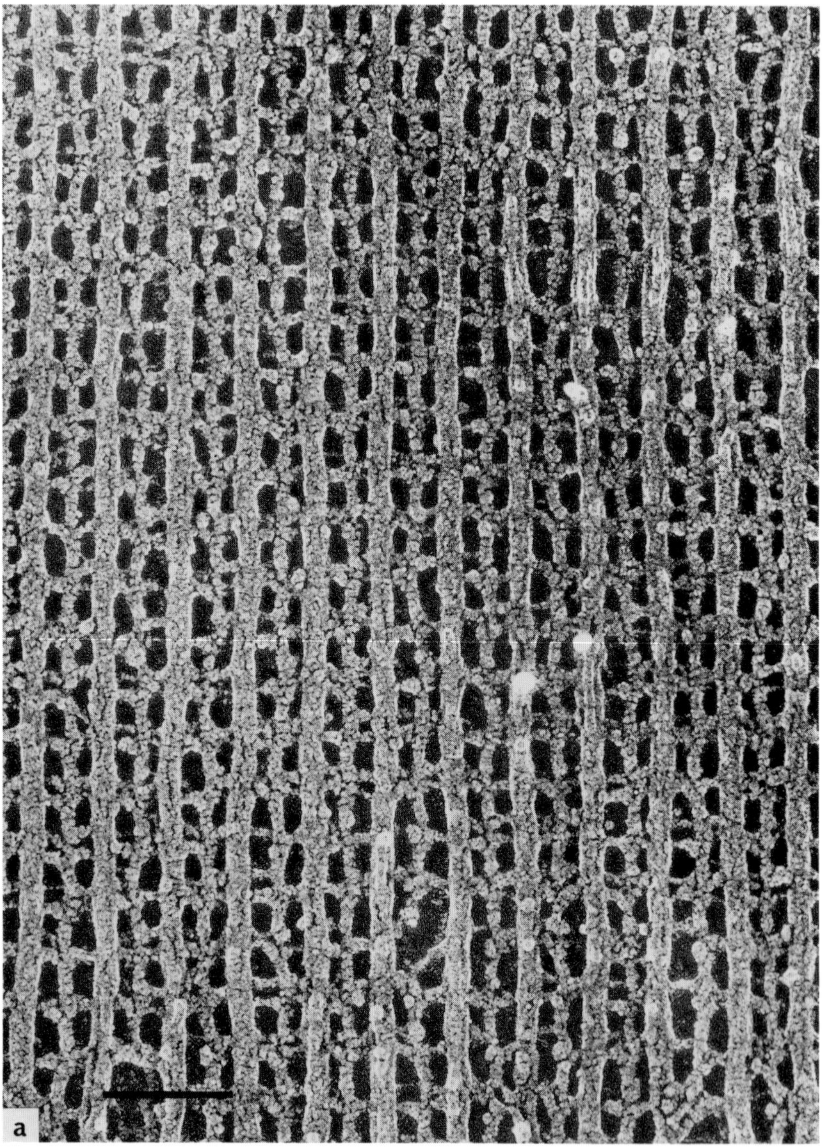

a

It is important to note that the two-filament interaction model is for muscle in rigor. A more difficult question is whether, in the rapid, dynamic, non-equilibrium state of contraction such two-filament interaction would occur, when proximity of myosin heads to actin, rather than minimum strain, might determine the mode of attachment. Indeed, one might question whether both heads attach simultaneously at all (in any fashion) during contraction.

Rigor insect flight muscle (both fixed and unfixed) has recently been observed using the technique of rapid freezing followed by etching and rotary shadowing (Heuser, 1981; Heuser and Cooke, 1983). The micrographs confirm the periodic attachment of cross-bridges to actin, often in paired groups (Fig. 49). However, the cross-bridges are not highly or consistently tilted and this presents an apparent discrepancy with the conventionally embedded material (Fig. 45), which needs to be resolved in the future. Relaxed

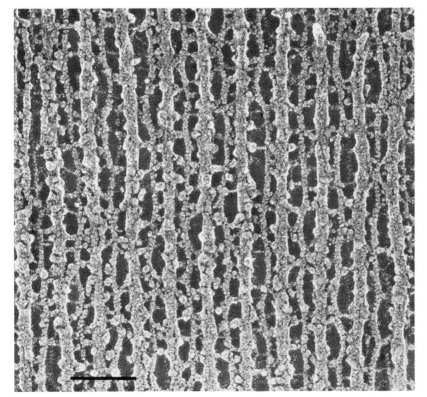

Fig. 49. Longitudinal views of insect flight muscle. Specimens prepared by rapid freezing, followed by freeze fracturing, etching and rotary shadowing. (Opposite) Rigor. The Z-line lies to the top of the field shown. (Above) Relaxed. Scale markers indicate 100 nm (from Heuser and Cooke (1983), courtesy of J. Heuser and Academic Press Inc.).

muscle treated in the same way shows a less regular structure with lumpier thick filaments and far fewer bridges attached to actin (Fig. 49b). It will be most interesting to see what can be learned of muscle caught during contraction by the same technique. It must be noted, however, that although rapid freezing can trap transient states, and is also free of many of the artifacts associated with conventional procedures (e.g. dehydration and embedding), it does not circumvent all problems (Heuser, 1981). Fixation is still generally necessary before freezing, although it can be avoided; and the removal of ice by etching, leaving the specimen unsupported in a vacuum, might allow movement of the structures of interest before shadowing has been carried out. Such effects could help to explain the differences between conventionally fixed and rapidly-frozen rigor insect flight muscle.

3. Actin-myosin Interaction in Contracting Muscle

When frog striated muscle contracts, the development of tension is accompanied by dramatic changes in the X-ray diffraction pattern. The myosin layer lines (including the 21·5 nm forbidden meridional reflection) diminish greatly in intensity (see Huxley et al., 1982) and the intensities of the equatorial reflections become similar to those of rigor muscle (see Huxley, 1979). The time course of these changes on the millisecond time-scale follows approximately that of tension development. The intensity of the 14·3 nm meridional reflection remains high during force production, but its spacing increases slightly to 14·4 nm (Huxley et al., 1982). No sign of any additional layer line reflections, for example those characteristic of rigor muscle, is seen during contraction. With muscles stretched beyond the point where the thick and thin filaments overlap, there is no change in myosin layer line intensity on stimulation (Yagi and Matsubara, 1980; Huxley et al., 1980). When an isometrically contracting muscle (at normal overlap) is suddenly stretched or allowed to shorten by ~ 10 nm per half sarcomere there is a sudden and dramatic drop in intensity of the 14·3 nm meridional reflection (Huxley et al., 1981).

These observations have been interpreted in the following way. The weakening of the myosin layer lines on contraction implies a disordering of the cross-bridges from their more regular arrangement in relaxed muscle. There is thought to be little change in the filament backbone, although the 1% change in the spacing of the 14·3 nm reflection could mean a slight lengthening of the filament. The changes in equatorial intensity suggest that the cross-bridges move away from the thick filament to the vicinity of the thin filament, where they are thought to attach to actin. The absence of any enhancement of the actin layer lines during contraction suggests that in contrast to rigor the cross-bridges have some freedom of rotation in an azimuthal direction about their attachment site on actin or that the actin helix itself becomes azimuthally disordered (Huxley et al., 1982). The large and rapid drop in intensity of the 14·3 nm meridional reflection that occurs on rapid stretch or release suggests a large change in conformation of the actomyosin complex, for example a tilt of attached cross-bridges (Huxley et al., 1981). The absence of any change in the myosin layer lines when a muscle, stretched to no-overlap, is stimulated implies that there is no movement of the cross-bridges in the absence of the thin filaments. This observation therefore provides no evidence for myosin-linked regulation in vertebrate striated muscle.

These results and those of muscle in rigor (Section VI.B.2), show that the thick filament, which, in its relaxed state, we have discussed as a fairly static entity (Section IV.A.1), is in fact able to undergo large changes in structure when a muscle contracts, or passes into rigor. This lability lies principally in the cross-bridges and their relatively weak and flexible attachment to the backbone, which enables them to form attachments to actin over a range of actin subunit orientations and over the changing interfilament spacings that occur as the sarcomere shortens (see Elliott et al., 1963; Huxley, 1969).

What is known about the detailed geometry of actin-myosin interaction in the cross-bridge cycle? The angle of head attachment at the beginning of the drive stroke has generally been taken to be the same as the cross-bridge angle in relaxed muscle. It has been assumed that the cross-bridge simply swings out from the thick filament and attaches in this orientation at the start of the drive stroke. X-ray patterns and longitudinal sections of relaxed insect flight muscle (Fig. 45a) have long been interpreted to show that this angle is 90° (Reedy et al., 1965).

More recent evidence casts doubt on the assumption that the initial angle of attachment is 90°. Both X-ray and electron microscope studies of invertebrate muscles suggest a variety of resting configurations (dependent on the muscle) involving considerable cross-bridge tilt (Section V.B). Detailed analysis of X-ray patterns has also recently suggested a tilted cross-bridge configuration for resting vertebrate striated muscle (Haselgrove, 1980). In addition to this, the angles observed in muscle sections are not necessarily the *in vivo* angles, since processing for electron microscopy may bring about changes in cross-bridge structure (Reedy and Barkas, 1974; Sjöström and Squire, 1977b). Most fundamentally, the angle in the relaxed state may not be the same as the initial angle of attachment to actin. As pointed out by Taylor (1979), the rotational relaxation time of a myosin head is ~ 1 μs, while tension development or change in intensity of the equatorial X-ray pattern occurs on a millisecond time-scale, so a model involving reorientation of the cross-bridge before attachment is perfectly possible.

Evidence for a tilted angle at the end of the power stroke (Fig. 32 bottom) comes from studies of rigor muscle discussed in Sections VI.A and VI.B.2. It is assumed that the rigor state (no ATP) is related to the end of the power stroke, where nucleotide has been displaced.

It is clear, then, that the hypothesis that muscle contracts by a change in the orientation of the cross-bridge on actin, whilst very attractive, has not been proved (see also Section VI.A.4). Although the recent X-ray studies of contracting muscle are suggestive of such a mechanism (Huxley *et al.*, 1981), it is clearly highly desirable that attached states other than rigor be observed by electron microscopy. There has consequently been much work with analogues of ATP, in particular AMP.PNP, in attempts to produce such states (see also Section VI.A.4). Mechanical and X-ray experiments have provided considerable but enigmatic data which have not proved to be easily interpretable (see Marston *et al.*, 1979, and Tregear and Marston, 1979, for references). The direct approach of electron microscopy has also given suggestive but inconclusive results (Marston *et al.*, 1976; Beinbrech *et al.*, 1976; Beinbrech, 1977; Meisner and Beinbrech, 1979). The best electron micrographs have been obtained by Reedy *et al.* (1983), but still no definite conclusions have been reached (Fig. 45c). In rigor insect flight muscle, treated with AMP.PNP, chevrons repeating every 38·5 nm are seen (but less clearly than in rigor) and a 14·5 nm striping of the thick filaments is also present; i.e. features characteristic of both rigor and relaxed states coexist in AMP.PNP. It is not known whether this is due to a mixture of two populations of cross-bridges, some attached to actin with the actin repeat and others detached with the myosin repeat, or to a single population, which are able to accommodate to both repeats (each cross-bridge having a distal rigor-like and a proximal relaxed-like component). The varied angles of the weakened chevrons suggest some change in conformation but it has not been determined exactly what this change is.

We have tried to show in this section, and throughout this chapter, that electron microscopy has played an essential role in elucidating the structures of the muscle filaments and their interaction. Nevertheless, it is not a technique that stands on its own and results must always be correlated with those of other approaches. Some questions have been answered fairly satisfactorily while other areas remain totally opaque. It is to be hoped that electron microscopy will continue, as techniques develop, to probe these areas successfully.

VII. ACKNOWLEDGEMENTS

We thank Drs M. Alhadeff, P. Bennett, H. Huxley, J. Kendrick-Jones, G. Offer, C. Shoenberg, J. Trinick and P. Vibert for reading and criticizing an earlier draft of the manuscript. We also thank Drs L. Amos, P. Flicker, J.

Heuser, H. Huxley, G. Offer, M. Reedy, J. Squire, M. Stewart, J. Trinick and P. Vibert for providing us with results or manuscripts prior to publication, and Ms J. Brightwell for typing the manuscript. Part of this work was carried out while the authors were at the Department of Biology, The Johns Hopkins University, Baltimore, Md (P.K.) and the Rosenstiel Basic Medical Sciences Research Center, Brandeis University, Waltham, Mass. and the MRC Laboratory of Molecular Biology, Hills Road, Cambridge (R.C.). The work was supported in part by NIH grant AM 04349 to Dr W. F. Harrington, grants from NIH (AM 17346), NSF (PCM 79-04396) and MDA to Dr C. Cohen, and by fellowships from the Medical Research Council to both authors.

VIII. REFERENCES

Adelstein, R. S. and Eisenberg, E. (1980). *Ann. Rev. Biochem.* **49**, 921–956.
Amos, L. A., Huxley, H. E., Holmes, K. C., Goody, R. S. and Taylor, K. A. (1982). *Nature* **299**, 467–469.
Ashton, F. T., Somlyo, A. V. and Somlyo, A. P. (1975). *J. Mol. Biol.* **98**, 17–29.
Auber, J. and Couteaux, R. (1963). *J. Microscopie* **2**, 309–324.
Bagshaw, C. R. (1977). *Biochemistry* **16**, 59–67.
Barrington-Leigh, J., Goody, R. S., Hofmann, W., Holmes, K., Mannherz, H. G., Rosenbaum, G. and Tregear, R. T. (1977). *In* "Insect Flight Muscle" (R. T. Tregear, ed.), pp. 137–146. Elsevier/North-Holland Biomedical Press, Amsterdam.
Bear, R. S. and Selby, C. C. (1956). *J. Biophys. Biochem. Cytol.* **2**, 55–69.
Beinbrech, G. (1977). *In* "Insect Flight Muscle" (R. T. Tregear, ed.), pp. 147–160. Elsevier/North-Holland Biomedical Press, Amsterdam.
Beinbrech, G., Kuhn, H. J., Herzig, J. W. and Rüegg, J. C. (1976). *Cytobiologie* **12**, 385–396.
Bennett, P. M. (1976). *Proc. 6th Eur. Congr. Elect. Microsc.* **2**, 517–519.
Bennett, P. M. (1977). PhD Thesis, University of London.
Bennett, P. M. (1981). *J. Mol. Biol.* **146**, 201–221.
Bennett, P. M. and Elliott, A. (1981). *J. Muscle Res. Cell Motil.* **2**, 65–81.
Bond, M., Somlyo, A. V., Butler, T. M. and Somlyo, A. P. (1981). In "Abstracts of the VII International Biophysics Congress and III Pan-American Biochemistry Congress", Mexico City, August 23–28, 1981. p. 46.
Brahms, J. and Brezner, J. (1961). *Arch. Biochem. Biophys.* **95**, 219–228.
Brann, L., Dewey, M. M., Baldwin, E. A., Brink, P. and Walcott, B. (1979). *Nature* **279**, 256–257.
Bullard, B., Hammond, K. S. and Luke, B. M. (1977). *J. Mol. Biol.* **115**, 417–440.
Burke, M., Himmelfarb, S. and Harrington, W. F. (1973). *Biochemistry* **12**, 701–710.
Caspar, D. L. D. and Klug, A. (1962). *Cold Spring Harb. Symp. Quant. Biol.* **27**, 1–24.
Chalovich, J. M. and Eisenberg, E. (1982). *J. Biol. Chem.* **257**, 2432–2437.
Chalovich, J. M., Chock, P. B. and Eisenberg, E. (1981). *J. Biol. Chem.* **256**, 575–578.
Chantler, P. D. and Szent-Györgyi, A. G. (1978). *Biochemistry* **17**, 5440–5448.
Chantler, P. D. and Szent-Györgyi, A. G. (1980). *J. Mol. Biol.* **138**, 473–492.
Chowrashi, P. K. and Pepe, F. A. (1977). *J. Cell Biol.* **74**, 136–152.

Clarke, M., Schatten, G., Mazia, D. and Spudich, J. A. (1975). *Proc. Natn. Acad. Sci. USA* **72**, 1758–1762.
Cohen, C. (1975). *Sci. Amer.* **233** (5), 36–45.
Cohen, C. (1982). *Proc. Natn. Acad. Sci. USA* **79**, 3176–3178.
Cohen, C. and Holmes, K. C. (1963). *J. Mol. Biol.* **6**, 423–432.
Cohen, C. and Szent-Györgyi, A. G. (1957). *J. Amer. Chem. Soc.* **79**, 248.
Cohen, C. and Szent-Györgyi, A. G. (1971). In "Contractility of Muscle Cells and Related Processes" (R. J. Podolsky, ed.), pp. 23–36. Prentice-Hall, Inc., New York and London.
Cohen, C., Lowey, S., Harrison, R. G., Kendrick-Jones, J. and Szent-Györgyi, A. G. (1970). *J. Mol. Biol.* **47**, 605–609.
Cohen, C., Szent-Györgyi, A. G. and Kendrick-Jones, J. (1971). *J. Mol. Biol.* **56**, 223–237.
Craig, R. W. (1975). PhD Thesis, University of London.
Craig, R. (1977). *J. Mol. Biol.* **109**, 69–81.
Craig, R. and Megerman, J. (1977). *J. Cell Biol.* **75**, 990–996.
Craig, R. and Megerman, J. (1979). In "Motility in Cell Function" (F. A. Pepe, J. W. Sanger and V. T. Nachmias, eds), pp. 91–102. Academic Press, New York.
Craig, R. and Offer, G. (1976a). *J. Mol. Biol.* **102**, 325–332.
Craig, R. and Offer, G. (1976b). *Proc. R. Soc. Lond. B.* **192**, 451–461.
Craig, R., Szent-Györgyi, A. G., Beese, L., Flicker, P., Vibert, P. and Cohen, C. (1980). *J. Mol. Biol.* **140**, 35–55.
Craig, R., Smith, R. and Kendrick-Jones, J. (1983). *Nature* **302**, 436–439.
Davis, J. S. (1981). *Biochem. J.* **197**, 309–314.
Davis, J. S., Buck, J. and Greene, E. P. (1982). *FEBS Lett.* **140**, 293–297.
DeRosier, D. J. and Klug, A. (1968). *Nature* **217**, 130–134.
Dewey, M. M., Levine, R. J. C. and Colflesh, D. E. (1973). *J. Cell Biol.* **58**, 574–593.
Dewey, M. M., Walcott, B., Colflesh, D. E., Terry, H. and Levine, R. J. C. (1977). *J. Cell Biol.* **75**, 366–380.
Dover, S. D. and Elliott, A. (1979). *J. Mol. Biol.* **132**, 340–341.
Draper, M. H. and Hodge, A. J. (1949). *Aust. J. Exp. Biol. Med. Sci.* **27**, 465–503.
Eaton, B. L. and Pepe, F. A. (1974). *J. Mol. Biol.* **82**, 421–423.
Ebashi, S. and Endo, M. (1968). *Prog. Biophys. Mol. Biol.* **18**, 123–183.
Eisenberg, E. and Hill, T. L. (1978). *Prog. Biophys. Mol. Biol.* **33**, 55–82.
Eisenberg, E. and Moos, C. (1967). *J. Biol. Chem.* **12**, 2945–2951.
Eisenberg, E., Zobel, C. R. and Moos, C. (1968). *Biochemistry* **7**, 3186–3194.
Elfvin, M., Levine, R. J. C. and Dewey, M. M. (1976). *J. Cell Biol.* **71**, 261–272.
Elliott, A. (1974). *Proc. R. Soc. Lond. B.* **186**, 53–66.
Elliott, A. (1979). *J. Mol. Biol.* **132**, 323–341.
Elliott, A. and Lowy, J. (1970). *J. Mol. Biol.* **53**, 181–203.
Elliott, A. and Offer, G. (1978). *J. Mol. Biol.* **123**, 505–519.
Elliott, A., Offer, G. and Burridge, K. (1976). *Proc. R. Soc. Lond. B.* **193**, 45–53.
Elliott, G. F. (1964). *J. Mol. Biol.* **10**, 89–104.
Elliott, G. F., Lowy, J. and Worthington, C. R. (1963). *J. Mol. Biol.* **6**, 295–305.
Elliott, G. F., Lowy, J. and Millman, B. M. (1967). *J. Mol. Biol.* **25**, 31–45.
Elzinga, M. and Collins, J. H. (1977). *Proc. Natn. Acad. Sci. USA* **74**, 4281–4284.
Emes, C. H. and Rowe, A. J. (1978). *Biochim. Biophys. Acta* **537**, 110–124.
Engelhardt, W. A. and Ljubimowa, M. N. (1939). *Nature* **144**, 668–669.
Flicker, P., Wallimann, T. and Vibert, P. (1983). *J. Mol. Biol.* **169**, in press.
Frank, G. and Weeds, A. G. (1974). *Eur. J. Biochem.* **44**, 317–334.

Franzini-Armstrong, C. (1970). *J. Cell Sci.* **6**, 559–592.
Franzini-Armstrong, C. and Porter, K. (1964). *J. Cell Biol.* **22**, 675–696.
Gadasi, H., Maruta, H., Collins, J. H. and Korn, E. D. (1979). *J. Biol. Chem.* **254**, 3631–3636.
Gazith, J., Himmelfarb, S. and Harrington, W. F. (1970). *J. Biol. Chem.* **245**, 15–22.
Gilëv, V. P. (1966a). *Biochim. Biophys. Acta* **112**, 340–345.
Gilëv, V. P. (1966b). In "Electron Microscopy 1966" (R. Uyeda, ed.), Vol. II, pp. 689–690. Maruzen Company, Ltd, Tokyo.
Gillis, J. M. and O'Brien, E. J. (1975). *J. Mol. Biol.* **99**, 445–459.
Godfrey, J. E. and Harrington, W. F. (1970a). *Biochemistry* **9**, 886–893.
Godfrey, J. E. and Harrington, W. F. (1970b). *Biochemistry* **9**, 894–908.
Goldman, R., Pollard, T. and Rosenbaum, J. (1976). *Cold Spring Harb. Conf. Cell Proliferation* **3**, "Cell Motility". Books A and B.
Gratzer, W. B. and Lowey, S. (1969). *J. Biol. Chem.* **244**, 22–25.
Greene, L. E. and Eisenberg, E. (1980). *J. Biol. Chem.* **255**, 543–548.
Hall, C. E., Jakus, M. A. and Schmitt, F. O. (1945). *J. Appl. Physics* **16**, 459–465.
Hall, C. E., Jakus, M. and Schmitt, F. O. (1946). *Biol. Bull.* **90**, 32–50.
Hanson, J. and Huxley, H. E. (1953). *Nature* **172**, 530–532.
Hanson, J. and Huxley, H. E. (1955). *Symp. Soc. Exp. Biol.* **9**, 228–264.
Hanson, J. and Huxley, H. E. (1957). *Biochim. Biophys. Acta* **23**, 250–260.
Hanson, J. and Lowy, J. (1964). *Proc. R. Soc. Lond.* B **160**, 523–524.
Hanson, J., O'Brien, E. J. and Bennett, P. M. (1971). *J. Mol. Biol.* **58**, 865–871.
Hardwicke, P. M. D. and Hanson, J. (1971). *J. Mol. Biol.* **59**, 509–516.
Harrington, W. F. (1971). *Proc. Natn. Acad. Sci. USA* **68**, 685–689.
Harrington, W. F. (1979). *Proc. Natn. Acad. Sci. USA* **76**, 5066–5070.
Harrington, W. F. and Burke, M. (1972). *Biochemistry* **11**, 1448–1455.
Harrington, W. F. and Himmelfarb, S. (1972). *Biochemistry* **11**, 2945–2952.
Harrington, W. F. and Josephs, R. (1968). *Dev. Biol. Suppl.* **2**, 21–62.
Harrison, R. G., Lowey, S. and Cohen, C. (1971). *J. Mol. Biol.* **59**, 531–535.
Haselgrove, J. C. (1975). *J. Mol. Biol.* **92**, 113–143.
Haselgrove, J. C. (1980). *J. Muscle Res. Cell Motil.* **1**, 177–191.
Hayashi, T., Silver, R. B., Ip, W., Cayer, M. L. and Smith, D. S. (1977). *J. Mol. Biol.* **111**, 159–171.
Herbert, T. J. and Carlson, F. D. (1971). *Biopolymers* **10**, 2231–2252.
Heuser, J. (1981). *Trends Biochem. Sci.* **6**, 64–68.
Heuser, J. E. and Cooke, R. S. (1983). *J. Mol. Biol.* **169**, 97–122.
Heuser, J. E., Reese, T. S., Dennis, M. J., Jan, Y., Jan, L. and Evans, L. (1979). *J. Cell Biol.* **81**, 275–300.
Highsmith, S., Akasaka, K., Konrad, M., Goody, R., Holmes, K., Wade-Jardetsky, N. and Jardetsky, O. (1979). *Biochemistry* **18**, 4238–4244.
Hinssen, H., D'Haese, J., Small, J. V. and Sobieszek, A. (1978). *J. Ultrastr. Res.* **64**, 282–302.
Hoh, J. F. Y. and Yeoh, G. P. S. (1979). *Nature* **280**, 321–323.
Holmes, K. C., Tregear, R. T. and Barrington-Leigh, J. (1980). *Proc. R. Soc. Lond.* B **207**, 13–33.
Holmes, K. C., Goody, R. S. and Amos, L. A. (1982). *Ultramicroscopy* **9**, 37–44.
Holt, J. C. and Lowey, S. (1977). *Biochemistry* **16**, 4398–4402.
Holtzer, A. and Lowey, S. (1959). *J. Amer. Chem. Soc.* **81**, 1370–1377.
Hoyle, G. and McNeill, P. A. (1968). *J. Exp. Zool.* **167**, 487–522.
Hoyle, G., McAlear, J. H. and Selverston, A. (1965). *J. Cell Biol.* **26**, 621–640.

Huxley, A. F. (1957). *Prog. Biophys. Biophys. Chem.* **7**, 255–318.
Huxley, A. F. (1980). "Reflections on Muscle". The Sherrington Lectures XIV. Liverpool University Press, Liverpool.
Huxley, A. F. and Niedergerke, R. (1954). *Nature* **173**, 971–973.
Huxley, A. F. and Simmons, R. M. (1971). *Nature* **233**, 533–538.
Huxley, H. E. (1953a). *Proc. R. Soc. Lond. B* **141**, 59–62.
Huxley, H. E. (1953b). *Biochim. Biophys. Acta* **12**, 387–394.
Huxley, H. E. (1957). *J. Biophys. Biochem. Cytol.* **3**, 631–648.
Huxley, H. E. (1960). In "The Cell" (J. Bracket and A. E. Mirsky, eds), Vol. 4, Chap. 7. Academic Press, New York and London.
Huxley, H. E. (1963). *J. Mol. Biol.* **7**, 281–308.
Huxley, H. E. (1965). *Sci. Amer.* **213** (6), 18–27.
Huxley, H. E. (1967). *J. Gen. Physiol.* **50** (Suppl.), 71–83.
Huxley, H. E. (1968). *J. Mol. Biol.* **37**, 507–520.
Huxley, H. E. (1969). *Science* **164**, 1356–1366.
Huxley, H. E. (1972). *Cold Spring Harb. Symp. Quant. Biol.* **37**, 361–376.
Huxley, H. E. (1973). *Nature* **243**, 445–449.
Huxley, H. E. (1979). In "Cross-Bridge Mechanism in Muscle Contraction" (H. Sugi and G. H. Pollack, eds), pp. 391–401. University of Tokyo Press, Tokyo.
Huxley, H. E. and Brown, W. (1967). *J. Mol. Biol.* **30**, 383–434.
Huxley, H. E. and Hanson, J. (1954). *Nature* **173**, 973–976.
Huxley, H. E. and Hanson, J. (1957). *Biochim. Biophys. Acta* **23**, 229–249.
Huxley, H. E., Faruqi, A. R., Bordas, J., Koch, M. H. J. and Milch, J. R. (1980). *Nature* **284**, 140–143.
Huxley, H. E., Simmons, R. M., Faruqi, A. R., Kress, M., Bordas, J. and Koch, M. H. J. (1981). *Proc. Natn. Acad. Sci. USA* **78**, 2297–2301.
Huxley, H. E., Faruqi, A. R., Kress, M., Bordas, J. and Koch, M. H. J. (1982). *J. Mol. Biol.* **158**, 637–684.
Ishikawa, H., Bischoff, R. and Holtzer, H. (1969). *J. Cell Biol.* **43**, 312–328.
Jahromi, S. S. and Atwood, H. L. (1969). *J. Exp. Zool.* **171**, 25–38.
Jeacocke, S. A. and England, P. J. (1980). *FEBS Lett.* **122**, 129–132.
Josephs, R. and Harrington, W. F. (1966). *Biochemistry* **5**, 3474–3487.
Josephs, R. and Harrington, W. F. (1968). *Biochemistry* **7**, 2834–2847.
Kaminer, B. (1969). *J. Mol. Biol.* **39**, 257–264.
Kaminer, B. and Bell, A. L. (1966). *J. Mol. Biol.* **20**, 391–401.
Katsura, I. and Noda, H. (1971). *J. Biochem. (Tokyo)* **69**, 219–229.
Katsura, I. and Noda, H. (1973a). *J. Biochem. (Tokyo)* **73**, 245–256.
Katsura, I. and Noda, H. (1973b). *J. Biochem. (Tokyo)* **73**, 257–268.
Kendrick-Jones, J. and Jakes, R. (1977). In "Myocardial Failure" (Riecker, G., Weber, A. and Goodwin, J. eds), pp. 28–40. Springer-Verlag, Berlin, Heidelberg.
Kendrick-Jones, J., Szent-Györgyi, A. G. and Cohen, C. (1971). *J. Mol. Biol.* **59**, 527–529.
Kendrick-Jones, J., Jakes, R., Tooth, P., Craig, R. and Scholey, J. (1982). In "Basic Biology of Muscles: A Comparative Approach". (Twarog, B. M., Levine, R. J. C. and Dewey, M. M., eds), pp. 255–272. Raven Press, New York.
Kensler, R. W. and Levine, R. J. C. (1982). *J. Cell Biol.* **92**, 443–451.
Kensler, R. W. and Stewart, M. (1983). *J. Cell Biol.* **96**, 1797.
Knappeis, G. G. and Carlsen, F. (1968). *J. Cell Biol.* **38**, 202–211.
Kominz, D. R., Carroll, W. R., Smith, E. N. and Mitchell, E. R. (1959). *Arch. Biochem. Biophys.* **79**, 191–199.

Kretzschmar, K. M., Mendelson, R. A. and Morales, M. F. (1978). *Biochemistry* **17,** 2314–2318.
Kuczmarski, E. R. and Spudich, J. A. (1980). *Proc. Natn. Acad. Sci. USA* **77,** 7292–7296.
Lamvik, M. K. (1978). *J. Mol. Biol.* **122,** 55–68.
Lehman, W. and Szent-Györgyi, A. G. (1975). *J. Gen. Physiol.* **66,** 1–30.
Levine, R. J. C. and Kensler, R. W. (1982). *Biophys. J.* **37,** 50a.
Levine, R. J. C., Elfvin, M., Dewey, M. M. and Walcott, B. (1976). *J. Cell Biol.* **71,** 273–279.
Lovell, S. J. and Harrington, W. F. (1981). *J. Mol. Biol.* **149,** 659–674.
Lovell, S. J., Knight, P. J. and Harrington, W. F. (1981). *Nature* **293,** 664–666.
Lowey, S. (1971). *In* "Subunits in Biological Systems, Part A". (S. N. Timasheff and G. D. Fasman, eds), Chapt. 5. Marcel Dekker Inc., New York.
Lowey, S. and Cohen, C. (1962). *J. Mol. Biol.* **4,** 293–308.
Lowey, S. and Risby, D. (1971). *Nature* **234,** 81–85.
Lowey, S., Kucera, J. and Holtzer, A. (1963). *J. Mol. Biol.* **7,** 234–244.
Lowey, S., Goldstein, L., Cohen, C. and Luck, S. M. (1967). *J. Mol. Biol.* **23,** 287–304.
Lowey, S., Slayter, H. S., Weeds, A. G. and Baker, H. (1969). *J. Mol. Biol.* **42,** 1–29.
Lowey, S., Benfield, P. A., Silberstein, L. and Lang, L. M. (1979). *Nature* **282,** 522–524.
Lowy, J. and Hanson, J. (1962). *Physiol. Rev.* **42,** (Suppl. 5), 34–47.
Lowy, J. and Small, J. V. (1970). *Nature* **227,** 46–51.
Lowy, J., Poulsen, F. R. and Vibert, P. J. (1970). *Nature* **225,** 1053–1054.
Lu, R. C. (1980). *Proc. Natn. Acad. Sci. USA* **77,** 2010–2013.
Luther, P. and Squire, J. (1978). *J. Mol. Biol.* **125,** 313–324.
Luther, P. K. and Squire, J. M. (1980). *J. Mol. Biol.* **141,** 409–439.
Luther, P. K., Munro, P. M. G. and Squire, J. M. (1981). *J. Mol. Biol.* **151,** 703–730.
Lymn, R. W. and Taylor, E. W. (1970). *Biochemistry* **9,** 2975–2983.
Margossian, S. S. and Lowey, S. (1973). *J. Mol. Biol.* **74,** 313–330.
Margossian, S. S., Stafford, W. F., III and Lowey, S. (1981). *Biochemistry* **20,** 2151–2155.
Marsh, D. J. and Lowey, S. (1980). *Biochemistry* **19,** 774–784.
Marston, S. B. and Taylor, E. W. (1980). *J. Mol. Biol.* **139,** 573–600.
Marston, S. B. and Tregear, R. T. (1972). *Nature New Biol.* **235,** 23–24.
Marston, S. B., Rodger, C. D. and Tregear, R. T. (1976). *J. Mol. Biol.* **104,** 263–276.
Marston, S. B., Tregear, R. T., Rodger, C. D. and Clarke, M. L. (1979). *J. Mol. Biol.* **128,** 111–126.
Maruyama, K. and Weber, A. (1972). *Biochemistry* **11,** 2990–2998.
Masaki, T. and Takaiti, O. (1974). *J. Biochem. (Tokyo)* **75,** 367–380.
Matsubara, I. (1974). *J. Physiol. (Lond.)* **238,** 473–486.
Matsubara, I. and Millman, B. M. (1974). *J. Mol. Biol.* **82,** 527–536.
Maw, M. C. and Rowe, A. J. (1980). *Nature* **286,** 412–414.
McLachlan, A. D. and Karn, J. (1982). *Nature* **299,** 226–231.
Meisner, D. and Beinbrech, G. (1979). *Eur. J. Cell Biol.* **19,** 189–195.
Mendelson, R. A. and Cheung, P. (1976). *Science* **194,** 190–192.
Mendelson, R. and Kretzschmar, K. M. (1980). *Biochemistry* **19,** 4103–4108.
Mendelson, R. A., Morales, M. F. and Botts, J. (1973). *Biochemistry* **12,** 2250–2255.
Mihályi, E. and Szent-Györgyi, A. G. (1953). *J. Biol. Chem.* **201,** 189–196.
Miller, A. and Tregear, R. T. (1972). *J. Mol. Biol.* **70,** 85–104.
Millman, B. M. (1979). *In* "Motility in Cell Function" (F. A. Pepe, J. W. Sanger and V. T. Nachmias, eds), pp. 351–354. Academic Press, New York.

Millman, B. M. and Bennett, P. M. (1976). *J. Mol. Biol.* **103**, 439–467.
Moore, P. B., Huxley, H. E. and De Rosier, D. J. (1970). *J. Mol. Biol.* **50**, 279–295.
Moos, C. (1972). *Cold Spring Harb. Symp. Quant. Biol.* **37**, 137–143.
Moos, C., Offer, G., Starr, R. and Bennett, P. (1975). *J. Mol. Biol.* **97**, 1–9.
Morimoto, K. and Harrington, W. F. (1972). *J. Biol. Chem.* **247**, 3052–3061.
Morimoto, K. and Harrington, W. F. (1974a). *J. Mol. Biol.* **88**, 693–709.
Morimoto, K. and Harrington, W. F. (1974b). *J. Mol. Biol.* **83**, 83–97.
Mornet, D., Bertrand, R., Pantel, P., Audemard, E. and Kassab, R. (1981). *Nature* **292**, 301–306.
Mueller, H. (1965). *J. Biol. Chem.* **240**, 3816–3828.
Nakamura, A., Sreter, F. and Gergely, J. (1971). *J. Cell Biol.* **49**, 883–898.
Needham, D. M. (1971). "Machina Carnis". Cambridge University Press, Cambridge.
O'Brien, E. J., Bennett, P. M. and Hanson, J. (1971). *Phil. Trans. R. Soc. Lond. Ser. B* **261**, 201–208.
O'Brien, E. J., Gillis, J. M. and Couch, J. (1975). *J. Mol. Biol.* **99**, 461–475.
O'Brien, E. J., Couch, J., Johnson, G. R. P. and Morris, E. P. (1983). *In* "Actin: Structure and Function in Muscle and Non-Muscle Cells" (dos Remedios, C. G., ed.), pp. 3–15. Academic Press, Australia.
Offer, G. (1972). *Cold Spring Harb. Symp. Quant. Biol.* **37**, 87–93.
Offer, G. (1974). *In* "Companion to Biochemistry: Selected Topics for Further Study". (A. T. Bull, J. R. Lagnado, J. O. Thomas and K. F. Tipton, eds), Chapt. 21. Longman Group Ltd, London.
Offer, G. and Elliott, A. (1978). *Nature* **271**, 325–329.
Offer, G., Moos, C. and Starr, R. (1973). *J. Mol. Biol.* **74**, 653–676.
Offer, G., Couch, J., O'Brien, E. and Elliott, A. (1981). *J. Mol. Biol.* **151**, 663–702.
Onishi, M. and Wakabayashi, T. (1982). *J. Biochem. (Tokyo)* **92**, 871–879.
Page, S. G. (1965). *J. Cell Biol.* **26**, 477–497.
Page, S. G. and Huxley, H. E. (1963). *J. Cell Biol.* **19**, 369–390.
Pepe, F. A. (1967a). *J. Mol. Biol.* **27**, 203–225.
Pepe, F. A. (1967b). *J. Mol. Biol.* **27**, 227–236.
Pepe, F. A. (1971). *Prog. Biophys. Mol. Biol.* **22**, 75–96.
Pepe, F. A. and Dowben, P. (1977). *J. Mol. Biol.* **113**, 199–218.
Pepe, F. A. and Drucker, B. (1972). *J. Cell Biol.* **52**, 255–260.
Pepe, F. A. and Drucker, B. (1975). *J. Mol. Biol.* **99**, 609–617.
Pepe, F. A. and Drucker, B. (1979). *J. Mol. Biol.* **130**, 379–393.
Pepe, F. A., Ashton, F. T., Dowben, P. and Stewart, M. (1981). *J. Mol. Biol.* **145**, 421–440.
Philpott, D. E. and Szent-Györgyi, A. G. (1954). *Biochim. Biophys. Acta* **15**, 165–173.
Pinset-Härström, I. and Truffy, J. (1979). *J. Mol. Biol.* **134**, 173–188.
Pollard, T. D. (1975). *J. Cell Biol.* **67**, 93–104.
Pollard, T. D. and Weihing, R. R. (1974). *CRC Crit. Rev. Biochem.* **2**, 1–65.
Pollard, T. D., Stafford, W. F., III and Porter, M. E. (1978). *J. Biol. Chem.* **253**, 4798–4808.
Potter, J. D. (1974). *Arch. Biochem. Biophys.* **162**, 436–441.
Pringle, J. W. S. (1957). "Insect Flight", Cambridge University Press, London and New York.
Pringle, J. W. S. (1965). *In* "Physiology of the Insecta" (M. Rockstein, ed.), Vol. 2, pp. 283–329. Academic Press, New York.
Pringle, J. W. S. (1972). *In* "The Structure and Function of Muscle" (G. H. Bourne, ed.), 2nd edn, Vol. I, Chapt. 10. Academic Press, New York and London.

Pringle, J. W. S. (1978). *Proc. R. Soc. Lond. B* **201**, 107–130.
Reedy, M. C., Reedy, M. K. and Goody, R. S. (1983). *J. Muscle Res. Cell Motil.* **4**, 55–81.
Reedy, M. K. (1968). *J. Mol. Biol.* **31**, 155–176.
Reedy, M. K. (1971). *In* "Contractility of Muscle Cells and Related Processes" (R. J. Podolsky, ed.), pp. 229–246. Prentice-Hall Inc., New York and London.
Reedy, M. K. and Barkas, A. E. (1974). *J. Cell Biol.* **63**, 282a.
Reedy, M. K., Holmes, K. C. and Tregear, R. T. (1965). *Nature* **207**, 1276–1280.
Reedy, M. K., Leonard, K. R., Freeman, R. and Arad, T. (1981). *J. Muscle Res. Cell Motil.* **2**, 45–64.
Reisler, E., Burke, M. and Harrington, W. F. (1974). *Biochemistry* **13**, 2014–2022.
Rice, R. V. (1961). *Biochim. Biophys. Acta* **52**, 602–604.
Rome, E. (1972). *Cold Spring Harb. Symp. Quant. Biol.* **37**, 331–339.
Rome, E., Offer, G. and Pepe, F. A. (1973). *Nature New Biol.* **244**, 152–154.
Rozsa, G., Szent-Györgyi, A. and Wyckoff, R. W. G. (1950). *Exp. Cell Res.* **1**, 194–205.
Safer, D. and Pepe, F. A. (1980). *J. Mol. Biol.* **136**, 343–358.
Saide, J. D. (1981). *J. Mol. Biol.* **153**, 661–679.
Scholey, J. M., Taylor, K. A. and Kendrick-Jones, J. (1980). *Nature* **287**, 233–235.
Scholey, J. M., Taylor, K. A. and Kendrick-Jones, J. (1981). *Biochimie* **63**, 255–271.
Seymour, J. and O'Brien, E. J. (1980). *Nature* **283**, 680–682.
Shoenberg, C. F. (1969). *Tissue Cell* **1**, 83–96.
Shoenberg, C. F. and Haselgrove, J. C. (1974). *Nature* **249**, 152–154.
Shoenberg, C. F. and Needham, D. M. (1976). *Biol. Rev. Camb. Phil. Soc.* **51**, 53–104.
Shoenberg, C. F. and Stewart, M. (1980). *J. Muscle Res. Cell Motil.* **1**, 117–126.
Silberstein, L. and Lowey, S. (1981). *J. Mol. Biol.* **148**, 153–189.
Sivaramakrishnan, M. and Burke, M. (1981). *J. Biol. Chem.* **256**, 2607–2610.
Sivaramakrishnan, M. and Burke, M. (1982). *J. Biol. Chem.* **257**, 1102–1105.
Sjöström, M. and Squire, J. M. (1977a). *J. Mol. Biol.* **109**, 49–68.
Sjöström, M. and Squire, J. M. (1977b). *J. Microsc.* **111**, 239–278.
Slayter, H. S. and Lowey, S. (1967). *Proc. Natn. Acad. Sci. USA* **58**, 1611–1618.
Small, J. V. (1977). *J. Cell Sci.* **24**, 329–349.
Small, J. V. and Squire, J. M. (1972). *J. Mol. Biol.* **67**, 117–149.
Small, P. A., Harrington, W. F. and Kielley, W. W. (1961). *Biochim. Biophys. Acta* **49**, 462–470.
Sobieszek, A. (1972). *J. Mol. Biol.* **70**, 741–744.
Sobieszek, A. and Small, J. V. (1972). *Cold Spring Harb. Symp. Quant. Biol.* **37**, 109–111.
Somlyo, A. P., Somlyo, A. V., Devine, C. E. and Rice, R. V. (1971). *Nature New Biol.* **231**, 243–246.
Somlyo, A. P., Devine, C. E., Somlyo, A. V. and Rice, R. V. (1973). *Phil. Trans. R. Soc. Lond. Ser. B* **265**, 223–229.
Somlyo, A. V., Ashton, F. T., Lemanski, L. F., Vallières, J. and Somlyo, A. P. (1977). *In* "The Biochemistry of Smooth Muscle" (N. L. Stephens, ed.), pp. 445–470. University Park Press, Baltimore.
Somlyo, A. V., Butler, T. M., Bond, M. and Somlyo, A. P. (1981). *Nature* **294**, 567–569.
Spudich, J. A., Huxley, H. E. and Finch, J. T. (1972). *J. Mol. Biol.* **72**, 619–632.
Squire, J. M. (1971). *Nature* **233**, 457–461.
Squire, J. M. (1972). *J. Mol. Biol.* **72**, 125–138.
Squire, J. M. (1973). *J. Mol. Biol.* **77**, 291–323.

Squire, J. M. (1975). *Ann. Rev. Biophys. Bioeng.* **4**, 137–163.
Squire, J. M. (1979). *In* "Fibrous Proteins: Scientific, Industrial and Medical Aspects" (D. A. D. Parry and L. K. Creamer, eds), Vol. 1, pp. 27–70. Academic Press, London.
Squire, J. (1981). "The Structural Basis of Muscular Contraction". Plenum Publishing Corporation, New York.
Squire, J. M., Sjöström, M. and Luther, P. (1976). *Proc. 6th Eur. Congr. Elect. Microsc.* pp. 91–95.
Squire, J. M., Harford, J. J., Edman, A. C. and Sjöström, M. (1982). *J. Mol. Biol.* **155**, 467–494.
Sreter, F. A., Gergely, J., Salmons, S. and Romanul, F. (1973). *Nature New Biol.* **241**, 17–19.
Sreter, F. A., Balint, M. and Gergely, J. (1975). *Dev. Biol.* **46**, 317–325.
Stafford, W. F. and Szent-Györgyi, A. G. (1978). *Biochemistry* **17**, 607–614.
Starr, R. and Offer, G. (1971). *FEBS Lett.* **15**, 40–44.
Starr, R. and Offer, G. (1978). *Biochem. J.* **171**, 813–816.
Stewart, M., Kensler, R. W. and Levine, R. J. C. (1981). *J. Mol. Biol.* **153**, 781–790.
Sutoh, K. and Harrington, W. F. (1977). *Biochemistry* **16**, 2441–2449.
Sutoh, K., Sutoh, K., Karr, T. and Harrington, W. F. (1978). *J. Mol. Biol.* **126**, 1–22.
Suzuki, H., Onishi, H., Takahashi, K. and Watanabe, S. (1978). *J. Biochem. (Tokyo)* **84**, 1529–1542.
Suzuki, H., Kamata, T., Onishi, H. and Watanabe, S. (1982). *J. Biochem. (Tokyo)* **91**, 1699–1705.
Szent-Györgyi, A. (1951). "Chemistry of Muscular Contraction". Academic Press, New York.
Szent-Györgyi, A. G., Cohen, C. and Kendrick-Jones, J. (1971). *J. Mol. Biol.* **56**, 239–258.
Szent-Györgyi, A. G., Szentkiralyi, E. M. and Kendrick-Jones, J. (1973). *J. Mol. Biol.* **74**, 179–203.
Szuchet, S. (1977). *Arch. Biochem. Biophys.* **180**, 493–503.
Takahashi, K. (1978). *J. Biochem. (Tokyo)* **83**, 905–908.
Takashi, R. (1979). *Biochemistry* **18**, 5164–5169.
Taylor, E. W. (1979). *CRC Crit. Rev. Biochem.* **6**, 103–164.
Taylor, K. A. and Amos, L. A. (1981). *J. Mol. Biol.* **147**, 297–324.
Thomas, D. D. and Cooke, R. (1980). *Biophys. J.* **32**, 891–906.
Thomas, D. D., Ishiwata, S., Seidel, J. C. and Gergely, J. (1980). *Biophys. J.* **32**, 873–890.
Tilney, L. G., DeRosier, D. J. and Mulroy, M. J. (1980). *J. Cell Biol.* **86**, 244–259.
Toyoshima, C. and Wakabayashi, T. (1979). *J. Biochem. (Tokyo)* **86**, 1887–1890.
Tregear, R. T. and Marston, S. B. (1979). *Ann. Rev. Physiol.* **41**, 723–736.
Tregear, R. T. and Squire, J. M. (1973). *J. Mol. Biol.* **77**, 279–290.
Trinick, J. A. (1981). *J. Mol. Biol.* **151**, 309–314.
Trinick, J. and Cooper, J. (1980). *J. Mol. Biol.* **141**, 315–321.
Trinick, J. and Elliott, A. (1979). *J. Mol. Biol.* **131**, 133–136.
Trinick, J. and Lowey, S. (1977). *J. Mol. Biol.* **113**, 343–368.
Trinick, J. and Offer, G. (1979). *J. Mol. Biol.* **133**, 549–556.
Trybus, K. M. and Taylor, E. W. (1980). *Proc. Natn. Acad. Sci. USA* **77**, 7209–7213.
Trybus, K. M., Huiatt, T. W. and Lowey, S. (1982). *Proc. Natn. Acad. Sci. USA* **79**, 6151–6155.

Tsao, T.-C., Tsou, Y.-S., Kung, T.-H., Pan, C.-H. and Lu, Z.-X. (1966). *Kexue Tongbao* **17**, 308–310.
Turner, D. C., Wallimann, T. and Eppenberger, H. M. (1973). *Proc. Natn. Acad. Sci. USA* **70**, 702–705.
Vibert, P. and Craig, R. (1982). *J. Mol. Biol.* **157**, 299–319.
Vibert, P. and Craig, R. (1983). *J. Mol. Biol.* **165**, 303–320.
Vibert, P., Szent-Györgyi, A. G., Craig, R., Wray, J. and Cohen, C. (1978). *Nature* **273**, 64–66.
Wachsberger, P. R. and Pepe, F. A. (1974). *J. Mol. Biol.* **88**, 385–391.
Wagner, P. D. and Giniger, E. (1981). *Nature* **292**, 560–562.
Wagner, P. D. and Weeds, A. G. (1977). *J. Mol. Biol.* **109**, 455–473.
Wakabayashi, T. (1980). *In* "Muscle Contraction: Its Regulatory Mechanisms" (S. Ebashi, K. Maruyama and M. Endo, eds), pp. 79–97. Japan Scientific Societies Press, Tokyo; Springer-Verlag, Berlin.
Wakabayashi, T., Huxley, H. E., Amos, L. A. and Klug, A. (1975). *J. Mol. Biol.* **93**, 477–497.
Wallimann, T., Turner, D. C. and Eppenberger, H. M. (1977). *J. Cell Biol.* **75**, 297–317.
Wallimann, T., Pelloni, G., Turner, D. C. and Eppenberger, H. M. (1978). *Proc. Natn. Acad. Sci. USA* **75**, 4296–4300.
Wallimann, T., Hardwicke, P. M. D. and Szent-Györgyi, A. G. (1982). *J. Mol. Biol.* **156**, 153–173.
Weber, A. and Murray, J. M. (1973). *Physiol. Rev.* **53**, 612–673.
Weeds, A. G. and Burridge, K. (1975). *FEBS Lett.* **57**, 203–208.
Weeds, A. G. and Lowey, S. (1971). *J. Mol. Biol.* **61**, 701–725.
Weeds, A. G. and Pope, B. (1977). *J. Mol. Biol.* **111**, 129–157.
Weeds, A. G. and Taylor, R. S. (1975). *Nature* **257**, 54–56.
Weeds, A. G., Trentham, D. R., Kean, C. J. C. and Buller, A. J. (1974). *Nature* **247**, 135–139.
Wells, J. A., Knoeber, C., Sheldon, M. C., Werber, M. M. and Yount, R. G. (1980). *J. Biol. Chem.* **255**, 11 135–11 140.
Whalen, R. G., Sell, S. M., Butler-Browne, G. S., Schwartz, K., Bouveret, P. and Pinset-Härström, I. (1981). *Nature* **292**, 805–809.
Winkelman, L. (1976). *Comp. Biochem. Physiol.* **55B**, 391–397.
Woods, E. F., Himmelfarb, S. and Harrington, W. F. (1963). *J. Biol. Chem.* **238**, 2374–2385.
Wray, J. S. (1979a). *Nature* **277**, 37–40.
Wray, J. S. (1979b). *Nature* **280**, 325–326.
Wray, J. S., Vibert, P. J. and Cohen, C. (1974). *J. Mol. Biol.* **88**, 343–348.
Wray, J. S., Vibert, P. J. and Cohen, C. (1975). *Nature* **257**, 561–564.
Wray, J., Vibert, P. and Cohen, C. (1978). *J. Mol. Biol.* **124**, 501–521.
Yagi, K. and Otani, F. (1974). *J. Biochem. (Tokyo)* **76**, 365–373.
Yagi, N. and Matsubara, I. (1977). *J. Mol. Biol.* **117**, 797–803.
Yagi, N. and Matsubara, I. (1980). *Science* **207**, 307–308.
Yagi, N. and Offer, G. W. (1981). *J. Mol. Biol.* **151**, 467–490.
Yagi, N., O'Brien, E. J. and Matsubara, I. (1981). *Biophys. J.* **33**, 121–138.
Young, D. M., Himmelfarb, S. and Harrington, W. F. (1965). *J. Biol. Chem.* **240**, 2428–2436.
Zobel, C. R. and Carlson, F. D. (1963). *J. Mol. Biol.* **7**, 78–89.

3. The Proteins of the Erythrocyte Membrane

DAVID M. SHOTTON

Department of Zoology, South Parks Road, Oxford, England

I. Introduction	206
A. Scope of the Chapter	206
B. General Properties of the Erythrocyte	207
1. Origin and lifespan	207
2. Physiological role, cell shape and membrane properties	208
3. Membrane composition	210
4. Electrophoretic analysis and molecular properties of the membrane proteins	211
II. Freeze-etch Studies of Erythrocyte Membranes	215
A. Historical Perspective	216
B. The Nature of the Intramembrane Particles	219
C. Particle Aggregation Studies	222
D. Studies of the Etched Membrane Surfaces	227
1. The extracellular surface	227
2. The cytoplasmic surface	233
III. The Major Integral Membrane Proteins	234
A. Band Three, the Anion Channel Protein	234
1. General properties and protein chemistry	234
2. Electron microscopy in reconstituted systems	235
B. Glycophorin A, the Major Sialoglycoprotein	237
1. General properties of the erythrocyte membrane sialoglycoproteins	237
2. Protein chemistry of glycophorin A	238
3. Electron microscopy in reconstituted systems	239
IV. The Membrane Skeleton	241
A. Historical Perspective	241
B. Comparison with the Cytoskeleton of Nucleated Cells	242
C. Protein Components of the Membrane Skeleton	243
D. Cell Shape Changes	244
1. Material properties of the membrane	244

       2. Molecular basis of metabolite-sensitive and experimentally-
          induced reversible shape changes 245
       3. Irreversible shape changes—blebbing, vesiculation and eversion . 247
    E. Membrane Skeletal Control of Translational and Rotational
       diffusion 249
    F. Electron Microscopic Observations 255
       1. Thin section 255
       2. Scanning electron microscopy 257
       3. Negative staining 259
       4. Freeze etching 260
  V. The Proteins of the Membrane Skeleton 260
    A. Spectrin 260
       1. General properties and protein chemistry 260
       2. Negative-staining studies 267
       3. Molecular shape as revealed by low-angle shadowing . . . 267
    B. Actin 279
    C. Band 4·1 282
    D. The Molecular Associations of Spectrin, Actin and Band 4·1 within
       the Membrane Skeleton 282
    E. The Attachment of the Membrane Skeleton to Integral Proteins:
       Ankyrin and Glycoconnectin 291
 VI. Other Major Membrane Proteins 294
    A. Polypeptides of Medium Molecular Weight 294
    B. Polypeptides of Low Molecular Weight: Torin and Cylindrin . . 295
       1. Early Studies 295
       2. Torin 296
       3. Cylindrin 298
VII. The Erythrocyte Membrane in Disease 303
    A. Hereditary Spherocytosis, Hereditary Elliptocytosis and Hereditary
       Pyropoikilocytosis . . . .
    B. Spectrin-Deficient Mutants in Mice 307
    C. Glycophorin-Deficient Mutants in Man 308
    D. Sickle Cell Anaemia 308
    E. Muscular Dystrophy 309
       1. Introduction 309
       2. Biochemical and biophysical differences 311
       3. Scanning electron microscopic studies 313
       4. Freeze-fracture morphology 314
       5. Conclusion 314
VIII. Conclusion: The Erythrocyte Membrane as a Paradigm . . .
 IX. Acknowledgements 320
  X. References 320

I. INTRODUCTION

A. Scope of the Chapter

In writing this chapter, it has been my purpose to bring together and discuss the best available electron micrographs of the major proteins of the erythrocyte membrane which show their appearance and location within the membrane, their molecular shapes after purification, or the nature of their intractions with other membrane components. A full review of the biochemistry and physiology of the erythrocyte membrane and its proteins is beyond the scope of this volume, but I have attempted to include sufficient biochemical information to allow the significance of the micrographs to be fully appreciated. The erythrocyte membrane and its proteins have been extensively studied during the last decade, and their biochemical properties are the subject of several excellent reviews (Steck, 1974; Kirkpatrick, 1976; Marchesi et al., 1976; Steck, 1978; Lux, 1979a, b; Marchesi, 1979a, b; Lux and Glader, 1980) to which the reader should refer for further information.

The scientific literature of the last decade is, unfortunately, littered with biochemical and ultrastructural studies on the erythrocyte membrane and its component proteins in which uninterpretable results have been presented, or false conclusions drawn. Reliable electron microscopic studies of the erythrocyte membrane, of which there are many, are characterized by the careful avoidance of conditions which might favour unintentional proteolysis or denaturation, by detailed biochemical analysis of the material under study, and by the absence of total reliance on a single preparative technique or electron microscopic method which might result in failure to recognize attendant artifacts. It is such studies which I have sought to present below, most of which have been of the proteins from normal human erythrocyte membranes from healthy donors. Very little is known about the effects of various diseases upon the individual membrane proteins (for reviews see Lux, 1979b; Lux and Glader, 1980), but wherever possible such information is also given. I have deliberately chosen to omit scanning electron micrographs of the many strange forms exhibited by abnormal and diseased erythrocytes, since at present we are unable to interpret these shapes in terms of the properties of the membrane proteins themselves, and since many of these shapes have already been well documented by Bessis (1973a, b, 1974).

B. General Properties of the Erythrocyte

1. Origin and Lifespan

The erythrocyte (red blood cell) has the simplest cellular structure imaginable, being a single limiting plasma membrane without regional specializations enclosing a homogeneous cytoplasm devoid of internal membraneous organelles, cytoskeletal elements, ribosomes and other insoluble components. This striking simplicity is the result of a lengthy pathway of cellular differentiation, erythropoiesis, which probably involves five mitotic divisions. In this, multipotent stem cells in the bone marrow develop under the influence of the hormone erythropoietin through stages of increasing specialization, during which the required components of the mature erythrocyte are first synthesized and then the supernumerary cellular organelles are disassembled, discarded or otherwise disposed of. The most dramatic event in the final stages of this process is the suicidal expulsion of the nucleus from reticulocytes to yield mature erythrocytes which are released into the peripheral circulation at the rate of 10^{11} cells per day. These cells are incapable of division or active movement and lack the ability to synthesize macromolecules or to generate ATP by oxidative phosphorylation. Nevertheless they are able to survive and function for some 120 days in the bloodstream, until desialation of their surface glycoproteins, and other poorly understood degradative changes to components which they are at this stage unable to replace, lead to their removal from the bloodstream in the spleen or the liver.

2. Physiological Role, Cell Shape and Membrane Properties

The cytoplasm of the circulating erythrocyte contains an extremely high concentration of the respiratory protein haemoglobin, and lower concentrations of many other proteins, including glycolytic enzymes which effect the synthesis of sufficient ATP to fuel various protein kinases and the transmembrane cation pumps by which the cell maintains ionic homeostasis and osmotic equilibrium. As such, in comparison with other cells, it is highly simplified and specialized for its physiological role of carrying dissolved oxygen and carbon dioxide (in the form of bicarbonate ions) between the lungs and the peripheral tisssues. Such cellular sequestration of haemoglobin within erythrocytes is essential if the overall haemoglobin concentration of the blood is to be high enough to support the high metabolic rate characteristic of mammals, while yet keeping the viscosity of the blood at a reasonably low value.

The cell's shape and viscoelastic properties, too, are highly adapted for its physiological role. The normal erythrocyte ("discocyte") has a characteristic shape (Fig. 1a), being a flattened biconcave disc approximately 8 μm diameter and 2 μm thick, in which the maximum intracellular diffusion path for the dissolved gases is thus limited to 1 μm. The cell's membrane is both extremely strong, enabling it to withstand the considerable sheer forces it experiences during the half million or so turbulent passages it makes through the heart during its 120-day lifespan, and also quite flexible, allowing the cell to distort and pass through splenic fenestrations and peripheral capillaries less than half its own diameter, from which it emerges undamaged, resuming its biconcave shape as soon as it has space to do so. The cell surface possesses a uniform high negative charge density, which is probably of importance in minimizing contact and adhesion with other cells, both those circulating in the blood and

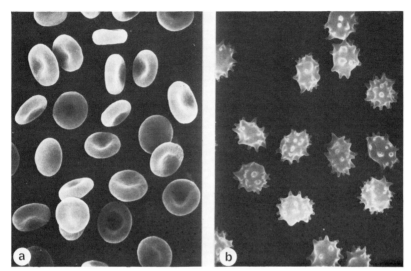

Fig. 1. Scanning electron micrographs of human erythrocytes: (a) Normal discocytes; (b) echinocytes (from Lovrien and Anderson (1980). J. Cell Biol. **85**, 534–548, with permission). Magnification approximately ×1250.

also those comprising the walls of the blood vessels. This property, together with the ease of deformability of the erythrocyte membrane, contributes greatly to the remarkably low viscosity of the blood, despite the fact that in a healthy person it contains approximately 10^{10} erythrocytes per ml, occupying more than 40% of the total blood volume.

Erythrocytes are exquisitely sensitive to osmotic stress. In hypertonic media they shrink, while in hypotonic conditions the cells swell, become spherical and then, if the internal turgor pressure exceeds the mechanical ability of their membranes to withstand rupture, they undergo hypotonic lysis and loss of their cytoplasmic contents. Further washing of the lysed cells with hypotonic buffers will complete the removal of the cytoplasm, resulting in a pure preparation of erythrocyte plasma membranes, usually referred to as "erythrocyte ghosts". Ghosts prepared from fresh, healthy erythrocytes retain the biconcave discoid shape of the cells from which they were derived, demonstrating that this physiologically-significant shape is an intrinsic property of the erythrocyte membrane itself, and represents the lowest free-energy state of this complex laminated molecular structure, which is described in detail below. Such freshly-prepared ghosts are said to be "leaky", being freely permeable to small molecules, and are thus insensitive to osmotic stress unless resealed by incubation at physiological ionic strength.

Incubation of intact erythrocytes under conditions which result in a decrease in their intracellular ATP concentration (for instance during prolonged blood bank storage) causes the cells to undergo a reversible shape change known as crenation, to form "echinocytes" (Fig. 1b). This change, which is accompanied by an increase in intracellular calcium due to failure of the calcium pump, a dephosphorylation of certain membrane proteins, and a marked and physiologically disastrous reduction in the deformability of the membrane, can be delayed by the addition of low concentrations of adenine to the citrate phosphate dextrose anticoagulant solution with which the blood is mixed at the start of storage, and is readily reversed by incubation of the crenated cells in the presence of glucose and adenosine, allowing the resynthesis of intracellular ATP, under which conditions the cells may become cup-shaped "stomatocytes". While the cause of the membrane stiffening which accompanies crenation during metabolic depletion is at present not fully understood, it is interesting to note that leaky ghosts prepared by hypotonic lysis of crenated erythrocytes are also crenated in shape, and that they too can be made to revert to flexible biconcave discs or stomatocytes either by incubation in the presence of ATP or by chelation of the calcium with EDTA.

3. Membrane Composition

The erythrocyte membrane is composed of approximately 52% protein, 40% lipid and 8% carbohydrate by weight, the carbohydrate being mainly in the form of short oligosaccharides attached via O- and N-glycosidic bonds to serine, threonine and asparagine residues within the polypeptide chains of glycoproteins, but with a small proportion (7%) as glycosphingolipids.

The membrane is remarkably rich in cholesterol, the lipids themselves being an almost equal mixture of cholesterol and phospholipids (cholesterol:phospholipid molar ratio = 0.8), together with a small proportion of glycolipids. These are arranged in a single bilayer approximately 140 μm^2 in surface area which is extremely asymmetric, phosphatidylinositol and the reactive aminophospholipids, phosphatidylserine and phosphatidylethanolamine, being largely confined to the cytoplasmic half of the bilayer, while most of the phosphatidylcholine and sphingomyelin and all the glycosphingolipids (which bear the majority of the A, B, H, Le^a, Le^b and P blood group antigenic determinants) are found in the extracellular half-bilayer. Furthermore the phospholipids which comprise the cytoplasmic half-bilayer are highly enriched in unsaturated fatty acyl side chains relative to the extracellular half-bilayer, rendering the cytoplasmic half-bilayer more fluid. Individual phospholipid molecules undergo spontaneous translocation from one half-bilayer to another ("flip-flop") at a finite although very slow rate, and their asymmetric equilibrium is maintained by preferential associations of the structural protein spectrin, located on the cytoplasmic surface of the bilayer, with the anionic phospholipids (which include phosphatidylethanolamine at physiological pH) but not with the neutral phospholipids bearing tertiary amino groups.

The cholesterol is present in both halves of the bilayer, although there is evidence that it is more abundant in the extracellular half, and being uncharged it can undergo flip-flop much more easily than the phospholipids. Both the cholesterol and the phospholipids in the outer half-bilayer can undergo ready exchange with unesterified cholesterol and phospholipids contained within plasma lipoproteins. The lipid molecules are free to diffuse within the plane of the bilayer, and exhibit very rapid rotational and translational mobility, although a minority may be in special association with specific membrane proteins and thus have their moblity restricted.

Following conventional wisdom, the proteins of the erythrocyte membrane may be conveniently divided into *the integral membrane proteins*, some parts of which penetrate the hydrophobic domain of the lipid bilayer, from which they cannot easily be solubilized without detergent disruption of the bilayer itself, and *the peripheral membrane proteins*, whose associations with one another, with the integral proteins and with the hydrophilic lipid headgroups are thought to involve primarily polar and ionic interactions (ion pairs, hydrogen bonds, etc.), since they can be solubilized without detergent treatment by suitable manipulations of pH and ionic concentrations. This division is largely an operational one, and certain membrane proteins exhibit particular properties characteristic of both classes. The elucidation of the molecular organization of the perpheral membrane proteins and the nature of their interactions with the integral proteins and the lipid bilayer is one of the most active and exciting areas of current membrane research.

4. Electrophoretic Analysis and Molecular Properties of the Membrane Proteins

The protein components of the erythrocyte membrane are usually defined as those present in erythrocyte ghosts prepared by hypotonic lysis following the procedures of Dodge *et al.* (1963), and have historically been identified numerically (Steck, 1974) by their electrophoretic mobility on one-dimensional polyacrylamide gels run in the presence of sodium dodecyl sulphate (SDS) under the conditions first described by Fairbanks *et al.* (1971), in which the proteins are separated with mobilities approximately proportional to the inverse logarithm of their molecular weights.

Such a one-dimensional fractionation has severe intrinsic limitations in resolution, and certain electrophoretic bands (for instance band 3; Steck, 1978; Lux and Glader, 1980) are now well known to contain several quite distinct polypeptides. Slight improvements in technique have occasionally resulted in the resolution of additional bands beyond those originally described (for instance the separation of band 4 into band 4·1 and band 4·2; Fairbanks *et al*., 1971; Steck and Yu, 1973), and, as research has progressed, other polypeptides (for instance band 2·1 and band 4·9) which were initially overlooked have been recognized to be functionally significant and have been assigned first numbers and later names.

Figure 2a shows the electrophoretic pattern typical of human erythrocyte membrane proteins prepared from fresh ghosts under conditions which prevent proteolytic degradation, run according to Steck and Yu (1973) in 5% polyacrylamide, 0·2% SDS, and 10% glycerol, and stained with Coomassie Brilliant Blue R250 (Shotton *et al*., 1979). The banding pattern observed is essentially identical with those reported in other careful studies (for instance Steck, 1974, shown here as Fig. 3a; Luna *et al*., 1979; Siegel *et al*., 1980) and the numbering scheme used harmonizes and extends the various partial numbering schemes previously used (Steck, 1974; Wang and Richards, 1974; Lux and John 1977; Luna *et al*., 1979; Sheetz, 1979a; White and Ralston, 1979; Fowler and Taylor, 1980).

The major sialoglycoproteins stain poorly with Coomassie Brilliant Blue, and are generally identified by staining their oligosaccharides using the periodic acid-Schiffs reagent ("PAS") procedure (Zacharius *et al*., 1969; Fairbanks *et al*., 1971), giving electrophoretic bands originally numbered PAS-1, PAS-2, etc., as shown in Fig. 3b, or by radiofluorography (Owens *et al*., 1980; see Fig. 2b).

Significant differences in the electrophoretic pattern given by the proteins of the erythrocyte membrane are obtained when the fractionation is carried out in a discontinuous buffer system of Laemmli (1970) rather than the continuous system of Fairbaks *et al*. (1971). The most obvious of these is the clear resolution of band 4·1 into two distanct bands, 4·1a and 4·1b (Edwards *et al*., 1979), and the altered patterns of migration of the sialoglycoproteins (Mueller *et al*., 1976; Furthmayr, 1978; Owens *et al*., 1980; Thompson *et al*., 1980). Until recently, there has been considerable confusion in the literature over the identity, oligomeric state and nomenclature of the erythrocyte sialoglycoproteins, both because glycophorin A and B can form homodimers and heterodimers which are resistant to dissociation in SDS, and also because several of the proteins are not resolved in the Fairbanks gel system. However, these recent studies have resulted in a more precise analysis of the nature of the individual sialoglycoproteins, leading to a revised and consistent nomenclature (Furthmayr, 1978a; Owens *et al*., 1980; Mueller and Morrison, 1980), as shown in Fig. 2b and detailed in Table 1B. It is likely, but not yet proven, that Furthmayr's glycophorin C is identical to Mueller and Morrison's glycoconnectin (PAS 2).

As a result of intense research over the last decade (reviewed in part by

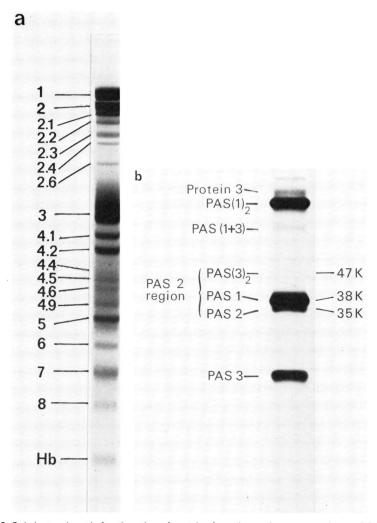

Fig. 2. Gel electrophoretic fractionation of proteins from the erythrocyte membrane: (a) fresh erythrocyte ghosts (Dodge et al., 1963) were fractionated in a 0·2% SDS–5% polyacrylamide tube gel using the continuous buffer system of Fairbanks et al. (1971) as modified by Steck and Yu (1973), and stained for protein with Coomassie Brilliant Blue R250. The individual polypeptides are numbered according to the scheme of Steck (1974), with the additions discussed in the text and Table I (from Shotton et al. (1979). J. Mol. Biol. **131**, 303–329, with permission). (b) Fresh erythrocyte ghosts, isolated from intact red cells labelled with periodate and [^{3}H]-NaBH$_4$, were fractionated in 0·1% sodium dodecyl sulphate on a 12–20% polyacrylamide gradient slab gel, using the discontinuous buffer system of Laemmli (1970). The labelled proteins were detected by radiofluorography. The electrophoretic positions of PAS 1, PAS 2 and PAS 3 monomers, homodimers and heterodimers (see text) are indicated (from Owens et al. (1980). Archs. Biochem. Biophys. **204**, 247–254, with permission).

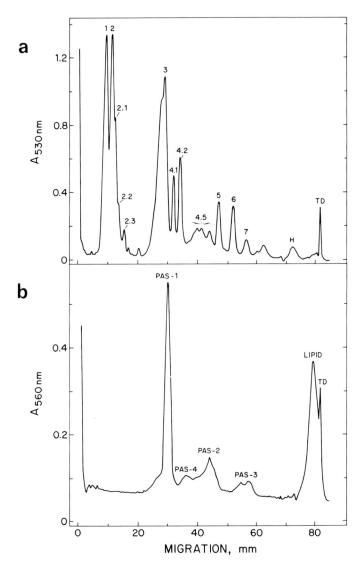

Fig. 3. Densitometer scans of erythrocyte membrane proteins fractionated on Fairbanks SDS-polyacrylamide tube gels as in Fig. 2a: (a) stained for protein with Coomassie Brilliant Blue R250 as in Fig. 2a; (b) stained for carbohydrate with PAS reagent (compare with Fig. 2b in which the same glycoproteins give a different pattern after being fractionated by the Laemmli gel electrophoretic system) (from Steck (1974). *J. Cell Biol.* **62**, 1–19, with permission, whose original nomenclature is retained).

Steck, 1974; Marchesi *et al.*, 1976; Lux and Glader, 1980), the molecular weights, natural abundance and topological arrangement of most of the major human erythrocyte membrane proteins are now fairly well established. I have summarized these data in Table I. The major integral proteins, with the possible exception of a component of the band 7 region of the gel, all span the lipid bilayer and are all glycoproteins, with their oligoscaccharide moieties confined to the extracellular surface, while the major peripheral membrane proteins are all located on the cytoplasmic surface of the bilayer, where they bind directly to integral membrane proteins and/or associate specifically with other peripheral membrane protein molecules.

In addition to the major proteins, which may be easily detected by such electrophoretic methods, the membrane contains a host of other proteins in smaller amounts, which often can only be recognized by functional assays on the intact membrane. These include transmembrane proteins such as the glucose carrier and the multicomponent sodium and calcium pumps, enzymes such as acetylcholinesterase which are confined to the extracellular surface of the membrane, and others like the protein kinases which are localized on the cytoplasmic surface (for review see Lux and Glader, 1980). Although very few copies of certain of these proteins are present within each cell membrane, their physiological significance is often tremendous. However, since little or no electron microscopic analysis of them has yet been undertaken, no further discussion of these minor proteins will be given here.

II. FREEZE-ETCH STUDIES OF ERYTHROCYTE MEMBRANES

A. Historical Perspective

The techniques of freeze fracture and freeze etching (Fischer and Branton, 1974; Southworth *et al.*, 1975; Sleytr and Robards, 1977b) are unique among electron microscopic methods in that they allow the study of the in-plane distribution of integral proteins spanning the lipid bilayer as a function of developmental stage, experimental conditions or onset of disease.

The erythrocyte membrane was one of the first biological membranes to be thoroughly studied by freeze fracture (Weinstein, 1969 and 1974) and provided much of the new data which led to the formulation of the fluid mosaic model of membrane structure by Singer and Nicolson in 1972. In particular, it was used in the classical freeze-etching studies of Pinto da Silva and Branton (1970) and Tillack and Marchesi (1970) which established that during freeze fracture of biological membranes the fracture plane passes between the terminal methyl groups of the opposed layers of fatty acyl chains, cleaving the lipid bilayer of the membrane into its two constituent mono-

TABLE I
Molecular Properties of the Major Human Erythrocyte Membrane Proteins[a]

A. Coomassie Blue Staining Proteins

Name	Properties of the SDS-denatured polypeptides				Properties of the native molecules				Other membrane polypeptides with which the molecule is known to interact
	Electrophoretic band[b]	Molecular weight	No. of polypeptide chains per ghost	% total protein[c]	Oligomerization state	Molecular weight	Sedimentation coefficient	Type[d]	
Spectrin	1	240 000	200 000	25	$\alpha_2\beta_2$ tetramer	960 000	12·5 S	P, MS	2·1, 4·1, 5 and probably 4·9
	2	220 000	200 000						
Ankyrin or syndeins[e]	2·1	210 000	100 000	5	Monomer	210 000		P	2 and 3
Anion channel protein	3[f]	93 000	1 200 000	25	Tetramer	373 000		I	2·1, 4·2 and 6. Coaggregates with glycophorin A
—	4·1	82 000	200 000	4–5	Monomer or dimer	82 000 or 164 000		P, MS	1 and 2, 5, PAS-2 and possibly 4·9
—	4·2	72 000	200 000	4–5	α_4 tetramer	288 000		P	3
—	4·9	50 000	100 000	1	Not known	—	—	P, MS	Probably 1, 2, 4·1 or 5
Actin	5	43 000	400 000	4–5	Probably α_{12} to α_{18} oligomers	Variable	Variable	P, MS	1 and 2, 4·1 and possibly 4·9
Glyceraldehyde 3-phosphate dehydrogenase	6	35 000	500 000	4–5	α_4 tetramer	140 000		P	3[g]
Hollow cylinder protein or cylindrin[h]	7·1	28 000	500 000	4–5	Probably $\alpha_{12}\beta_6/_{12}$ cylindrical complex	750 000	19·3 S	P	Not known
	7·2	25 000							
	8	22 500							
Torus protein or torin[h]	8·1	20 000	150 000	1	Probably α_{10} torus-shaped complex	200 000	9·0 S	P	Not known

Properties of the SDS-denatured polypeptides | | | | | Properties of the native molecules | | | |

Name	Electro-phoretic bands[j]	Apparent molecular weight[l]	No. of poly-peptide chains per ghost[k]	% total protein[k]	Oligomeric state	Molecular weight[l]	% total membrane sialic acid[m]	Type[d]	Other membrane poly-peptides with which the molecule is known to interact
Glycophorin A	PAS (1)₂ PAS 1	76 000 38 000	420 000	1·6	Monomer or dimer	62 000 31 000	80	I	Dimer stable in SDS. Monomer can form a PAS (1+3) heterodimer with glycophorin B. Can be co-aggregated with band 3
—	PAS 2' PAS 2	45 000 35 000	? 35 000		? Monomer	? ?	Trace 8	I I	4·1
Glycoconnectin Glycophorin B	PAS 3	24 000	70 000		Monomer or dimer	?	10	I	Can form a PAS (1+3) heterodimer with glyco-phorin A, as well as a PAS (3)₂ homodimer

[a] Best available data, based mainly on Steck (1974) and Lux and Glader (1980) and references therein unless otherwise indicated. All numerical values are approximations.
[b] After electrophoresis in 5% polyacrylamide gels containing 0·2% SDS and 10% glycerol (Fairbanks et al., 1971; Steck and Yu, 1973) as shown in Fig. 2a.
[c] Based on Coomassie Brilliant Blue staining intensity, by making the simplistic assumption that the colour yield for all proteins is identical.
[d] P: peripheral membrane protein located on the cytoplasmic surface of the lipid bilayer; MS: component of the membrane skeleton; I: integral glycoprotein spanning the lipid bilayer, with oligosaccharide moieties confined to the extracellular surface.
[e] In addition, bands 2·2 to 2·6 are a family of sequence-related peptides derived from band 2·1 which are present even after the most careful preparative conditions designed to avoid proteolysis (see Fig. 2a), implying that all of them are also present in the native erythrocyte (Luna et al., 1979; Yu and Goodman 1979; Siegel et al., 1980).
[f] Minor components of band 3, co-migrating with the anion exchange protein, include among other proteins acetylcholine esterase (approximately 0·25% of total membrane protein, i.e. 10 000 copies per ghost) and the Na⁺, K⁺-ATPase (approximately 0·005% of total membrane protein, i.e. 200 copies per ghost).
[g] Shares with phosphofructokinase and aldolase a common saturable binding site on the anion exchange protein.
[h] Data from Harris and Naeem (1978). White and Ralston (1979) and Harris (1980). It is reported that a second major water-soluble component from the band 7·1 region exists as a monomer, while one from the band 8 region appears to be a dimer (White and Ralston, 1979; and personal communication of their unpublished results). Another component in the band 7·1 region, originally called band 7, appears to resist extraction from the membrane and may be an integral protein (Steck and Yu, 1973; Steck, 1974). However, considerable uncertainty remains about the exact number, abundance and polymeric states of the small polypeptides in the band 7 and 8 regions, and the data given should at this stage be regarded as unestablished (see Section VI).
[i] Nomenclature of Furthmayr (1978a), Owens et al. (1980) and Mueller and Morrison (1980) (see also Fig. 2b).
[j] Apparent molecular weights after SDS-gel electrophoresis on 12·5% gels in the discontinuous buffer system of Laemmli (1970) as shown in Fig. 2b (Owens et al., 1980). The sialoglycoproteins are known to migrate anomalously slowly, giving spuriously high molecular weights, and these figures should only be used to compare relative mobilities.
[k] Furthmayr (1978a).
[l] True molecular weights of the glycophorin A homodimer and monomer, established from its amino acid and oligosaccharide composition (Tomita and Marchesi, 1975; Tomita et al., 1978). The molecule is approximately 40% polypeptide and 60% carbohydrate by weight.
[m] Owens et al. (1980).

layers. While the quality of the data presented in these original studies and the conclusions drawn from them have been criticized (Sjöstränd, 1979), the majority of subsequent investigations have provided overwhelming supportive evidence for the then revolutionary hypothesis that freeze-fracture splits the bilayer.

The freeze-fractured erythrocyte membrane exhibits two distinct appearances when replicas of the membrane's two fracture faces are viewed in the transmission electron microscope (Fig. 4). The protoplasmic or P fracture face (nomenclature of Branton *et al.*, 1975) is heavily studded with circa 8·5 nm diameter P-face intramembrane particles (IMP_P), which appear almost randomly distributed and which occupy about 20% of the surface area, interrupting the otherwise smooth fracture face formed by the terminal portions of the fatty acyl chains and cholesterol molecules in the protoplasmic (P) half of the split lipid bilayer. The complementary extracellular (E) fracture face contains a lower number of E-face intramembrane particles (IMP_E) which are rather more heterogeneous in appearance, and has a high density of

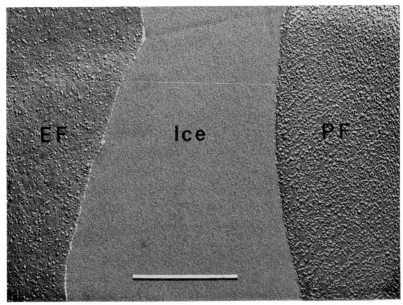

Fig. 4. Freeze fracture of two erythrocyte ghosts showing the typical apperance of the erythrocyte membrane's extracellular fracture face (E face) (left) and protoplasmic fracture face (P face) (right), separated by smooth ice (Shotton, unpublished work). This and subsequent freeze fracture, freeze etch and low-angle shadowing micrographs are all positive prints (platinum black, shadows white), with (except in those which have been rotary shadowed) the direction of platinum deposition from below. Magnification ×57 000; bar=500 nm.

less obvious depressions or pits formed by the removal of the P-face particles during the fracture process. Lack of complete complementarity between these fracture faces is attributed to infilling of depressions and to plastic distortion and "decoration" of particles during the fracturing and replication steps (Sleytr and Robards, 1977a).

The ability to separate the inner and outer halves of membranes by freeze fracture led Fisher (1975) to propose a novel way to exploit the technique to achieve half membrane enrichment. In his model system, erythrocytes attached to a polylysine-coated glass coverslip are frozen conventionally against a copper disc. When this frozen sandwich is cleaved under vacuum or liquid nitrogen, the fracture plane passes along the erythrocyte membranes where they are adjacent to the glass, thus removing most of each cell with the copper disc and leaving primarily E-half membrane fragments attached to the glass coverslip. As well as permitting replication of extensive flat fracture planes of P or E faces at a precisely known shadow angle, this method's primary importance is that it also allows the physical separation and subsequent chemical characterization of the components of the fractured half-membranes. This has since been exploited by Fisher (1976) for cholesterol analysis, by Fisher (1978) and subsequently Nermut and Williams (1980) for autoradiographic feasibility studies, and by Edwards *et al*. (1979) for the characterization of the proteins which partition with each half-membrane (see below).

B. The Nature of the Intramembrane Particles

There is now abundant evidence that the intramembrane particles seen in replicas of freeze-fractured erythrocyte membranes are the shadowed images of integral proteins exposed by the fracture process. This was first elegantly demonstrated by the application of the proteolytic enzyme pronase to erythrocyte ghosts before fracture, when it was found that the disappearance of the particles could be correlated with the degree of proteolytic destruction of the membrane proteins (Engstrom, 1970; Branton, 1971). This conclusion was strengthened by the subsequent demonstration that protein-free lipid blebs which could be induced to bud from the surface of fresh erythrocyte ghosts were also free of intramembrane particles (Elgsaeter and Branton, 1974; Elgsaeter *et al*., 1976), by the freeze-etch labelling studies described below, and by the various more specific reconstitution experiments mentioned in Section III.

Because of its abundance, the obvious candidate for the protein which constitutes the majority of these particles is the anion channel protein, band 3, particularly since it is possible to elute most of the other major proteins from the membrane while leaving the IMPs unchanged, and since the appearance of

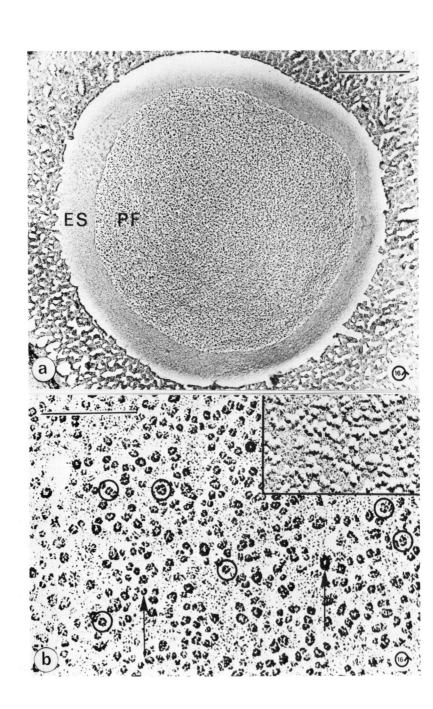

the IMPs is normal in En(a-) mutants (discussed in Section VII) which lack the other principal integral membrane protein, glycophorin A (Bächi et al., 1979; Gahmberg et al., 1978). However, there is extensive evidence from freeze-etch labelling and fluorescence mobility studies, discussed below, that glycophorin-A molecules normally associate with band 3 *in vivo* and redistribute with the intramembrane particles when these are aggregated.

The precise number of particles observable on the fracture faces depends crucially upon the exact method of replicating the fractured surface, the absence of contamination before and during replication, and the subjective criteria used by the investigator to define an intramembrane particle for the purpose of counting. Typical early published figures (Weinstein, 1969, 1974) are 2600 IMP_P per μm^2 (82% of total) and 575 IMP_E per μm^2 (18% of total), giving a total of approximately 460 000 IMP_P plus IMP_E per erythrocyte.

Rotary shadowing with platinum, rather than conventional unidirectional shadowing, during replication of the fractured membrane surface has been used to reveal the structure of the IMP_P more clearly (Margaritis et al., 1976, 1977). Some of the particles appear to be clearly tetrameric in structure, while a minority are larger than the others (Fig. 5). Using this method, which also decreases ambiguities in IMP identification for counting purposes, Weinstein *et al.* (1980) have recently re-estimated the density of IMP_P as 2430 ± 220 per μm^2 and their total number as $3 \cdot 2 \times 10^5$ per erythrocyte, and have concluded from considerations of their size, appearance and natural abundance that the majority of IMP_P are tetramers of the globular integral protein, band 3, approximately $1 \cdot 2 \times 10^6$ copies of which are present per cell. In support of this claim, Edwards *et al.* (1979) have shown, by chemical analysis of the polypeptides present in the E half of the membrane after freeze fracture, that band 3 remains predominantly on the P half membrane, and Pinto da Silva and Torrisi (1981) have obtained the same result by cytochemical labelling after freeze fracture. The heterogeneity seen among IMP_P can be explained both by the occurrence of subpopulations of minor proteins forming dissimilar particles, and also by the plastic distortion which the fracture process itself is known to induce in protein molecules (Sleytr and Robards, 1977a).

The nature of the IMP_E is at present less clear. They always appear more heterogeneous than the IMP_P, and in the best replicas a proportion of them

---

Fig. 5. Freeze-etched erythrocyte ghost after rotary shadowing at 16° angle. (a) Central P face surrounded by an annulus of etched E surface, embedded in rough etched ice. The overall contrast is higher than with conventional unidirectional shadowing at 45° angle (Fig. 4). Magnification ×40 000; bar=500 nm. (b) At high magnification many intramembrane particles exhibit tetrameric subunit structure (circles) normally not visible after unidirectional shadowing (inset). However, heterogeneity in size and substructure is also evident (arrows). Magnification ×250 000; bar=100 nm (from Margaritis *et al.* (1977). *J. Cell Biol.* **72**, 47–56, with permission).

are rod- or strand-like in appearance. Some may simply be the "tail views" of a minority of the band 3 particles which have by chance partitioned with the E half membrane. However, from their freeze-fracture cytochemical labelling studies and chemical analyses, respectively, Pinto da Silva and Torrisi (1981) and Edwards et al. (1979) conclude that the predominant polypeptides which partition with the E half membrane are not band 3 but rather monomers (but not dimers) of glycophorin A and large fragments of this protein formed by breakage of covalent bonds within the polypeptide chain during the fracture process. These molecules might thus constitute the smaller members of the IMP_E population, since glycophorin A reconstituted into synthetic bilayers gives similar small rugosities, particles and particle aggregates as discussed in Section III below.

Using a liquid propane jet freezer to give more rapid freezing, Elgsaeter and his colleagues (Elgsaeter et al., 1980; Espevik and Elgsaeter, 1981) obtained highly unusual images of the P and E faces of erythrocyte ghosts, with fibrous strands on the E face and grooves on the P face, although the replicas from intact erythrocytes frozen under identical conditions had conventional appearances. This finding has as yet no satisfactory explanation (Elgsaeter, personal communication) and has not at present been substantiated by other investigators.

C. Particle Aggregation Studies

In the pronase digestion experiment mentioned above (Engstrom, 1970; Branton, 1971), it was observed that during the initial stages of proteolysis of the erythrocyte membrane extensive aggregation of the IMPs occurred prior to any significant decrease in particle density. This suggested that proteolysis was loosening constraints upon the lateral mobility of the integral proteins imposed by the more readily accessible and thus preferentially digested peripheral proteins. In subsequent studies Pinto da Silva (1972) and Pinto da Silva and Branton (1972) were able to induce similar particle aggregation without adding proteolytic enzymes, and to demonstrate that it was both pH dependent and reversible. The molecular explanation of these phenomena was provided in subsequent studies by Elgsaeter and Branton (1974) and Elgsaeter et al. (1976) who showed that the normal constraints on IMP mobility could be removed by the selective elution of much of the spectrin and actin from the cytoplasmic surface of the membrane, and that, following this pretreatment, the IMPs could easily be induced to aggregate by a variety of conditions, all of which were effective in precipitating spectrin from solution. It was proposed that in these experiments particle aggregation was caused by the microprecipitation of residual spectrin molecules on the cytoplasmic surface of the membrane, to which the integral proteins were in some way attached.

Example of freeze-etched membranes after such particle aggregation are shown in Fig. 6. In support of this, Shotton et al. (1978) demonstrated by ferritin-conjugated anti-spectrin labelling that spectrin molecules underlay the aggregates of IMPs. More recently the protein ankyrin mediating the binding of the integral proteins to spectrin has been identified and is discussed in Section V below.

Similar aggregation has been induced by lectin binding to the external oligosaccharides of the integral glycoproteins (see below) and by the direct binding of divalent antispectrin antibodies to the cytoplasmic surface of the membrane under conditions of low ionic strength (Nicolson and Painter, 1973). This latter finding has been confirmed by Asane and Sekiguchi (1978), who also observed that similar treatment of ghosts with antispectrin antibodies under isotonic conditions failed to result in a redistribution of intramembrane particles and indeed inhibited the particle clustering normally seen by them during membrane fusion with low levels of *Sendai* virus under these isotonic conditions. Bächi et al. (1973) and Lalazar et al. (1976) had previously observed particle clustering during fusion with high levels of *Sendai* virus, but on the basis of extensive freeze-fracture studies on fusion of the viral envelope with human erythrocytes, Knutton (1977) claimed that particle redistribution did not always occur with low levels of *Sendai* virus. The reason for this discrepancy is unclear.

Occasionally experimental treatments have resulted in the aggregation of IMPs into striking linear arrays, in contrast to the random aggregates normally observed. For instance Lutz et al. (1977) found that sheep erythrocytes which were "aged" *in vivo* vesiculated to form long tube-like protrusions. These were depleted in spectrin but retained the integral membrane proteins, which were observed to be regularly arranged in paired particle rows (Fig. 7). In my own unpublished studies of particle aggregation in human erythrocyte ghosts after limited tryptic proteolysis, I observed that many of the IMPs crystallized into regular straight "avenues" about 8 particles wide. While plastic distortion obscured the original periodicity of the particle aggregates on the P faces (Fig. 8a), this periodicity could be clearly discerned in the depressions left by the IMP_P on the opposed E faces (Fig. 8b), a phenomenon subsequently noted after freeze fracture of other periodic systems (Branton and Kirchanski, 1977). The P_1 unit cell of this lattice has approximate dimensions $a=7\cdot3$ nm, $b=10\cdot4$ nm, $\alpha=86\cdot5°$, and is thus of similar area to a single 8·5 nm diameter IMP_P, as one would expect. While experiments of this sort obviously involve gross distortions of the membrane from its native structure, they show that the integral proteins comprising a majority of the IMP_P are sufficiently uniform in composition that they can spontaneously associate into two-dimensional crystalline arrays, once freed from the restraints normally exercised by the membrane skeleton.

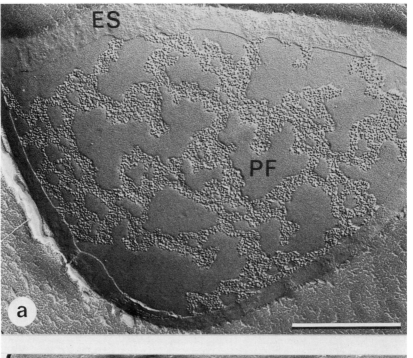
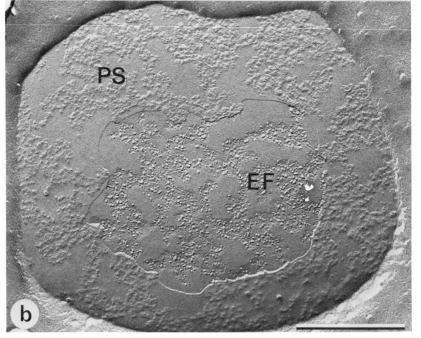

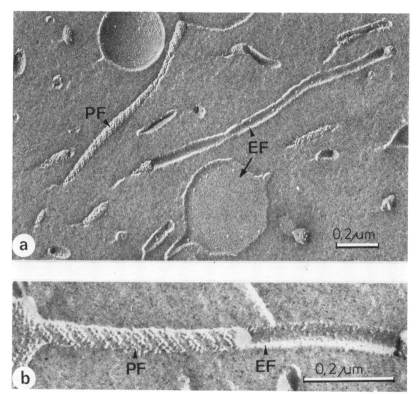

Fig. 7. Freeze fracture of stromalytic protrusions from "aged" sheep erythrocyte ghosts, showing helical double rows of P face intramembrane particles (from Lutz et al. (1977). *J. Cell Biol.* **74**, 389–398, with permission). Magnifications (a) ×56 000, (b) ×120 000; bars=200 nm.

Fig. 6. Freeze-etched erythrocyte ghosts after particle aggregation. (a) P face, with particle aggregates which have slightly collapsed during etching (Pinto da Silva, 1973; Pinto da Silva et al., 1973) and intervening smooth areas of lipid tails, surrounded by an annulus of etched E surface showing slight surface rugosities due to surface glycoproteins contiguous with the P face particle aggregates, and intervening smooth areas of lipid head groups (Shotton, unpublished work). (b) E face, with the pits from which most of the IMPs have been removed aggregated to form extensive shallow depressions containing the remaining E face particles, and intervening smooth areas of lipid tails, surrounded by an annulus of etched P surface showing prominent aggregates of peripheral membrane proteins, contiguous with the E face pits, and intervening smooth ares of lipid head groups (from Shotton et al. (1978). *J. Cell Biol.* **76**, 512–531, with permission). Magnifications ×60 000; bar=500 nm.

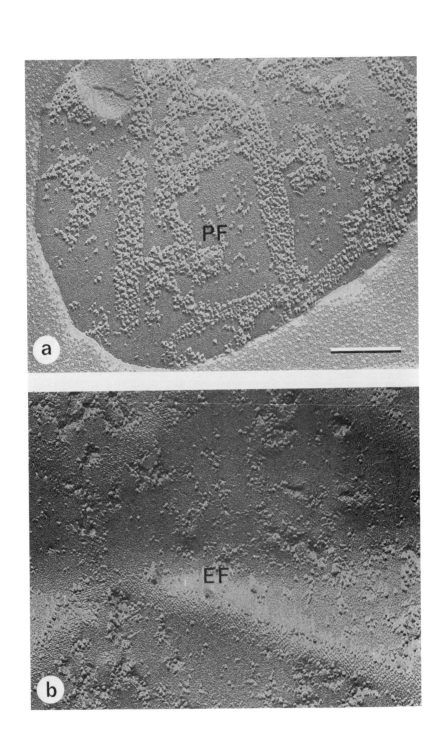

D. Studies of the Etched Membrane Surfaces

1. The Extracellular Surface

While it is known that the oligosaccharide-bearing regions of the band 3 polypeptide and of the sialoglycoproteins extend for a considerable distance from the external surface (ES) of the erythrocyte membrane into the extracellular space, these structures are not readily visualized after freeze etching to expose this membrane surface, presumably because they tend to collapse back onto the lipid bilayer. In the untreated erythrocyte membrane only the faintest indications of substructure are visible on the ES (Fig. 9a), and it is only after particle aggregation that the protruding protein components can be clearly seen as areas of increased surface texture (Fig. 6a). Weinstein *et al.* (1978) have shown that if the external surface of the erythrocyte membrane is subjected to proteolyis by chymotrypsin or pronase, which preferentially releases the N-terminal domains of the sialoglycoproteins, the appearance changes subtly to reveal 3 to 5 nm diameter ES particles, which coaggregate with the IMP_P and are presumed to be surface protrusions of the band 3 molecules (Fig. 10).

The chemical identity of the intramembrane particles has been widely investigated using a variety of ferritin conjugates whose location can be clearly seen on the external surface of the membrane either after freeze etching or thin sectioning. At the most basic level, the identification of the protrusions on the etched surface of the membrane as components of the integral membrane proteins has been confirmed by labelling the membrane with a ferritin-conjugated anti-hapten, after first coupling the hapten nonspecifically to the surface proteins of intact erythrocytes and subsequently inducing particle aggregation in the ghosts prepared from these hapten-labelled cells (Shotton *et al.*, 1978). Figure 11 shows the results of such an experiment, in which the contiguity between the areas of aggregated IMP_P on the PF and of aggregated ferritin-conjugated anti-hapten antibody on the ES can be clearly seen. Not only did these experiments demonstrate that the intramembrane particles contained protein, but, more importantly, they also ruled out the possibility that any significant proportion of the labelled surface proteins existed independently of the particles, able to maintain a dispersed distribution under conditions where the particles were aggregated.

Fig. 8. Freeze-fractured erythrocyte ghosts in which particle aggregation has been induced at pH 5·0 following limited proteolysis with trypsin. (a) P face, showing broad, straight "aveneus" of IMP_P without apparent internal order. (b) E face, showing particle-free "aveneus" of highly ordered pits, from which the IMP_Ps were removed by the fracture process (Shotton, unpublished work). Magnifications ×96 000; bar=200 nm.

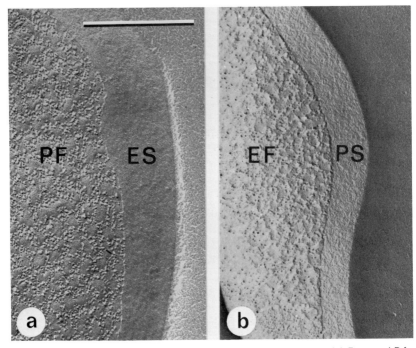

Fig. 9. Freeze-etched erythrocyte ghosts without particle aggregation. (a) Fractured P face (left) and etched E surface (Shotton, unpublished work). (b) Fractured E face (left) pitted by etching (Pinto da Silva et al., 1973) and etched P surface (from Shotton et al. (1978). J. Cell Biol. **76**, 512–531, with permission). Magnifications ×60 000; bar=500 nm.

Of the specific ferritin conjugates which have been employed, the simplest is cationized ferritin, which binds primarily to the anionic, sialic acid-rich, N-terminal portions of the glycophorin molecules. The use of this conjugate was pioneered by Danon et al. (1972), who were able to demonstrate by thin sectioning that the density of cationized ferritin binding sites was less in old rabbit erythrocytes than in young ones (Fig. 12) indicating a progressive desialation with advancing age. More recently Gahmberg et al. (1978) used the same method to study En(a−) erythrocyte membranes, in which glycophorin A is genetically absent (see Section VII, below), and found a low density of cationized ferritin binding sites. The first freeze etch studies employing cationized ferritin were conducted by Pinto da Silva et al. (1973), who induced intramembrane particle aggregation in normal erythrocyte ghosts by incubation at pH 5·0 and 35°C in the presence of 1 mM calcium chloride, in order to make visible the strong positive correlation which exists between the location of the cationized ferritin on the etched external surface of the ghosts and the intramembrane particle clusters on the adjacent P fracture

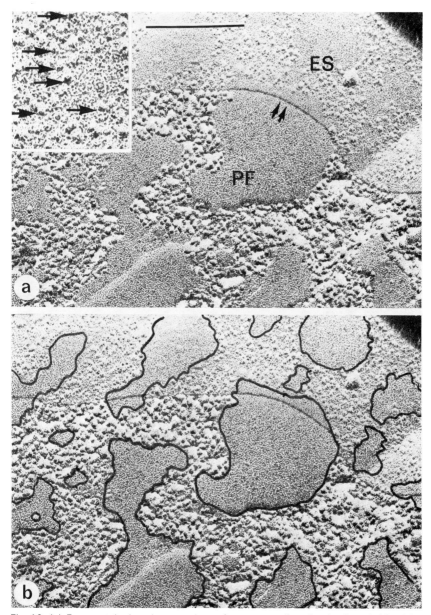

Fig. 10. (a) Freeze-etched erythrocyte ghost after pronase digestion (at 1 mg/ml for 4 h at 0 °C), showing aggregated IMP$_P$ and contiguous aggregates of small particles on the etched E surface. Double arrow shows the P face/E surface boundary. Magnification ×135 000; bar=200 nm. Inset: high magnification of ES particles (arrows) which are small and cast long shadows. Magnification ×270 000. (b) As (a), with boundaries between particle-rich and particle-free areas drawn with a pen, to illustrate that the IMP$_P$ and the ES particles share the same topographical distribution (from Weinstein et al. (1978). *J. Supramol. Struct.* **8**, 325–335, with permission).

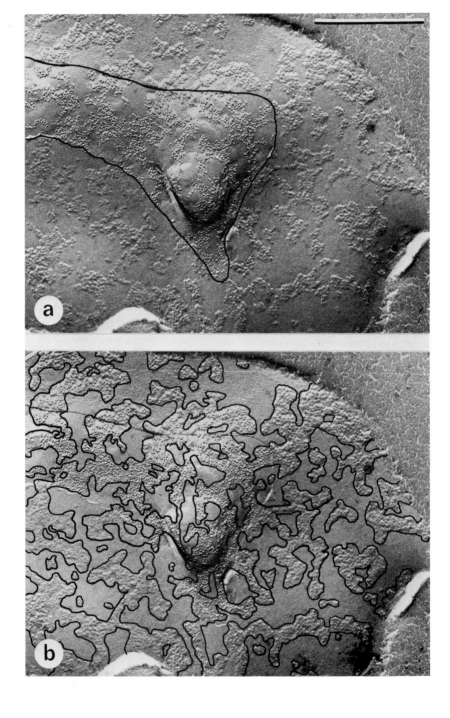

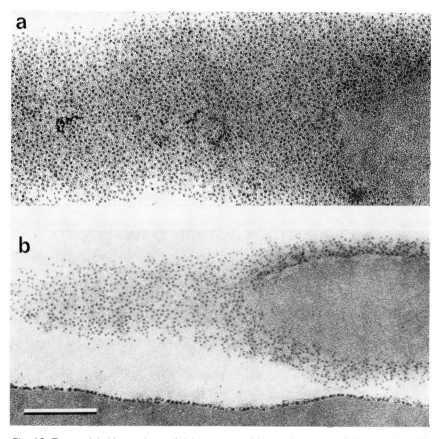

Fig. 12. Tangential thin sections of (a) a young rabbit erythrocyte and (b) an old rabbit erythrocyte, both labelled with cationized ferritin (small black dots) prior to fixation with glutaraldehyde. The older cell shows a lower ferritin density, reflecting the desialation of some of its surface sialoglycoproteins (from Dannon et al. (1972). J. Ultrastr. Res. **38**, 500–510, with permission). Magnifications ×100 000; bar=200 nm.

Fig. 11. (a) Freeze-etched erythrocyte ghosts in which the hapten p-diazonium-phenyl-β-D-lactoside (*lac*) has been covalently attached to the surface glycoproteins, which have then been subjected to particle aggregtion and labelled with ferritin-conjugated anti-*lac* immunoglobulin. The ferritin molecules appear as closely-packed spheres on the etched E surface. The P face/E surface boundary has been drawn in with a pen for clarity. (b) As (a), with the boundaries between particle- and ferritin-rich areas drawn in to illustrate that they share a common topographical distribution (from Shotton et al. (1978). J. Cell Biol. **76**, 512–531, with permission). Magnifications ×60 000; bar=500 nm.

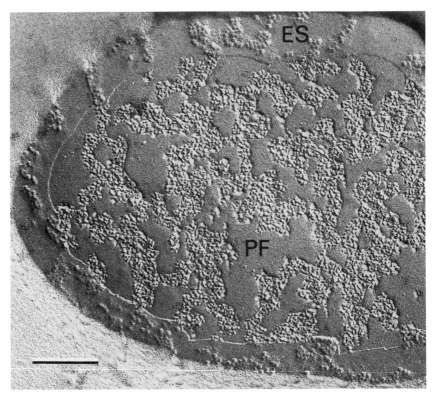

Fig. 13. Freeze-etched human erythrocyte ghost after particle aggregation and labelling with cationized ferritin at pH 5·5. Ferritin aggregates attached to sialoglycoproteins on the etched E surface are contiguous with IMP_P aggregates (from Pinto da Silva et al. (1973). *Exp. Cell Res.* **81**, 127–138, with permission). Magnification ×90 000; bar=200 nm.

face (Fig. 13), proving that the sialoglycoproteins are associated with the structural units which after freeze fracture give rise to the IMP_P. Marikovsky et al. (1978) have found that hypertonic crenation of erythrocytes leads to a reversible decrease in the binding of cationized ferritin molecules, particularly at the bases of the evaginated echinocytic spicules so formed.

Other workers have used conjugates to specific antisera or lectins, in every case obtaining a positive relationship between the distribution of the ligand on the etched membrane surface and the intramembrane particles on the fracture face. For instance, Tillach et al. (1972) and Pinto da Silva and Nicolson (1974) employed ferritin conjugates of the lectins phytohaemagglutinin from *Phaseolus vulgaris* and concanavalin A from jack beans, respectively, and thus identified glycophorin A and band 3 as components of the structural units which after freeze fracture give rise to the intramembrane particles (Fig. 14).

2. The Cytoplasmic Surface

By contrast with the external surface, the cytoplasmic surface (PS) of the human erythrocyte ghost membrane, after etching from low ionic strength buffers, always has a distinctly textured appearance covering the entire membrane surface (Fig. 9b), due both to the cytoplasmic protrusions of the integral membrane proteins and also to the dense underlying coat of peripheral membrane proteins forming the membrane skeleton. After aggregation of the IMPs, the PS shows rough areas of surface protein which are contiguous with the intramembrane particle clusters and intervening smooth areas of lipid head groups contiguous with the areas of smooth lipid tails from the opposing half-membrane on the adjacent EF (Fig. 6b). Ferritin-conjugated antibodies, directed either against spectrin or against a hapten coupled nonspecifically to accessible cytoplasmic surface protein molecules, label these rough PS areas exclusively, confirming them to be both composed of protein and also spectrin-rich (Shotton *et al.*, 1978). These areas thus contain the collapsed remains of the aggregated membrane skeleton left after etching, together with the cytoplasmic protrusions of the integral membrane proteins.

If fresh ghosts are resuspended in isotonic phosphate buffer before freeze

Fig. 14. Freeze-etched human erythrocyte ghost after particle aggregation, labelled with ferritin-conjugted concanavalin A. This binds to the extracellular C-terminal domain of the anion channel, band 3, whose topographical distribution is thereby revealed as being identical to and contiguous with that of the IMP$_P$ aggregates (from Pinto da Silva and Nicolson (1974). *Biochim. Biophys. Acta* **363**, 311–319, with permission). Magnification ×80 000; bar=200 nm.

etching, there is a subtle improvement in the appearance of the etched cytoplasmic surface (Shotton, unpublished data) which results in the spectrin molecules of the membrane skeleton being more clearly identifiable. The appearance under these conditions bears a strong resemblance to the images of the membrane skeleton obtained after tannic acid fixation or negative staining, described in Section IV below.

Weinstein et al. (1978) have shown that if all the peripheral membrane proteins are eluted from the cytoplasmic surface of the membrane, the cytoplasmic N-terminal 40 000 dalton domains of the integral band 3 molecules are revealed as a population of irregular "granofibrillar components", which can be selectively decorated with glyceraldehyde-3-phosphate dehydrogenase (G3PD; band 6). Proteolysis, to which this N-terminal domain of band 3 is exquisitely sensitive, removes these "granofibrillar components" and abolishes G3PD binding, leaving a low density of protease-resistant 9 nm diameter ES particles of unknown composition.

III. THE MAJOR INTEGRAL MEMBRANE PROTEINS

A. Band Three, the Anion Channel Protein

1. General Properties and Protein Chemistry

Band 3 is the most abundant polypeptide and the purported mediator of anion transport in the human erythrocyte membrane. Several major proteins co-electrophorese in the band 3 region, and band 3 itself shows oligosaccharide heterogeneity (Tsuji et al., 1981a) which broadens the electrophoretic band it forms (Mueller et al., 1979). However, this protein, which has not yet been given a specific name, is a unique species with well-defined molecular properties (reviewed by Cabantchik et al., 1978; Steck, 1978) although variants are known (Mueller and Morrison, 1977). It is a large globular glycoprotein, which spans the membrane asymmetrically and shows the hydrophobic behaviour typical of an integral membrane protein, being best solubilized and purified in non-ionic detergent in which it exists as non-covalent dimers or tetramers of identical polypeptides of approximately 93 000 daltons (Yu and Steck, 1975a, b; Kiehm and Ji, 1977; Nigg and Cherry, 1979b; Nakashima and Makino, 1980). The labelling studies which demonstrated its asymmetric transmembrane location and function are reviewed by Steck (1974) and Bretscher and Raff (1975), its protein chemistry and general properties are discussed in a review by Steck (1978), the structure of its oligosaccharides have been determind by Tsuji et al. (1980, 1981b), and their changes during development have been characterized by Fuduka and Fuduka (1981).

Unlike the situation with glycophorin-A and certain viral envelope proteins, the amino-terminus of band 3 is at the cytoplasmic surface of the membrane. The polypeptide chain probably traverses the bilayer several times in folding to form the globular band 3 molecule. The N-terminal part of the protein forms a large ~40 000-dalton hydrophilic globular segment on the cytoplasmic surface of the membrane, which can readily be released by proteolytic cleavage. Glyceraldehyde 3-phosphate dehydrogenase (band 6), phosphofrucktokinase, and aldolase bind reversibly in a metabolite sensitive fashion to this region of the molecule (Kliman and Steck, 1980), which appears not to be involved in anion transport (England *et al.*, 1980). Tetrameric band 4·2, a protein of unknown function, also binds strongly to this region (Yu and Steck, 1975b). Ferritin-conjugated antibodies to peptides prepared from this N-terminal region of band 3 exclusively label the cytoplasmic surface of the erythrocyte membrane (Fukuda *et al.*, 1978). The middle portion of the molecule traverses the membrane at least once in an outward N to C direction, and the C-terminal domain, which is variably glycosylated on an external portion, bearing receptors to lectins, has been shown by Markowitz and Marchesi (1981) to traverse the bilayer at least once in an inward N to C direction, having a lactoperoxidase accessible site on the membrane's cytoplasmic surface.

2. *Electron Microscopy in Reconstituted Systems*

Yu and Branton (1976 and 1977) demonstrated that when fully-purified or partially purified band 3 dimers were reconstituted into lipid vesicles and freeze fractured, circa 8·5 nm diameter intramembrane particles were seen on both convex and concave fracture faces (Fig. 15a), similar in size and appearance to many of the intramembrane particles seen in the native human erythrocyte membrane. Similar results have been reported by Gerritson *et al.* (1978, 1980) using alternative reconstitution procedures. If the partially-purified band 3 preparation of Yu and Branton (which contained traces of ankyrin) was reconstituted in the presence of a crude extract of spectrin and actin, these particles remained dispersed at pH 7·6 (Fig. 15b) but could be induced to aggregate reversibly by subsequent incubation at pH 5·5 (Fig. 15c), mimicking the pH 5·5-induced aggregation of intramembrane particles observed in pretreated erythrocyte membranes (Elgsaeter and Branton, 1974; Elgsaeter *et al.*, 1976) and demonstrating that the physiologically important binding of ankyrin to both band 3 and spectrin, described below, could occur in an *in vitro* reconstituted system.

No other electron microscopic studies of the purified protein have been reported, but the *in situ* freeze-etch appearance of the band 3 polypeptide in the erythrocyte membrane have been well documented and various electron

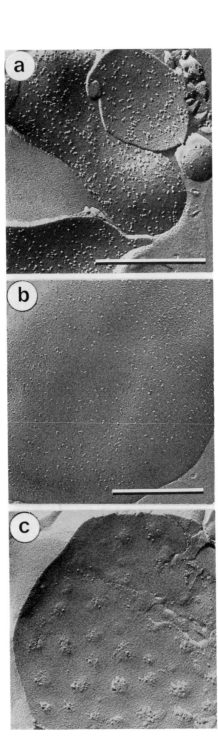

microscopic studies of the lateral distribution of the protein after ferritin labelling have been undertaken, as described in Section II above. Weinstein *et al.* (1980) have carefully compared the size, shape and density of intramembrane particles with the abundance of the band 3 polypeptide in the erythrocyte membrane, and have concluded that it is the tetramer which is the normal oligomeric form of this polypeptide in the erythrocyte membrane, in agreement with the detergent solubilization results of Nakashima and Makino (1980), and not the dimer as was originally supposed.

In an elegant application of freeze-fracture cytochemistry, Pinto da Silva and Torrisi (1981) have directly confirmed that during freeze fracture of intact human erythrocytes band 3 preferentially partitions with the protoplasmic half of the fractured membrane.

B. Glycophorin A, the Major Sialoglycoprotein

1. *General Properties of the Erythrocyte Membrane Sialoglycoproteins*

The sialoglycoprotein composition of the erythrocyte membrane, while complex and still not fully characterized, is now becoming intelligible (see Fig. 2b, and Table I and footnotes). The most abundant of these proteins is the major MN sialoglycoprotein, which bears the MN blood group antigens and receptors for influenza virus and the lectins phytohaemagglutinin and wheat germ agglutinin, and is now commonly known as glycophorin A (or PAS 1). It exists in two allelic forms, M and N, differing only in residues 1 and 5 of the polypeptide chain (Marchesi *et al.*, 1976; Furthmayer, 1978a, b). The minor sialoglycoprotein glycophorin B (or PAS 3) has the same N-terminal sequence as glycophorin A^N, explaining the low levels of blood group N activity expressed by glycophorin A^M homozygotes, and in many ways it appears to be a very similar although somewhat smaller protein. It is able to form PAS(1+3) heterodimers with glycophorin A which resist SDS dissociation (Furthmayer, 1978b) and which run in the region labelled PAS 4 in Fairbanks gels (see Figs

---

Fig. 15. Freeze fracture of unilamellar egg lecithin vesicles into which band 3 molecules have been reconstituted from a crude Triton X-100 extract of erythrocyte ghosts (also containing small amounts of ankyrin, band 2·1), by dialysis from cholate solutions. The reconstituted vesicles were stored as a pellet before freezing to promote fusion and thus give larger fracture faces. (a) Uniform, circa, 8·5 nm particles are seen on both concave and convex vesicle fracture faces of the untreated reconstituted vesicles. (b) And (c) reconstituted vesicles to which spectrin has been bound, presumably via bound ankyrin molecules, by the addition of a fresh spectrin-actin extract to the cholate solution before dialysis, subsequently incubated at 4 °C in isotonic phosphate-buffered saline at either pH 7·6 (b) or at pH 5·5 (c). Under the former conditions spectrin is soluble and the band 3 IMPs remain dispersed, while at pH 5·5 spectrin precipitates and the band 3 IMPs in the reconstituted vesicles are aggregated into small clusters (from Yu and Branton (1976). *Proc. Natn. Acad. Sci. USA* **73**, 3891–3895, with permission). Magnifications (a) ×60 000, (b) and (c) ×50 000; bars=500 nm.

2b and 3b). Both glycophorin A and glycophorin B can also exist as SDS-stable homodimers, PAS $(1)_2$ and PAS$(2)_2$ (Fig. 3b), and it is possible that they span the membrane as dimers *in vivo*.

The minor sialoglycoprotein, known as glycophorin C (Furthmayer, 1978a) or glyconnectin (PAS 2) (Owens *et al.*, 1980; Mueller and Morrison, 1980), is dissimilar from glycophorins A and B in that although it spans the lipid bilayer it is not readily labelled by lactoperoxidase-catalysed iodination of intact red cells, it contains tryptophan, and it does not exist as a dimer in SDS. Furthermore it is not readily solubilized in Triton X-100, because it binds to the membrane skeleton protein band 4·1 and is thus retained in the residual "Triton shell" after extraction of the other integral proteins (Mueller and Morrison, 1980; Owens *et al.*, 1980).

The nature of the high salt form of PAS 1 described by Potema and Garvin (1976), which runs in the "PAS 4" region (Fig. 3b), remains obscure. It is unlikely to be the PAS (1 + 3) heterodimer identified by Owens *et al.* (1980) both because it is too abundant and also because the heterodimer is reported by Owen *et al.* not to form in the phosphate-buffered Fairbanks gel system employed by Potema and Garvin. There is, in addition, a further minor sialoglycoprotein PAS 2' (Mueller and Morrison, 1974; Owens *et al.*, 1980) which is even less well characterized. No electron microscopy has yet been performed on these minor sialoglycoproteins.

2. Protein Chemistry of Glycophorin A

The polypeptide chain of glycophorin A, whose sequence of 131 amino acid residues has been determined (Tomita and Marchesi, 1975; Tomita *et al.*, 1978), may be divided into three clearly-defined structural domains. The extracellular N-terminal 72 residues bear 15 tetrasaccharide moieties with the sequence sialic acid $\alpha 2 \rightarrow 3$ galactosamine $\alpha 1 \rightarrow 3$ (sialic acid $\alpha 2 \rightarrow 6$) N-acetyl galactosamine or desialated versions thereof (Thomas and Winzler, 1969) linked via O-glycosidic bonds to seven serine and eight threonine residues, plus one complex sialo-oligosaccharide which contains 13 sugar residues including two sialic acids and is linked via an N-glycosidic bond to asparagine-26 (Yoshima *et al.*, 1980).

Residues 73 to 95 form a hydrophobic domain devoid of charged residues which spans the hydrophobic lipid bilayer. The isolated tryptic peptide containing these residues exhibits a high α-helical content (Segrest and Kohn, 1973), and there is strong circumstantial evidence that the transmembrane hydrophobic domain of glycophorin A adopts such an α-helical conformation *in vivo*, in which case the 23 residues would form 6·4 turns of α-helix extending 3·5 nm, just sufficient to span the hydrophobic region between the fatty acyl ester linkages of opposed phospholipids (Shulte and Marchesi, 1979). When this tryptic peptide, identified as T(is) (Sergrest *et al.*, 1974) is incorporated into egg lecithin liposomes it forms large intramembrane particles (diameter 7–8 nm) visible on the freeze-fractured faces of the lipid bilayers, whose numbers are proportional to the concentration of the peptide once a threshold concentration of one T(is) peptide to approximately 100 lecithin molecules is exceeded. Below this concentration, no large particles are seen. These large particles are interpreted as being due to parallel aggregates of circa 23 T(is) peptides spanning the phospholipid bilayer which form by spontaneous noncovalent association in a fashion similar to micelle formation, once the "critical multimer concentration" threshold for T(is) in egg lecithin is exceeded. They probably bear no structural relationship to the intramembrane particles seen in freeze-fractured erythrocyte membranes and are thus not shown here. Individual α-helical T(is) peptides would have a diameter of about 1 nm, too small to be resolved by the freeze-fracture technique.

The C-terminal region of glycophorin A (residues 96 to 131) contains both charged and hydrophobic amino acid residues and lies on the cytoplasmic surface of the bilayer, where it is probably folded into a compact globular domain. The transmembrane disposition of the polypeptide, originally proposed on the basis of labelling studies on intact cells and leaky ghosts, has subsequently been confirmed immunochemically in intact erythrocytes by electron microscopic localization of ferritin conjugates to antibodies directed against an antigenic site between residues 102 and 118 in the C-terminal domain of band 3 (Cotmore *et al.*, 1977).

3. Electron Microscopy in Reconstituted Systems

When intact purified glycophorin A molecules are incorporated into phospholipid vesicles by a variety of methods, the freeze-fracture faces of the vesicles show small ill-defined particles or rugosites, each probably composed of several glycophorin molecules, which are often aggregated into small ridges (Grant and McConnell, 1974; Branton and Kirchanski, 1977; Van Zoelen *et al.*, 1978a, b; Romans *et al.*, 1979; Gerritsen *et al.*, 1980). When these vesicles are composed of dissimilar phospholipids, and the sample is frozen from conditions in which phase separation exists in the synthetic bilayer, the glycophorin particles are localized in the fluid regions of the bilayer, and are excluded from the immobile crystalline regions (Fig. 16; Grant and McConnell, 1974). In vesicles of a single phospholipid frozen from below its phase transition, the glycophorin particles occur in the solid lipid phase (Van Zoelen, *et al.*, 1978b). The glycophorin molecules in these preparations have been shown to span the bilayers asymmetrically, particularly in very small sonicated vesicles (Van Zoelen *et al.*, 1978a), where their presence enhances membrane permeability, greatly increases the rate of transbilayer movement

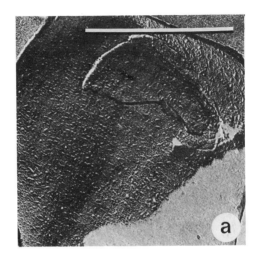

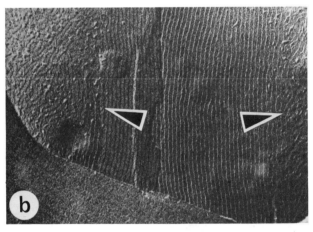

Fig. 16. (a) Freeze fracture of a vesicle of equimolar amounts of bovine brain phosphatidylserine and dimyristoylphosphatidylcholine into which glycophorin molecules have been reconstituted at a sialoglycoprotein:lipid molar ratio of 1:120. Frozen from 23 °C. The presence of glycophorin causes small bumps and irregularities in the fracture face. (b) As (a), but with equilmolar amounts of dielaidoylphosphatidylcholine and dipalmitoyl phosphatidyl-choline. Frozen from 23 °C at which roughly equal areas of fluid and crystalline lipid domains coexist. The glycophorin-related particles are confined to the fluid lipid domains (left and right of micrograph), which are separated by a crystalline lipid domain (centre, with ordered lines), the boundaries of which are indicated by the arrows (from Grant and McConnell (1974). *Proc. Natn. Acad. Sci. USA* **71**, 4653–4657, with permission). Magnifications ×80 000; bar=500 nm.

of the phospholipids ("flip-flop") and causes discontinuities in the lipid bilayer which makes the phospholipid molecules more susceptible to lipase attack (Van Zoelen et al., 1978b; de Kruijff et al., 1978; Gerritsen et al., 1980).

However, Lintz et al. (1979) were unable to detect any small particles in the highly curved membranes of freeze-fractured reconstituted vesicles containing glycophorin-A, which had been prepared by the removal of Triton X-100 from glycophorin-enriched Triton extracts of erythrocyte ghosts, although 8 nm diameter particles could be seen if band 3 was included in their extracts by increasing the Triton concentration.

While these reports clearly demonstrate that the incorporation of glycophorin A may in certain cases cause the apperance of surface irregularities or small particles on the fracture faces of freeze-fractured synthetic phospholipid vesicles, and while the interactions between glycophorin-A and band 3 in the native membrane are well established (see Sections II and IV), the contribution made by the sialoglycoproteins to the large 8·5 nm diameter particles observed on the P face of the freeze-fractured erythrocyte membrane seems minimal, since their absence in En(a-) mutants (Section VII) does not change the appearance of the IMP_P.

Pinto da Silva and Torrisi (1981) have shown by freeze-fracture cytochemistry that the wheat-germ agglutinin binding site associated with the N-terminal domain of glycophorin A preferentially partitions with the extracellular half of freeze-fractured human erythrocyte membranes, and Edwards et al. (1979) have demonstrated by direct polypeptide analysis that after fracture this extracellular half contains both intact glycophorin A molecules and also N-terminal fragments formed by breakage of covalent bonds within the polypeptide main chain. It is not known where in the amino-acid sequence such breaks occur, or what proportion of the molecules are thus destroyed, but one may assume that such fractured molecules would not protrude from the fracture face of the cleaved bilayer sufficiently to form intramembrane particles upon shadowing.

IV. THE MEMBRANE SKELETON

A. Historical Perspective

Over the past decade, understanding of biological membrane structure has been greatly influenced by the fluid mosaic model proposed in 1972 by Singer and Nicolson. Not only did this model bring together many previously unrelated data in a thermodynamically sensible scheme, but more importantly it provided a new conceptual framework within which to design further experiments. Unfortunately this model, at least in its initial and popularized forms, has led many people to think of biological membranes almost exclusively in terms of their constituent lipid bilayers and "floating" integral proteins, under emphasizing the importance of the peripheral proteins. Perhaps even the psychological connotations of the word "peripheral" have prevented many people from appreciating the fundamental structural and functional importance of these peripheral proteins as basic constituents of the membrane.

Nowhere is this importance better illustrated than in the erythrocyte membrane. Recent evidence reviewed below shows the erythrocyte membrane to be essentially a laminated structure, in which the membrane's lipid bilayer is underlain on its cytoplasmic surface by a thick complex meshwork of structural proteins, firmly attached to it by close specific associations between certain of these peripheral structural proteins and the integral proteins which span the bilayer. The meshwork of peripheral membrane proteins is of crucial importance to the cell, imparting to the membrane as a whole its characteristic biconcave shape, strength and flexibility, as well as limiting the translational mobility of the integral membrane proteins. Because of its functional role, this protein meshwork is best described as *the membrane skeleton*.

B. Comparison with the Cytoskeleton of Nucleated Cells

This erythrocyte *membrane skeleton* may be contrasted with the three-dimensional *cytoskeleton* found in other cell types (Goldman et al., 1976) in several respects. While the cytoskeleton is truly three-dimensional, penetrating all regions of the cytoplasm but making only limited terminal contacts with the plasma membrane of its cell, the erythrocyte membrane skeleton is a fundamental component of the cell's membrane, forming a two-dimensional meshwork probably no more than 20 nm thick, closely applied to the cytoplasmic surface of the membrane's phospholipid bilayer and making extensive contacts with the membrane's integral proteins and with the phospholipid molecules themselves. While the functions of the cytoskeleton are both structural and contractile, and while it is capable of undergoing functionally significant polymerization-depolymerization reactions crucial for cellular motility, the erythrocyte membrane skeleton as a whole has an essentially static, structural role, although the individual component molecules are involved in metabolically-sensitive dynamic binding interactions. Furthermore, the cytoskeletons of other cells are both heterogeneous and essentially linear in their molecular organization, involving three distinct systems of extremely long homopolymers (of actin to form microfilaments, of tubulin to form microtubules and of prekeratin, desmin and/or vimentin to form intermediate filaments), each system with its own set of specific accessory proteins involved in structural organization, crosslinking, motility and control. The erythrocyte membrane skeleton, by contrast, is essentially a structurally undifferentiated two-dimensional meshwork of interacting molecules, with little overall linear polarity, in which long homopolymers, either of spectrin or actin, are absent. However, the occurrence of actin in both systems, and the structural similarities discussed below between erythrocyte spectrin on the one hand and various non-erythroid spectrin-like molecules on the other, suggest certain common molecular binding mechanisms, and support the hope shared by many investigators that a detailed analysis of the proteins of the erythrocyte membrane skeleton will help our understanding of the nature of the interaction between cytoskeletal microfilaments and the plasma membrane in other cell types.

C. Protein Components of the Membrane Skeleton

If the majority of the lipid and integral protein molecules are removed from the erythrocyte membrane using Triton X-100, a non-ionic detergent (Yu *et al.*, 1973), the membrane skeleton survives intact and is left as a "Triton shell" which is easily seen in the light microscope if stained with uranyl acetate, although almost invisible by phase contrst microscopy without staining because of the removal of the highly refractile lipid bilayer. In comparative studies, Lux and his colleagues (Lux *et al.*, 1976; Lux and John, 1977; Lux 1979b) have shown that the Triton shells prepared from erythrocytes or their ghosts, while somewhat shrunken, retain the specific shape of the membrane before Triton extraction, whether this be biconcave, echinocytic or, in the case of hereditary spherocytosis and sickle cell anaemia, spherocytotic or sickled (as discussed more fully in Section VII below), indicating that the membrane skeleton is primarily responsible for determining the cell shape.

Yu *et al.* (1973) showed that increasing Triton X-100 concentrations progressively removed more and more of the integral glycoproteins of the erythrocyte membrane, leaving Triton shells containing primarily spectrin, actin and band 4·1. In subsequent careful studies, Sheetz and Sawyer (1978) and Sheetz (1979a) described conditions for obtaining the cleanest Triton shells, containing only spectrin, actin, band 4·1 and band 4·9, which must therefore be regarded as the essential components of the membrane skeleton. Depending upon the conditions, the two proteins now known to anchor the membrane skeleton to the lipid bilayer, namely ankyrin and glycoconnectin, may also be retained in the Triton shell (Sheetz, 1979; Mueller and Morrison, 1980).

D. Cell Shape Changes

1. Material Properties of the Membrane

The material properties of the erythrocyte membrane have been lucidly analysed and described by Evans and Lacelle (1975) and Evans and Hochmuth (1977). Because of its laminated structure, the membrane behaves under a deforming force or shear not as a liquid, as might be expected from the fluid mosiac model, but as an elastic solid, in which resistive forces are built up (i.e. energy is stored in the molecular structure of the membrane skeleton) as a function of the deformation or strain applied, allowing it to recoil elastically to its original shape when the deforming force is removed. The fluid lipid bilayer itself offers no resistance to such lateral distortion, but its molecular structure ensures that during such distortions the membrane surface area and thickness remain essentially constant. The membrane's elasticity is thus anisotropic in nature, being limited to the plane of the membrane. Within this plane the laminated molecular structure enables the membrane to behave like a weak elastomer (a loose polymeric meshwork) able to undergo large elastic deformations easily with a low and nearly constant shear modulus, while exhibiting the noncompresible properties of a fluid (i.e. having a very large bulk modulus). Very little force is thus required to induce erythrocytes to undergo the large deformations necessary for their passage through the narrow splenic fenestrations and peripheral capillaries, as demonstrated by Weed and Lacelle (1969), who showed that normal erythrocytes could be easily sucked into capillaries of 3 μm internal diameter by a negative pressure of only 4 mm H_2O.

Prolonged laboratory application of a local distorting strain to a region of an erythrocyte membrane results in a permanent deformation (i.e. plastic flow) of the membrane, which is probably the result of local molecular rearrangements between the non-covalently linked components of the membrane skeleton.

The laterally homogeneous, closed-surface laminated structure of the membrane, in which the outer phospholipid bilayer is noncompressible and the inner membrane skeleton is under a slight tension, is directly responsible for the overall biconcave shape characteristic of the erythrocyte. Very simple physical models built on these principles spring back into this biconcave shape when a distorting force is removed. The shape thus represents the lowest free-energy state of the membrane under normal physiological conditions (Evans, 1973; Evans and Hochmuth, 1977, 1978).

2. Molecular Basis of Metabolite-sensitive and Experimentally-induced Reversible Shape Changes

In a manner analogous to the one-dimensional curvature induced by cooling or warming a bimetallic strip, small changes in the relative surface area occupied by the two layers which make up the laminated membrane of the normal erythrocyte will profoundly influence the membrane's two-dimensional curvature and hence the cell's overall shape.

Since the proteins of the erythrocyte membrane are exquisitely sensitive to their metabolic environment, conditions such as metabolic depletion which alter protein-protein interactions within the membrane skeleton, or lectin binding which leads to crosslinking of integral glycoproteins, induce dramatic changes in the erythrocyte membrane shape, leading either to crenation or stomatocyte formation observable by scanning electron microscopy (Sheetz and Singer, 1977; Birchmeier and Singer, 1977; Lovrein and Anderson, 1980). Hence storage of erythrocytes under conditions of ATP-depletion (e.g. prolonged blood bank storage) causes the cells to become less deformable, highly crenated echinocytes, while incubation in adenosine and glucose, which allows the intracellular resynthesis of ATP, almost totally reverses this affect (Weed and Lacelle, 1969; Weed et al., 1969). Similarly, echinocytic ghosts, prepared by lysing fresh erythrocytes once in 10 mM Tris buffer pH 7·4 and resuspending without further washes in an isotonic salt solution, return to the discocyte shape and eventually become cup-shaped stomatocytes when incubated at 37 °C in the presence of Mg^{++}-ATP (Sheetz and Singer, 1977; Birchmeier and Singer, 1977; Sheetz et al., 1978). These authors proposed that a specific protein kinase-catalysed phosphorylation of a component of the membrane skeleton, probably band 2 of spectrin, is necessary to induce this change from echinocyte to discocyte, by promoting a slight relative *expansion* of the inner membrane skeleton from its contracted echinocytic state, relative to the constant surface area of the outer phospholipid bilayer. They found that the transition to discocyte was promoted by low concentrations of divalent antispectrin antibody (but not by F_{ab} fragments of this) suggesting that spectrin crosslinking was an important factor in the phosphorylation-induced transition, and was inhibited by 10^{-5} M Ca^{++}. The manner in which such phosphorylation may induce the shape change is not at present understood. Since these workers did not observe similar shape transitions in haemoglobin-free but spectrin-rich ghosts prepared according to Dodge et al. (1963), they suggested that the extensive washing involved in preparing those ghosts had in some way altered the normal *in vivo* protein-protein interactions of the membrane skeleton, a factor which may go towards explaining the complete lack of differences in behaviour observed in *in vitro* experiments with phosphorylated and dephosphorylated spectrin (see Section V below).

However, more recently it has become clear that the protein interactions of the membrane skeleton are sensitive to 2,3-diphosphoglycerate and other polyphosphate metabolites at concentrations which might permit metabolic control of membrane deformability and lateral mobility. These studies have relied largely upon measurements of lateral protein mobility rather than observations of shape changes to detect the effects of these polyphosphates, and are described in the next section (IV E).

The erythrocyte membrane is somewhat more complex than the simple laminated model described above, since the lipid bilayer component of the membrane is itself an asymmetric laminated coupled structure. Thus lipid soluble reagents which can insert into and hence alter the surface area of either the inner or the outer half bilayer can also induce similar shape changes, both in intact cells and in ghosts. Local anaesthetics and other amphipathic or lipid-soluble drugs (Sheetz and Singer, 1974, 1976), lysolecithin (Mohandas et al., 1978), alkyl sulphate detergents such as SDS (Lovrein and Anderson, 1980), and even increased quantities of normal membrane components such as cholesterol (Cooper, 1978) have all been shown to have such effects.

The nature of the shape changes obtained, i.e. whether from normal discocytes (Fig. 1a) to echinocytes (Fig. 1b) or to cup-shaped stomatocytes, has been explained by the bilayer couple hypothesis of Sheetz and Singer (1974) to depend upon the topology of insertion of the hydrophobic portion of the adduct. Thus anionic drugs, anionic fatty acids and alkyl sulphates, which cannot easily cross the hydrohobic domain of the erythrocyte membrane, insert their hydrophobic portions from the outside of intact erythrocytes, causing a very slight increase in surface area of the external half-bilayer, resulting in crenation of the cells. In contrast cationic compounds which are able to diffuse across the membrane as deprotonated neutral species preferentially partition into the cytoplasmic half-bilayer where they are stabilized by the high concentration of the anionic phospholipid phosphatidylserine, and there induce the cells to become cup-shaped stomatocytes. The initial bilayer couple hypothesis proposed that the extent of the shape changes induced was in direct proportion to the per cent change in surface area of the effected half-bilayer. However, from their quantitative study of lysolecithin insertion into the external and cytoplasmic half-bilayer, Mohandas et al. 1978) concluded that this simple quantitative explanation is insufficient. The situation is more complex, and presumably reflects the different environments of the two lipid half-bilayers, the external one being relatively exposed while the cytoplasmic one is in close association with the proteins of the membrane skeleton.

3. Irreversible Shape Changes—Blebbing, Vesiculation and Eversion

More severe shape changes, for instance the stromalytic extensions studied by Lutz et al. (1977) (Fig. 7) have been induced by a variety of experimental manipulations of erythrocytes or their ghosts, but rarely has it been possible to interpret these observed structural changes in terms of the molecular properties of the membrane components. An exception to this is the effect of polycations such as polylysine and basic proteins such as protamine on fresh ghosts. These proteins, which are very effective at precipitating spectrin (isoelectric point pH 4·8) from aqueous solution, cause the membrane to undergo irreversible blebbing, as shown in Fig. 17, releasing protein-free lipid vesicles from the membrane surface (Elgsaeter et al., 1976). This effect seems to be the result of a "precipitation" of spectrin *in situ*, throwing the membrane skeleton into a contraction. The force of this contraction apparently exceeds the collapse pressure of the noncompressible lipid bilayer, which responds by blebbing off protein-free lipid vesicles, as depicted in Fig. 18, thereby reducing its surface area to accommodate to that of the contracted membrane skeleton.

The importance of the peripheral membrane proteins in maintaining cell shape is also nicely demonstrated by the changes which occur when intact fresh biconcave erythrocyte ghosts are incubated at low ionic strength, a procedure which results in the selective solubilization of spectrin and actin from the membrane skeleton, with a corresponding loss of its elastomeric behaviour. As this happens the physical properties become more and more determined by the lipid bilayer alone, and a characteristic sequence of shape changes is observed. Initially the biconcave shape is lost and the ghost assumes a spherical or oblate spheroid shape, resembling a giant liposome. The membrane at this stage is very weak and if subjected to any mechanical agitation will vesiculate into a complex mixture of inside-out and right-side-out vesicles. By the use of appropriate ionic conditions and high mechanical shear forces, more homogeneous populations of small inside-out or right-side-out vesicles may be obtained and purified by centrifugation (Steck and Kant, 1975). However, if care is taken to avoid warming and mechanical agitation, the normally closed surface of the spectrin-depleted ghost will rupture to form an open bell-shaped figure in which the cytoplasmic membrane surface is patent to the surrounding medium. These open ghosts eventually evert, through omega-forms to form inside-out toroidal scrolls (Fig. 19; Shotton, unpublished observations). The existence of free membrane edges in such open ghosts implies the stabilizing contributions of the residual membrane proteins, since such edges are not found in synthetic bilayers of erythrocyte lipids.

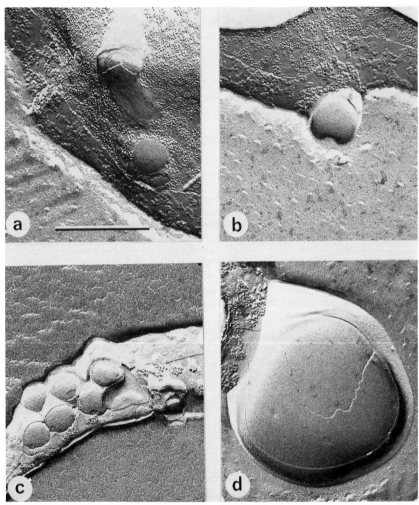

Fig. 17. Freeze-etched erythrocyte ghosts illustrating the formation of particle-free lipid vesicles induced by incubation at pH 7·6 in protamine sulphate (0·05 mg/ml), without prior pretreatment to release spectrin and actin (which would have resulted in massive particle aggregation but no blebbing). (a), (b) And (d), *lac*-modified ghosts labelled after protamine treatment with ferritin-conjugated anti-*lac* antibodies (see Fig. 11 legend), showing progressive stages in vesicle formation and blebbing. Note that both the IMP$_P$ and the ferritin markers stop at the neck of the large vesicle in (d). (c) Unmodified ghosts showing an early stage of blebbing (from Elgsaeter *et al.* (1976). *Biochim. Biophys. Acta* **426**, 101–122, with permission). Magnifications ×52 000; bar=500 nm.

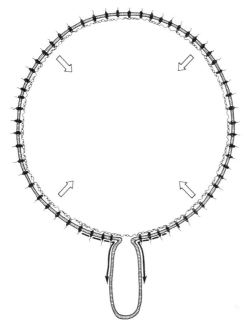

Fig. 18. Diagrammatic illustration of the behaviour of the erythrocyte membrane in producing protein-free lipid blebs when a fresh leaky ghost is exposed to strong spectrin-precipitating agents such as protamine sulphate, causing the membrane skeleton (wavy lines) to contract (open arrows). Since the membrane skeleton is intimately connected to the integral proteins of the membrane (black ovals) the compressive force transmitted to the non-compressible lipid bilayer will cause bulk lateral flow of lipid molecules (black arrows) past the immobilized integral proteins to form protein-free lipid vesicle blebs (compare with Fig. 17) (from Elgsaeter et al. (1976). Biochim. Biophys. Acta **426**, 101–122, with permission).

E. Membrane Skeletal Control of Translational and Rotational Diffusion

Molecules of band 3 and to a lesser extent glycophorin A are the primary protein components to be labelled when intact erythrocytes are incubated with fluorescein isothiocyanate. By fusing labelled with unlabelled cells using *Sendai* virus or polyethylene glycol, and then observing the subsequent redistribution of fluorescent label in the hybrids so formed, Fowler and Branton (1977) were able to show that the minimal translational diffusion coefficient of the FITC-labelled protein molecules in the membranes of fresh erythrocytes or ghosts suspended in isotonic phosphate-buffered saline at pH 7·6 ranged from approximately 4×10^{-12} cm^2/s at 37 °C to approximately 8×10^{-12} cm^2/s at 23 °C, with complete immobilization of the label at 0 °C. Even lower values were obtained from "aged" (ATP depleted) cells. These values were significantly lower than those obtained for plasma membrane

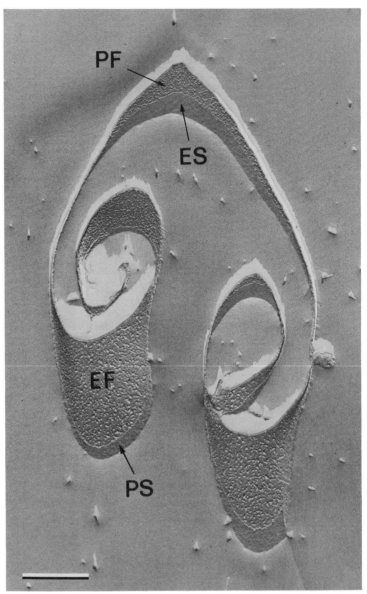

Fig. 19. Freeze-etch electron microgaph of an inverted membrane scroll formed after dialysis of fresh erythrocyte ghosts against 0·5 mM sodium phosphate buffer, pH 8·1 at 0 °C for 18 h. The surrounding ice surface is interrupted by spectrin molecules, exposed by etching, which have dissociated from the membrane (Shotton, unpublished work). Magnifiction ×36 000; bar=500 nm.

proteins of other cells and suggested that the translational mobility of the integral protein molecules in the erythrocyte membrane might be restricted by dynamic interactions with components of the membrane skeleton. In a subsequent study, Fowler and Bennett (1978) demonstrated that incubation of leaky labelled ghosts with a 72 000-dalton proteolytic fragment of ankyrin, which contained that protein's binding site for spectrin, lead to competition with the native ankyrin binding sites and resulted in the selective dissociation of spectrin from the membrane and a concomitant two-fold increase in the lateral diffusion coefficient of the labelled integral protein molecules, thus providing direct evidence that the lateral mobility of the integral proteins is restricted by ankyrin-mediated interactions with the membrane skeleton (see Section V).

Schindler *et al.* (1980) refined this technique of fluorescence redistribution after fusion (FRAF), directly quantifying the fluorescence intensity across pairs of labelled and unlabelled erythrocyte or ghost membranes fused with polyethylene glycol by scanning with the focused beam of a laser emitting at the excitation wavelength. Their determination of the average diffusion coefficients at 30 °C for the major glycoproteins, labelled on free amino groups with dichlorotriazinyl aminofluorescein, were $9 \cdot 2 \times 10^{-11}$ cm^2/s for the fused ghosts of fresh cells and $2 \cdot 2 \times 10^{-11}$ cm^2/s for those of ATP-depleted erythrocytes. They found that the addition of 12·5 mM ATP to the ATP-depleted erythrocytes increased the lateral diffusion coefficient by four-fold, restoring it almost to the value for fresh cells, and that 12·5 mM 2,3-diphosphoglycerate increased it by 2·5-fold, both of these metabolites being at higher concentrations in oxygenated blood, thus opening the possibility that the protein interactions within the membrane skeleton are under metabolic control. These findings are in accord with studies of Wolfe *et al.* (1980), Casaly and Sheetz (1980) and Sheetz and Casaly (1980), who demonstrated that polyphosphates disrupt isolated skeletons derived from intact erythrocytes using Triton X-100. In contrast, 0·6 mM concentrations of the polyamines neomycin or spermine were found to totally inhibit redistribution of the fluorescent label, an effect which is likely to be related to the strong interactions of polycations and basic proteins with spectrin noted by Elgsaeter *et al.* (1976) and Shotton *et al.* (1978) and discussed above.

Golan and Veatch (1980) have more recently investigated lateral protein mobility using eosin isothiocyanate-labelled ghost membranes, in which about 80% of the label is localized on band 3 molecules. Using the technique of fluorescence recovery after photobleaching (FRAP) which permits a more detailed kinetic analysis, they were able to identify two populations of band 3 molecules, one immobilized and the other mobile. The latter represented only 10% of the total at 21 °C and 45 mM ionic strength of sodium phosphate buffer, but rose to 90% when the temperature was increased to 37 °C and the ionic strength dropped to 13 mM sodium phosphate. This rise, which was not fully reversible, was accompanied by a separate, dramatic and fully-reversible fifty-fold increase in the translational diffusion coefficient of the mobile fraction of band 3 molecules from 4×10^{-11} cm^2/s to 200×10^{-11} cm^2/s, a value approaching that observed for the smaller integral protein rhodopsin in retinal rod discs (Poo and Cone, 1974), in which the protein molecules are believed to be unrestricted in their translational mobility. However, these two phenomena seemed distinct, since the increase in the proportion of mobile molecules occurred more rapidly and over a higher ionic strength range than the reversible increase in the translational diffusion coefficient of that mobile population. Neither bore a direct correlation with the dissociation of spectrin into free solution from the labelled ghost membrane which occurs at even lower ionic strengths!

There has always been some suspicion that the physical effects of such laser photobleaching might damage the membrane in such a way as to alter the lateral diffusion coefficients of the labelled proteins, but in an elegant comparison of the FRAF and FRAP techniques on the same fused erythrocyte specimens, Koppel and Sheetz (1981) have shown that the experimental results obtained from these two methods are indistinguishable, thereby validating conclusions obtained by the FRAP method.

Using this method, Koppel et al. (1981) have studied the increase in lateral mobility of the erythrocyte membrane glycoproteins consequent upon the absence of spectrin in spectrin-deficient spherocytic mouse mutants (discussed in Section VII below), have confirmed the results of Golan and Veatch (1980) mentioned above, and have developed a mathematical description for the lateral mobility of the band 3 molecules. The total immobilization of a proportion of the band 3 molecules is understood to be due to direct binding via ankyrin to the spectrin molecules of the membrane skeleton, while the low translational diffusion coefficient of those band 3 molecules which *are* mobile is thought to be caused by a "corralling" effect of the membrane skeletal components, restraining the free lateral movement of the integral proteins by steric hindrance of their protruding cytoplasmic domains. While rotational diffusion is not hindered in these circumstances (see below), the speed of lateral diffusion is controlled and limited by the dynamics of dissociation and rearrangement of the membrane skeletal components necessary to allow an integral protein to move out of one "corral" and enter the next. Koppel and his colleagues have termed this effect "matrix control".

In support of these ideas, Schekman and Singer (1976) and Tokuyasu et al. (1979) have demonstrated a greater translational mobility of concanavalin A receptors, in comparison with adult cells, in discrete membrane domains of neonatal human erythrocytes and reticulocytes where spectrin is absent, while adjacent areas of the immature cells which have a normal complement of spectrin do not show this increase.

Nigg and Cherry (1979a, b, 1980) have conducted similar investigations of the rotational diffusion of band 3 in eosin isothiocyanate- and eosin 5-maleimide-labelled erythrocyte membranes by polarized fluorescence bleaching and the subsequent measurement, within the next few milliseconds, of the rapid decay of the transient fluorescence anisotrophy thereby induced. They too found two populations of band 3 molecules (Nigg and Cherry, 1979a), one free to rotate rapidly upon its own axis within the plane of the membrane, and one able to do so only very much more slowly, and observed that the latter became increasingly more abundant at lower temperatures, although the exact values obtained for the proportions of mobile and restricted molecules differed somewhat from those obtained for translational mobility by Golan and Veatch (1980). Both fast and slow rotations could be abolished by crosslinking the membrane proteins with 1% glutaraldehyde for 30 min at 22 °C or by aggregation of the intramembrane particles, but alteration of the cholesterol content of the membrane between 38% and 185% of its normal value, or increasing the viscosity of the external buffer by the inclusion of 70% glycerol were surprisingly without effect upon the observed decay times of the anisotrophy (Nigg and Cherry, 1979a). However, proteolytic cleavage with trypsin of the 40 000-dalton N-terminal fragment containing the ankyrin-binding site from the band 3 molecules at the cytoplasmic surface of the membrane greatly enhanced the rate of decay of the anisotropy of the eosin probe attached to these molecules at their extracellular surface (Nigg and Cherry, 1980), by converting into rapidly rotating molecules that population of band 3 whose mobility had formerly been restricted by interactions between this N-terminal domain and ankyrin. Similarly, while low ionic strength elution of spectrin from the cytoplasmic surface of the labelled membrane was without effect upon the relaxation times (Nigg and Cherry, 1979a), the subsequent high ionic strength elution of ankyrin and band 4·1 resulted in an equally strong enhancement of the rate of decay of the induced anisotropy (Nigg and Cherry, 1980), providing strong evidence for the normal immobilization of a significant fraction (up to 40% in their hands) of band 3 molecules by virtue of their binding via ankyrin to the membrane skeleton. Nigg and Cherry (1979b) found no change in the rate of decay of the induced anisotropy of eosin-labelled band 3 upon crosslinking into covalent dimers with copper and *o*-phenanthroline (Steck, 1972), indicating that the natural asociation state of the band 3 molecules is at least dimeric. However, in other experiments, Nigg *et al.* (1980a) showed that the rotational diffusion of the freely mobile population of band 3 in eosin-labelled membranes could be greatly reduced by crosslinking of glycophorin A with a divalent anti-glycophorin A antibody, while this effect was not obtained with

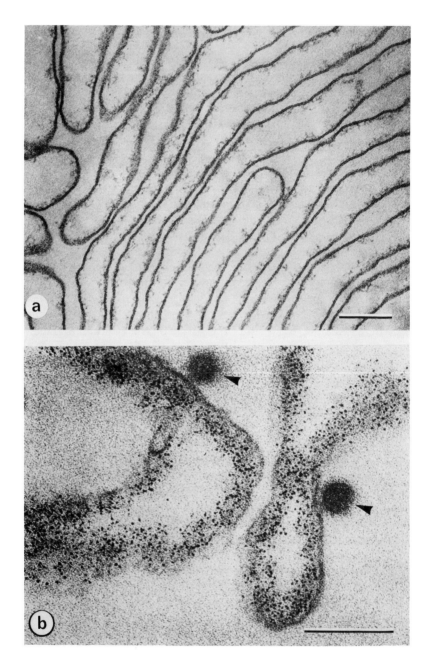

monovalent anti-glycophorin A F_{ab} fragments, antispectrin or non-specific antibodies. This was taken as evidence for a pre-existing integral protein complex between band 3 and glycophorin A in the erythrocyte membrane, which could thus be caused to aggregate using the divalent anti-glycophorin A antibody, in agreement with the particle aggregation studies reviewed in Section II above. However, in En(a-) membranes, in which glycophorin A is entirely absent, and in normal membranes from which the sialic acid residues of glycophorin A had been removed by neuraminidase digestion, the rotational diffusion properties of eosin-labelled band 3 were essentially unaltered, indicating that the glycophorin A molecules were contributing no more than 15% to the overall cross-sectional diameter of the band 3 glycophorin A complex in the plane of the membrane (Nigg et al., 1980b). These results are entirely consistent with current concepts of the transmembrane conformations of the α-helical glycophorin A polypeptide chains and the globular band 3 oligomers, and their relative proportions within the intramembrane particles discussed above.

F. Electron Microscopic Observations

1. Thin Section

Awareness of the existence and importance of the erythrocyte membrane skeleton grew slowly during the 1970s, largely from indirect evidence. Most early thin section studies of erythrocyte ghosts (for example Marchesi and Palade 1967a; Tilney and Detmers, 1975) showed wispy strands of material extending for variable distances from the cytoplasmic surface of the membrane (Fig. 20a) but gave little structural information about the membrane skeleton. The most important contribution in this period was the classic study of Nicholson et al. (1971), who demonstrated by labelling fixed leaky ghosts with ferritin-conjugated rabbit anti-spectrin that spectrin was localized on the cytoplasmic surface of the membrane, extending up to 60 nm from the lipid bilayer (Fig. 20b). Recently the application of the tannic acid fixation technique to the erythrocyte membrane by Tsukita et al. (1980) has permitted direct visualization in thin-section electron micrographs of the membrane's laminated structure with a clarity previously unobtainable, and

---

Fig. 20. (a) Thin sections of human erythrocyte ghosts showing fibrillar material extending from their cytoplasmic surfaces (from Marchesi and Palade (1967a). J. Cell Biol. **35**, 385–404, with permission). Magnification ×72 000; bar=200 nm. (b) As (a), after incubation with ferritin-conjugated anti-spectrin (small black dots), which specifically labels the fibrillar material on the cytoplasmic surface of each membrane. The extracellular membrane surfaces have been identified by the specific binding of influenza viruses (arrowheads) to the glycosylated domains of the glycophorin-A molecules (from Nicolson et al. (1971). J. Cell Biol. **51**, 265–272, with permission). Magnification ×120 000; bar=200 nm.

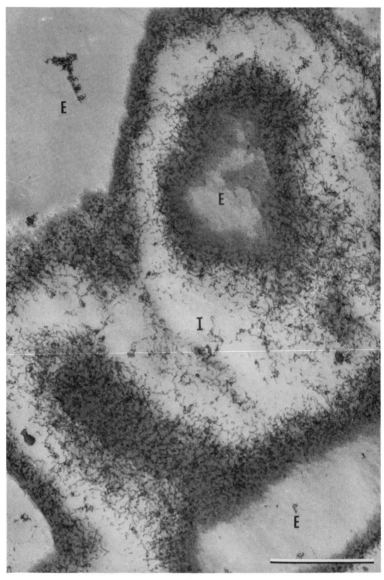

Fig. 21. Thin section of an erythrocyte ghost after initial fixation in 2% tannic acid–2·5% glutaraldehyde. The membrane skeletal network underlying the phospholipid bilayer (grey areas) is clearly visible in obliquely or tangentially sectioned areas of the membrane. E, extracellular space. I, intracellular space (from Tsukita et al. (1980). J. Cell Biol. **85**, 567–576, with permission). Magnification ×56 000; bar=500 nm.

has shown some details of the molecular nature of the membrane skeleton *in situ*. Their results are shown in Figs 21 and 22. The lipid bilayer, seen in cross-section (Fig. 22a) after osmic acid fixation and tannic acid staining as two parallel dark lines, is underlain on its cytoplasmic surface by a filamentous network of proteins composed of two layers: the first, a layer of vertical elements with granular appearance, which are seen to be directly associated with bilayer, and the second a horizontally disposed anastomosing meshwork of long filamentous components, seen more clearly in tangential section (Figs 21 and 22b), situated between 10 and 30 nm from the lipid bilayer and linked to it by the vertical densities. From what is now known of the structure and associations of the individual components of the membrane skeleton, described below, the major filamentous components can be identified with confidence as spectrin molecules, while the vertical granular densities linking the membrane skeleton to the bilayer may be ascribed to ankyrin molecules associated with the N-terminal cytoplasmic protrusions of the band 3 integral proteins. The diameter and appearance of isolated spectrin molecules fixed and stained similarly resembles the filamentous components. Extraction of spectrin and actin from the ghosts by dialysis against 0·1 mM EDTA, pH 8·0, resulted, as expected, in the disappearance of the filamentous meshwork, leaving only the granular components (Tsukita *et al.*, 1980; micrographs not shown here).

Because of the high protein density in the cytoplasm, thin section studies of intact erythrocytes show no features of the membrane skeleton. However, the location of the spectrin molecules *in situ* has been determined by Ziparo *et al.* (1978), who labelled frozen thin sections of intact erythrocytes with ferritin-conjugated rabbit anti-spectrin IgG antibodies (Fig. 23). As can be seen, the ferritin molecules in these preparations lie in a very well defined layer between 10 and 15 nm from the cell surface. Spectrin retains this sub-bilayer location in the spiny processes of echinocytes produced by ATP depletion (micrographs not shown here).

Taken together, these reports suggest first that in the intact erythrocyte, the spectrin tetramer molecules of the membrane skeleton, although individually about 200 nm in length (see below) do not project to any significant extent into the cytoplasm but are localized in a distinct and well defined layer on the cytoplasmic surface of the bilayer, and second that upon making ghosts the membrane skeleton is slightly perturbed from its normal location, but does not suffer gross disruption.

2. Scanning Electron Microscopy

While scanning electron microscopy has been used to study the overall shape of cells, ghosts and Triton shells under various conditions, its use in studying

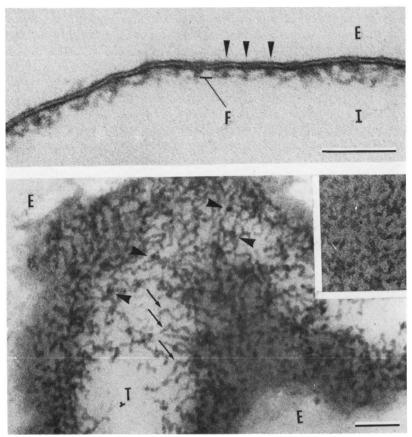

Fig. 22. High power micrographs of thin sections of erythrocyte ghosts after tannic acid-glutaraldehyde fixation. E, extracellular space. I, intracellular space. (a) In cross-section the membrane skeleton is clearly visible on the cytoplasmic surface of the lipid bilayer, which after osmic acid postfixation has a characteristic trilaminar appearance. The skeletal structure is resolved into two layers: horizontally-arranged fibrous components of the membrane skeleton (F) are linked to the lipid bilayer by vertical densities (arrowheads) which may be ankyrin molecules. Note that the middle lamella of the lipid bilayer is denser at the point of attachment of these vertical densities than elsewhere, perhaps because of the presence of integral band 3 molecules to which the ankyrin molecules bind. (b) In an obliquely cut section the appearance of the layered structure of the membrane skeleton's protein meshwork gradually changes at different depths. Next to the homogeneous grey layer of the lipid bilayer is seen the granular layer (also shown in the inset), then the filamentous meshwork is seen superimposed on the granular layer, and finally only the filamentous meshwork is observed. The diameter and appearance of the filamentous components (small arrows) are very similar to those of purified spectrin tetramers. Note the scattered occurrence of round spots which may represent actin protofilaments (arrowheads), to each of which several filamentous components appear to attach (from Tsukita et al. (1980). J. Cell Biol. **85**, 567–576, with permission). Magnifications (a) ×200 000, (b) ×120 000; bars=100 nm.

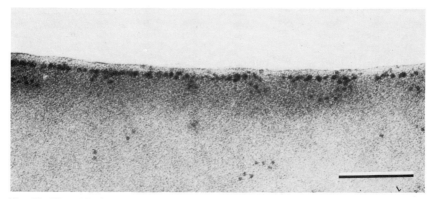

Fig. 23. Ultra-thin frozen section of an intact human erythrocyte after incubation with ferritin-conjugated anti-spectrin antibodies, but without any heavy metal staining. Almost all the ferritin molecules (black dots) are located in a very narrow band immediately subjacent to the periphery of the cell (from Ziparo et al. (1978). J. Cell Sci. **34**, 91–101, with permission). Magnification ×200 000; bar=100 nm.

the substructure of the membrane skeleton has been limited. Hainfeld and Steck (1977) examined glutaraldehyde-fixed ghosts and Triton shells in the scanning electron microscope after freeze-drying and coating with gold/palladium. Their images of the ghosts are difficult to interpret, but those of the Triton shells show a reticular pattern (Fig. 24) similar to but considerably coarser than that seen after tannic acid fixation and thin sectioning.

3. Negative Staining

The earliest electron microscopic images of erythrocyte membrane skeletons were obtained by the negative staining of Triton shells (Yu et al., 1973; Fig. 25). These early micrographs were at low magnification and showed little detail of the skeletal substructure, but later studies by Sheetz and Sawyer (1978), by Lux (1979) and particularly the recent work by Pinder et al. (1981; Fig. 26) show not only spectrin tetramers forming the fibrous strands which constitute the bulk of the membrane skeleton (images which closely resemble those of negtively-stained purified spectrin; see Fig. 28), but in addition show the actin branchpoints at which the spectrin molecules are crosslinked to form a two-dimensional protein network. While in no cases have filaments resembling F-actin been identified by negative staining in unmodified ghosts or Triton shells, there is evidence from cytochalasin binding studies (Lin and Lin, 1979) that the erythrocyte actin normally exists as short double-helical oligomeric strands which form the branchpoints for the membrane skeletal structure by offering attachment sites for several spectrin molecules, as discussed in Section V below.

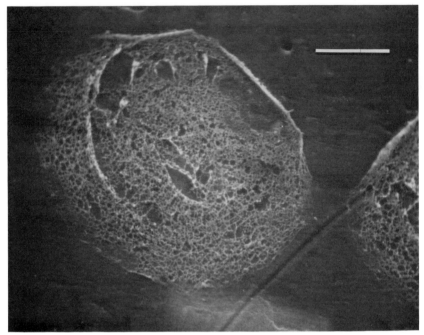

Fig. 24. Scanning electron micrograph of the residual membrane skeleton after Triton X-100 extraction of a freshly prepared erythrocyte ghost (from Hainfeld and Steck (1977). *J. Supramol. Struct.* **6,** 301–311, with permission). Magnification ×10 000; bar=2 μm.

4. *Freeze etching*

The appearance of the membrane skeleton after freeze etching has been described in Section II.D.2 above.

V. THE PROTEINS OF THE MEMBRANE SKELETON

A. Spectrin

1. *General Properties and Protein Chemistry*

Spectrin was first characterized and named by Marchesi and Steers (1968), and it is Marchesi (1979) who has most recently reviewed the status of this protein. The early work on this protein (reviewed by Steck, 1974; Singer, 1974; Kirkpatrick, 1976) showed it to consist of two very large dissimilar polypeptide chains, with approximate molecular weights of 240 000 (band 1)

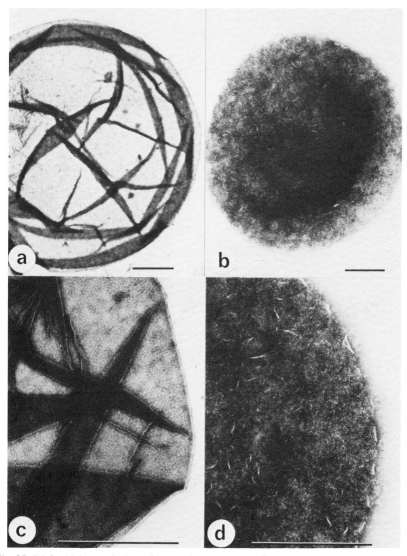

Fig. 25. (a) And (c) negatively-stained erythrocyte ghosts. The dark lines are stain-enclosing folds caused by the collapse of the erythrocyte membrane as it dried flat onto the electron microscope grid support film. (b) And (d) negatively-stained erythrocyte membrane skeletons ("Triton shells") produced by extraction of erythrocyte ghosts with 0·5% Triton X-100 (from Yu et al. (1973). J. Supramol. Struct. **1**, 233–248, with permission). Stain: 1% unbuffered uranyl acetate. Magnification (a) and (b) ×11 000, (c) and (d) ×33 000; bars=1 μm.

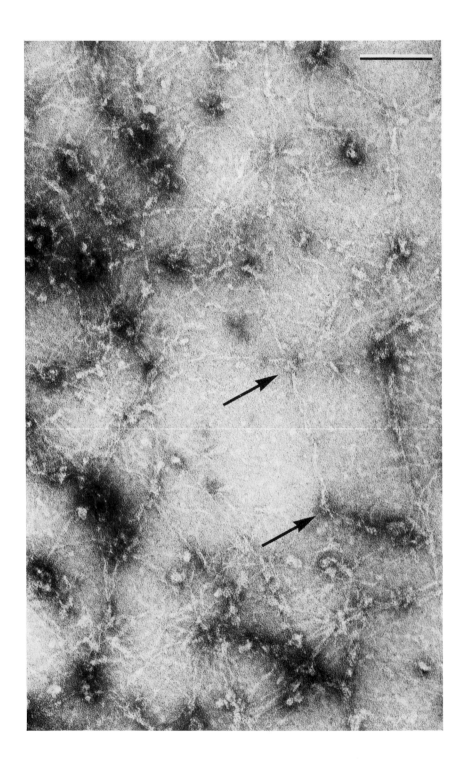

and 220 000 (band 2), having a high α-helical content and being capable of forming aperiodic fibrous aggregates, especially in the presence of divalent cations. Its location on the cytoplasmic surface of the erythrocyte membrane, from which it is readily eluted by low ionic strength buffers, has already been discussed.

Modern understanding of the molecule was initiated by the work of Ralston and his colleagues (Ralston, 1975, 1976; Ralston et al., 1977), who showed that spectrin could be extracted from ghosts in a variety of oligomeric forms. The $\alpha\beta$ heterodimer, composed of one band 1 polypeptide and one band 2 polypeptide, with a molecular weight of approximately 460 000, is the predominant species when spectrin is extracted at 37 °C, while the $\alpha_2\beta_2$ tetramer, made from two heterodimers and having a total molecular weight of approximately 920 000, is the major form when spectrin is extracted at 4 °C. These two forms can be separated by gel filtration in physiological ionic strength buffers, in which they exhibit sedimentation coefficients of approximately 9S and 12S, respectively. Ungewickell and Gratzer (1978) demonstrated that interconversion between heterodimer and tetramer could occur at 29 °C, obeying simple equilibrium thermodynamics, but that both species become kinetically "trapped" when cooled at 4 °C, no interconversion occurring over a period of several days. These findings suggested that, while in dynamic equilibrium with dimers, it is the tetramer which is the predominant association state *in vivo* on the cytoplasmic surface of the erythrocyte membrane. Photochemical crosslinking studies by Ji et al. (1980) and Middaugh and Ji (1980) have supported this idea.

Despite the formidable size of the component polypeptide chains of spectrin, significant advances have recently been made in the protein chemical analysis of the protein. Anderson (1979) demonstrated that the spectrin chains (or at least band 2) can be split by trypsin at low ionic strength into a series of very large fragments differing from one another by approximately 8 000 daltons, while at high ionic strength access to these trypsin-sensitive sites is restricted, and relatively trypsin-insensitive fragments of 170 000, 150 000 and 80 000 daltons are the primary digestion products. In an elegant study, Speicher et al. (1980) have extended this work by showing that prolonged

Fig. 26. High resolution electron micrograph showing an area of Triton X-100 extracted erythrocyte membrane skeleton after negative staining. The skeleton consists of a number of branchpoints provided by actin protofilaments, from which fibrous spectrin molecules radiate, bound at their distal ends to actin, band 4·1 and possibly band 4·9 molecules (see text, Section V). Two such branchpoints are marked (arrows) and are directly linked by a single extended spectrin tetramer whose structure closely resembles that of isolated spectrin molecules after low angle shadowing (Figs 32 and 33), even to the extent of revealing the plectonemic winding of the constituent polypeptide chains (from Pinder et al. (1981). In "The Red Cell" (G. J. Brewer, ed.), *Prog. Clin. Biol. Res.* **55**, 343–354, A. R. Liss, New York, with permission. Magnifications ×195 000; bar=100 nm.

tryptic digestion of the separately purified spectrin polypeptides in 40 mM NaCl, 10 mM sodium phosphte buffer, pH 8·0, at 0 °C results in the production of a set of proteolytically-resistant fragments, those from band 1 having apparent molecular weights of 80 000, 52 000, 46 000 and 41 000, and those from band 2 of 74 000, 65 000, 40 000 and 33 000. By examining larger partial digestion products they were able unambiguously to determine the linear order of these fragments within each chain, and proposed that each fragment represents a stable trypsin-insensitive tertiary structural domain, linked to its neighbouring domains by small proteolytically-sensitive segments. In an accompanying paper, Morrow *et al.* (1980) were able to establish which domains are involved in ankyrin binding, dimer stabilization and dimer-dimer interactions to form tetramers, their results being in excellent agreement with the published electron microscopic studies of these phenomena described below.

Calvert *et al.* (1980a) also studied the separated polypeptide chains of spectrin. They found that after denaturation from urea each chain was monomeric and yet resembled the native spectrin dimer in its solubility, α-helical content and elongated shape (as indicated by negative staining and sedimentation coefficients), and demonstrated that when these separate subunits were combined they were able to form normal spectrin heterodimers. They determined the ankyrin binding site to be on band 2 and showed that the formation of a ternary complex with actin and band 4·1, discussed below, requires the presence of both spectrin subunits. In a related study, Calvert *et al.* (1980b) found evidence from circular dichroism measurements during urea denaturation for the existence of several independently-folded structural domains within each spectrin subunit, and from protein magnetic resonance studies concluded that the flexible regions of the polypeptide chains linking these domains are highly hydrophobic in character. They suggested that these regions might be involved in the direct interactions of spectrin with lipid molecules, for which Sweet and Zull (1970), Juliano *et al.* (1971), Mombers *et al.* (1977, 1979, 1980), Johnson and Kirkwood (1978), Haest *et al.* (1978) and Williamson *et al.* (1982) have provided good experimental evidence (but see also Cassoly *et al.*, 1980), to suggest that it is the preferential interaction of spectrin with the anionic phospholipids, and its failure to bind the neutral phospholipids phosphatidylcholine and sphingomyelin which determines the asymmetric distribution of phospholipids in the native erythrocyte membrane.

Spectrin is normally phosphoryolated, and can be readily labelled with [^{32}P] if intact erythrocytes are incubated with [^{32}P] orthophosphate or if ghosts are given [γ-^{32}P] ATP as a substrate for the specific cAMP-independent spectrin kinase (Roses and Appel, 1973a; Avruch and Fairbanks, 1974; Hosey and Tao, 1977; Plut *et al.*, 1978; Fairbanks *et al.*, 1978). Wolfe and Lux (1978) showed the [^{32}P] label to be present on multiple sites, which Anderson (1979) demonstrated to be within approximately 8 000 daltons of one or both ends of the band 2 polypeptide, since it was absent from the largest tryptic fragment of band 2. More recently Harris and Lux (1980) have undertaken a more detailed characterization of the phosphorylated peptide, showing the native spectrin heterodimer to have four exchangeable phosphorylation sites on band 2, three phosphoserines and one phosphothreonine, which are normally almost fully phosphorylated *in vivo* (Harris *et al.*, 1980). They also demonstrated that these phosphorylation sites are clustered in the carboxyl terminal end of the band 2 polypeptide, which contains a high proportion of proline residues and probably occupies an exposed surface position.

While experiments with ghosts (Sheetz and Singer, 1977; Birchmeier and Singer, 1977) and crude spectrin extracts (Pinder *et al.*, 1975, 1977, 1978) suggested spectrin phosphorylation regulates membrane shape changes and membrane skeletal crosslinking, possibly by modulating spectrin-actin interactions, more recent studies with erythrocyte membranes and also with purified proteins *in vitro* have failed to confirm this, and have revealed no differences in behaviour between phosphorylated and dephosphorylated spectrin. Thus Patel and Fairbanks (1981) found no correlation between the phosphorylation state of spectrin and Mg^{++}-ATP-dependent shape changes, Ungewickel and Gratzer (1978) showed the spectrin dimer-tetramer equilibrium to be unaffected by its phosphorylation state, and Anderson and Tyler (1980) demonstrated that the phosphorylation state of spectrin alters neither the affinity nor the saturation levels of its rebinding to spectrin high-affinity sites on ankyrin molecules attached to spectrin-depleted inside-out vesicles. They further observed that no appreciable spectrin dephosphorylation occurred during the first four hours of glucose starvation of intact erythrocytes, during which extensive metabolic depletion of ATP and cell crenation occurred, nor during the crenation and deformability loss induced by calcium entry after erythrocytes were incubated with the calcium ionophore A 23187. Neither does the phosphorylation state of spectrin seem to affect its interactions with actin and band 4·1 (Ungewickell *et al.*, 1979; Cohen and Branton, 1979). Thus, to-date, no causal relationship has been established between spectrin phosphorylation levels and either erythrocyte shape and deformability or spectrin binding to other peripheral membrane proteins, despite the evidence for active metabolic turnover of the spectrin phosphates and the widespread use elsewhere of phosphorylation as a mechanism for regulation of protein function (Rubin and Rosen, 1975; Greengard, 1978). Even the mechanism of spectrin phosphorylation and dephosphorylation seems to be in doubt, since the evidence of Fairbanks *et al.* (1978) for the importance of a specific membrane-bound cAMP-independent spectrin kinase and for a cytoplasmic phosphospectrin phosphatase has been chal-

lenged by Imhof *et al.* (1980), who suggest that both the phosphorylation and dephosphorylation of spectrin are autocatalytic.

More recently Sobue *et al.* (1981) and Glenney *et al.* (1982) have shown that band 1 of erythrocyte spectrin binds the calcium sequestering protein calmodulin. While the importance of this observation is at present unclear, it is possible that the bound calmodulin confers a calcium sensitivity to the association between spectrin and the other components of the membrane skeleton (see below) which could affect membrane deformability.

2. Negative-Staining Studies

In many early publications (reviewed by Kirkpatrick, 1976) different workers have shown by negative staining that spectrin is capable of forming non-periodic fibrous aggregates, especially in the presence of divalent cations. These structures, an example of which is shown in Fig. 27, suggested that spectrin molecules are elongated, but revealed little further in the way of specific molecular details. Individual unaggregated spectrin molecules, like those of myosin (Huxley, 1973; Takahashi, 1978), have proved difficult to visualize by negative staining. Figure 28 shows a micrograph in which spectrin tetramers may be seen with reasonable clarity after negative staining in the absence of divalent cations, and should be compared with Fig. 26 in which the spectrin molecules are visible after negative staining of an intact membrane skeleton prepared by Triton extraction, as discussed in Section IV above. However, even this favourable preparation would be difficult to interpret were it not for the results from low angle shadowing studies of spectrin described below, which have provided a clear conceptual image of this molecule's conformation, and it is fair to say that the earlier negative-staining studies failed to produce any general agreement among workers in this field as to the shape of this large and important structural protein.

3. Molecular Shape as Revealed by Low-angle Shadowing

Prior to its use with spectrin, low-angle shadowing (reviewed by Slayter, 1976) had been used to elucidate the structures of several macromolecules including collagen (Hall and Doty, 1958), fibrinogen (Hall and Slayter, 1959) and myosin (Slayter and Lowey, 1967; Lowey *et al.*, 1969; Elliott *et al.*, 1976). The inclusion of glycerol to prevent rapid air drying of aerosol droplets of the

---

Fig. 27. Fibrous aggregates of spectrin molecules formed by incubation for 10 min at 0 °C in 1 mM $MgCl_2$-100 mM KCl, and subsequently negative stained with unbuffered 1% uranyl acetate (Shotton, unpublished work). Magnification ×170 000; bar=100 nm.

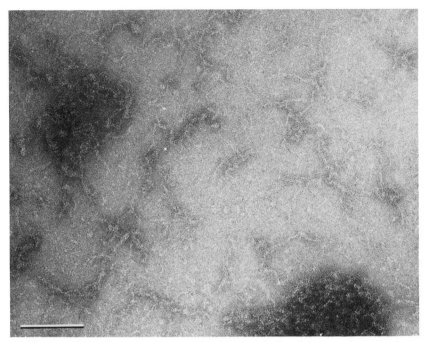

Fig. 28. Spectrin tetramers negatively stained with unbuffered 1% uranyl acetate in the absence of divalent cations. Although some aggregation has occurred, many molecules may be seen lying separately in regions of thin stain (Shotton, unpublished work). Magnification ×170 000; bar=100 nm.

protein solution sprayed on freshly-cleaved mica prior to low-angle shadowing, which avoids the damaging shear forces resulting from the rapid meniscus movements of such air drying, was found by Elliott et al. (1976) to be important in obtaining clear low-angle shadowed images of myosin, and it was this methodological improvement which was employed by Shotton et al. (1978, 1979) to produce interpretable shadowed images of individual spectrin dimers and tetramers, in which the general conformation of the molecule and the relative positioning of the band 1 and band 2 polypeptide chains within the heterodimer could for the first time be clearly seen.

Figure 29 shows a preparation of spectrin heterodimers after unidirectional and rotary shadowing, while Fig. 30 shows enlargements of selected molecules to illustrate the range of conformations which these flexible proteins can adopt. In their most extended form, they appear as straight compact rods, but

Fig. 29. Fields of spectrin heterodimer molecules after (a) unidirectional and (b) rotary shadowing (from Shotton et al. (1979). J. Mol. Biol. **131**, 303–329, with permission). Magnification ×90 000; bar=100 nm.

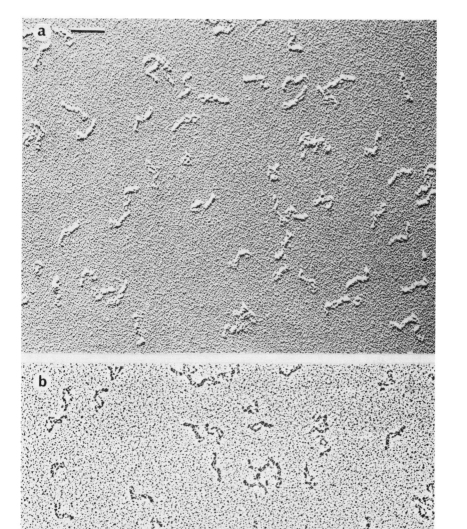

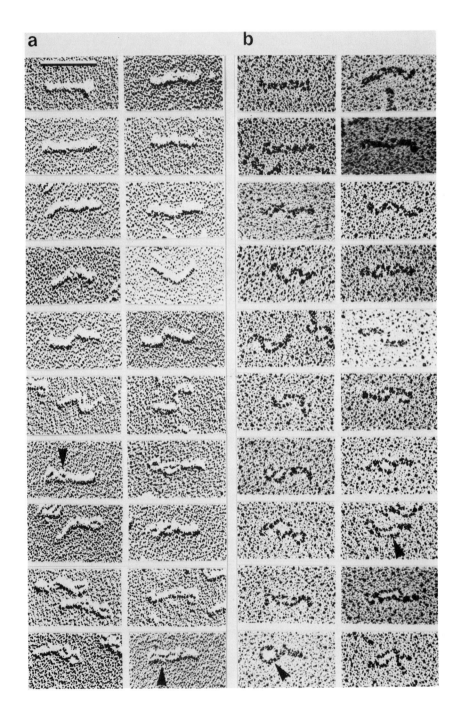

most molecules are curved into a variety of L, C or S shapes or lie collapsed upon themselves. As is most clearly seen in the rotary-shadowed specimens, many molecules appear somewhat unfolded and can be seen to consist of two separate strands. No obvious structural differentiation is visible between or within these two strands which appear to extend the whole length of the molecule, lying partially separated from one another, side-by-side or twisted round one another in a loose double helix. Both left- and right-handed forms can be seen (Figs 29a and 30a), suggesting that this coiling is a non-specific phenomenon resulting from the flexible nature of the strands. The most straightforward interpretation is that these strands are the shadowed images of the elongated tertiary structures of the individual monomer polypeptides (bands 1 and 2) lying side-by-side. Even in cases where these two strands lie separated along most of their lengths, they are normally joined at their ends. Images in which one end of the heterodimer is splayed into two free polypeptide strands (bottom right image of Fig. 30b) are rare. The mean length of the heterodimer molecules after correction for shadow thickness, as described by Shotton *et al.* (1979), was 97 nm, the individual measured lengths having the distribution shown in Fig. 31.

From these studies and the protein chemical investigations described above, a picture emerges of the spectrin heterodimer molecule as being formed by the lateral non-covalent association of its two component subunits, each of which is an elongated strand composed of stable and probably fairly rigid closely-packed structural domains, linked by flexible proteolytically sensitive regions, giving the whole molecule a high degree of structural flexibility. The spin labelling results of Cassoly *et al.* (1980) are consistent with this structure and show that the conformation of the spectrin molecule is stabilized both by the divalent calcium and magnesium cations and also by rebinding to spectrin-depleted inside-out erythrocyte membrane vesicles. Further discussion of the conformation of the polypeptides within each physical strand is given by Shotton *et al.* (1979).

The appearance of spectrin tetramers after unidirectional and rotary shadowing is shown in Figs 32 and 33. Measurements of these shadowed molecules gives a mean length after correction for shadow thickness of 194 nm, exactly twice that for the heterodimers, the individual measured lengths having the distribution shown in Fig. 31. It is thus clear that spectrin tetramers are formed by the end-to-end association of pairs of heterodimers, without

Fig. 30. Spectrin heterodimer molecules after (a) unidirectional and (b) rotary shadowing. In this Fig. and in Fig. 32, the images have been selected to show the typical variety of extended, curved, helical and partially unfolded configurations which the molecules adopt. In many cases the two strands which comprise each heterodimer molecule are clearly visible (arrows). Molecules exhibiting similar configurations are grouped on corresponding rows in the two halves of the Fig. (from Shotton *et al.* (1979). *J. Mol. Biol.* **131**, 303–329, with permission). Magnification ×153 000; bar=100 nm.

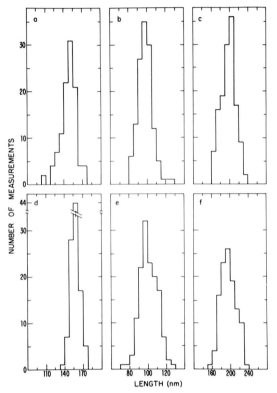

Fig. 31. Histograms showing the distribution of measured molecular lengths, prior to correction for shadow thickness, after [(a) to (c)] unidirectional and [(d) to (f)] rotary shadowing. (a) And (d) myosin tails (myosin molecules were used as control specimens for the shadowing procedure); (b) and (e) spectrin heterodimers; and (c) and (f) spectrin tetramers. Histograms are plotted in nm, at intervals of approximately 0·05 of the molecular lengths (from Shotton et al. (1979). J. Mol. Biol. **131**, 303–329, with permission, in which statistics and further discussion are given).

measurable overlap, to give molecules which when extended have axial ratios clearly exceeding 10:1. Figure 34b to 34e shows stereoscopic enlargements of spectrin tetramers in which the separation and coiling of the individual polypeptides can be clearly discerned.

Topological arguments led Shotton et al. (1979) to propose that this association between two heterodimers to form the tetramer was of a head-to-head rather than a head-to-tail nature. In the latter case there is no

Fig. 32. Fields of spectrin tetramer molecules after (a) unidirectional and (b) rotary shadowing (from Shotton et al. (1979). J. Mol. Biol. **131**, 303–329, with permission). Magnification ×90 000; bar=100 nm.

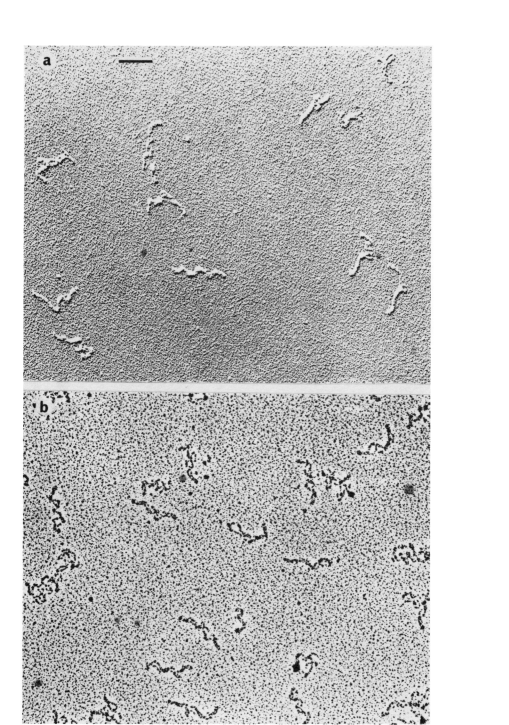

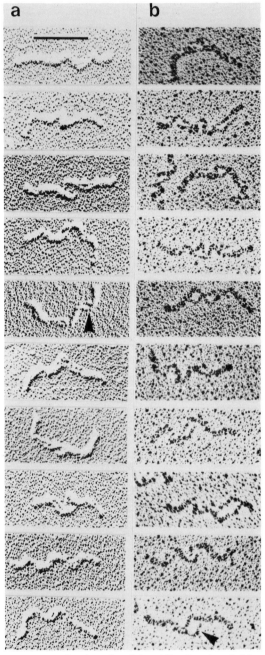

Fig. 33. Spectrin tetramer molecules after (a) unidirectional and (b) rotary shadowing. Details as for Fig. 30 (from Shotton et al. (1979). J. Mol. Biol. **131**, 303–329, with permission). Magnification ×153 000; bar=100 nm.

reason to expect the process of oligomerization to stop at the tetramer. Linear hexamers, octomers and higher oligomers might all be expected to occur under appropriate conditions. If, in contrast, tetramers are formed by the head-to-head association of heterodimers, linear elongation beyond the tetramer is impossible without a second distinct binding site or a second molecular component to crosslink individual tetramers into the molecular meshwork of the membrane skeleton. No evidence for such linear higher oligomers was found in the absence of other components of the membrane skeleton.

Such head-to-head binding must logically involve reciprocal binding at two identical pairs of binding sites, if all binding valencies are to be satisfied. The micrographs of shadowed tetramers (Figs 32 and 33, particularly the lower right image of Fig. 33b) show that interdimer binding occurs between *both* of the paired monomer strands. These observations were explained by Shotton *et al.* (1979) by postulating a binding site A at one end of a band 1 monomer which can bind *either* to a complementary site B on its sister band 2 *or* to a chemically identical B site on band 2 of a second heterodimer (Fig. 35). This would free site B on the sister monomer of the first heterodimer to form a reciprocal association with site A on band 1 of the second heterodimer. The spectrin molecules are obviously sufficiently flexible to permit this type of alternative binding. In this model the intradimer binding interactions between the distal ends of sister monomer strands remain undisturbed by the process of tetramer formation. Direct labelling studies described below have subsequently verified this head-to-head binding hypothesis.

At the time of these initial shadowing studies we noticed the additional occurrence, in high molecular weight gel filtration column fractions, of spectrin hexamers formed by the head-to-head binding of *three* spectrin heterodimers, such that the quasi-equivalence of the individual binding sites (Caspar and Klug, 1962) was conserved, as illustrated in Fig. 35c. These observations were subsequently confirmed by Tyler *et al.* (1980) and independently by Morrow and Marchesi (1981). The apperance of spectrin hexamers after rotary shadowing is illustrated in Fig. 34a, while Fig. 34b shows a stereoscopic enlargement of one such hexamer together with three spectrin tetramers, in which all 18 individual polypeptide chains can be seen with great clarity (Burke, Burnip and Shotton, unpublished observations).

Direct estimates of molecular thickness from the shadowed images of elongated molecules (in contrast to estimates of their length) are unreliable, because of the proportionately greater errors involved in correcting for the thickness of the shadow material. Some estimate of the thickness of the individual strands of spectrin molecules may, however, be gained from Fig. 36 by direct comparisons with the tails of myosin molecules shadowed and observed under identical conditions. These show that the width of one

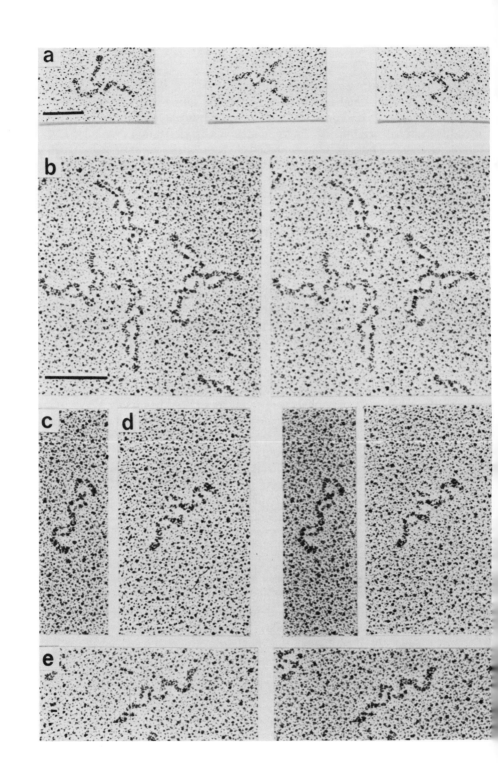

3. Proteins of the Erythrocyte Membranes 277

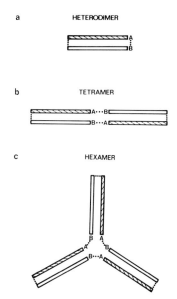

Fig. 35. Diagrammatic illustration of the inter-chain binding in spectrin (a) heterodimers, (b) tetramers and (c) hexamers. The hatched and empty bars represent band 1 and band 2, respectively. The three dots represent quasi-equivalent binding interactions at sites A and B between sister strands in a heterodimer and between paired strands in a tetramer and hexamer, and additional binding interactions at the distal ends of each component dimer (after Shotton et al. (1979). J. Mol. Biol. **131**, 303–329).

spectrin molecule slightly exceeds that of two myosin tails lying side-by-side, while the width of a single spectrin polypeptide strand, seen where individual molecules lie partially unfolded, appears somewhat narrower than the two myosin tails and a little wider than a single tail. Since the diameter of a single myosin tail, formed by the supercoiling of two α-helices, is approximately 2 nm, this suggests that an individual spectrin polypeptide is between 2 and 3 nm in thickness, while the spectrin molecule as a whole has a maximum width of between 4 and 6 nm.

In a project designed to correlate the physical structure of the spectrin molecule with its protein chemistry and functional binding interactions, we

---

Fig. 34. (a) And (b) spectrin hexamers and (b–e) spectrin tetramers after rotary shadowing. (b–e) Are stereoscopic pairs (which should be viewed with a stereoscopic viewer with interocular distance 6·5 cm) in which the individual polypeptide strands within each molecules can be clearly seen, some of them coiling in a helical conformation above the surface of the mica substrate. The hexamers are formed by the head-to-head-to-head binding of three spectrin heterodimers, as shown diagrammatically in Fig. 35c (Burke, Burnip and Shotton, unpublished work). (a) Magnification ×110 000. (b–e) Specimen tilt +12° and −12° (total tilt 24°). Magnification ×170 000. Bars=100 nm.

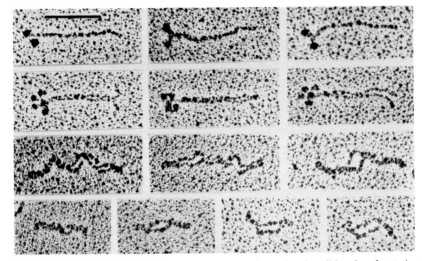

Fig. 36. Rotary shadowed images of (a) single myosin molecules, (b) pairs of entwined myosin molecules, (c) spectrin tetramers and (d) spectrin heterodimers. All the molecules were shadowed in the same experiment, receiving identical exposure to the stream of platinum vapour. The shadow thickness on all the molecules can thus be assumed to be the same (mean platinum shadow thickness on the mica=0·78 nm) so that direct visual comparisons of molecular thicknesses are possible, as discussed in the text (from Shotton *et al.* (1979). *J. Mol. Biol.* **131**, 303–329, with permission). All magnifications ×153 000; bar=100 nm.

have raised monoclonal antibodies to specific antigenic sites along the spectrin molecule and have mapped these by radioimmune labelling of tryptic fragments of spectrin fractionated by SDS slab gel electrophoresis and blotted onto nitrocellulose filters (Burke and Shotton, 1982; Scott, Pittman and Shotton, work in progress). The binding of one of these antibodies, which is specific for the terminal 80 000 dalton domain of band 1 (Speicher *et al.*, 1980), to spectrin dimers, visualized directly in the electron microscope after low-angle shadowing of the antibody-spectrin complex, is shown in Fig. 37. Since the antibody binds to the mid-point but not the ends of spectrin tetramers (micrograph not shown) the binding site is identified as being at that end of the heterodimer involved in tetramer formation, in agreement with the protein chemical conclusions of Morrow *et al.* (1980) derived from the associations between tryptic fragments in non-denaturing gels. This work provides the first firm evidence for the gross colinearity of the amino-acid sequence of band 1 with the physical strand it forms, although it is not yet known whether this binding site is at the N- or the C- terminal end of the polypeptide chain.

The ability to visualize, by electron microscopy after low-angle shadowing,

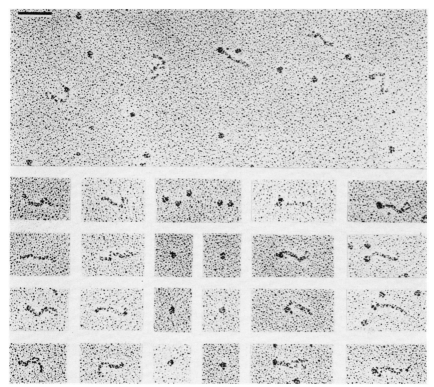

Fig. 37. Labelling of spectrin heterodimers with a purified monoclonal IgG preparation specific for an antigenic site on the 80 000 dalton terminal domain of band 1. Top: general field after incubation of the monoclonal antibody with spectrin heterodimers. Left two columns: spectrin heterodimers. Centre columns: IgG molecules. Right two columns: molecular complexes in each of which a single IgG molecule has bound to a unique site at one end of a spectrin heterodimer (from Burke and Shotton (1982). *EMBO J.* **1**, 505–508, with permission). All magnifications ×100 000; bar=100 nm.

a single immunoglobulin molecule bound to its antigenic site on an isolated spectrin molecule demonstrates the feasibility of using this approach more widely for the more detailed immunological dissection of this and other large protein molecules.

B. Actin

The earliest suggestion that the erythrocyte membrane contained actin came from Ohnishi (1962), who added muscle myosin to an aqueous extract of an acetone powder of erythrocyte ghosts and demonstrated by viscometry that the myosin interacted with a protein in the extract, observations which he has more recently confirmed and extended (Ohnishi, 1977). While the first electron microscopic observation of actin from erythrocytes was made by Marchesi and Palade (1967b), who demonstrated that when guinea pig erythrocyte ghosts were trypsinized in the presence of ATP and Mg^{++}, filaments appeared which had the double helical morphology of F-actin. Unfortunately the interpretation of this latter discovery was confused by the subsequent report of Marchesi and Steers (1968), published after studying an extract from guinea pig ghosts made with distilled water containing ATP and β-mercaptoethanol, in which the main protein component was spectrin. Negative staining of this extract without addition of divalent cations showed "amorphous material" (presumably poorly resolved spectrin molecules), while filaments similar to those seen by Marchesi and Palade were observed in addition to the amorphous material after the inclusion of 1 mM $MgCl_2$. However, the authors assumed these filaments to be of spectrin, formed (as they thought) by polymerization of the amorphous material, rather than of actin, which was an unnoticed minor component in their extract, particularly since their rabbit antiserum raised against the extract (presumably chiefly anti-spectrin) failed to cross-react with guinea pig muscle actin. Despite this, the idea that actin might be present in erythrocyte membranes persisted, associated with the mistaken but then widespread notion that spectrin was an erythrocyte myosin and that together these two proteins formed a contractile actomyosin ATPase system (Guidotti, 1972; Juliano, 1973; Singer, 1974; Avissar et al., 1975).

The first positive identification of band 5 (Figs 2 and 3, and Table I) as erythrocyte actin was achieved by Tilney and Detmers (1975) using classical actin extraction, polymerization and heavy meromyosin (HMM) decoration techniques. Figure 38a shows one of their erythrocyte actin filaments decorated with subfragment one (S-1) of rabbit HMM, in which the polarity of the filament is revealed by the characteristic "arrowhead pattern" formed by the projected images of the helically arranged negatively stained myosin heads. Similar images of undecorated (Fig. 38b) and decorated erythrocyte F-actin were published by Sheetz et al. (1976a, b), who confirmed this identification of band 5 as actin. Subsequently, Nakashima and Beutler (1979) have shown by isoelectric focusing that erythrocyte actin is mainly of the β form characteristic of platelet and other non-muscle actins, rather than the α form of muscle. In all other properties tested, isolated erythrocyte band 5 behaved similarly to actins from other sources. The appearance after rotary shadowing of F-actin filaments associated with other membrane skeletal proteins is described below.

In their report, Tilney and Detmers (1976) argued strongly that erythrocyte actin is present *in vivo* in a non-polymerized form in a molecular complex with spectrin, analogous to the complex formed between actin and profilin in the undischarged acrosomal process of the sperm of the echinoderm *Thyone*

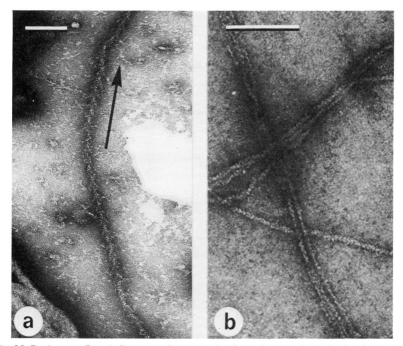

Fig. 38. Erythrocyte F-actin filaments after extraction from ghosts, polymerization, incubation with heavy meromyosin subfragment 1 (HMM S-1) and negative staining with uranyl acetate. In (a) the polarity of the characteristic "arrowhead" pattern of HMM S-1 molecules on the decorated F-actin filament (see text) is indicated. In (b), where 20 mM pyrophosphate was additionally included with the HMM S-1 to prevent the attachment of the myosin heads to the F-actin filaments, the double helical morphology of the naked filaments can be clearly seen. (a) From Tilney and Detmers (1975). *J. Cell Biol.* **66**, 508–520 with permission; (b) from Sheetz, Painter and Singer (1976). "Cell Motility" (R. Goldman, J. L. Rosenbaum and T. D. Pollard, eds) Vol. B, pp. 651–664, Cold Spring Harbor Laboratory, New York, with permission). Magnifications (a) ×115 000; (b) ×200 000; bars=100 nm.

(Tilney, 1974, 1975). They based this argument on the undisputed fact that while muscle actin filaments can be clearly seen after freeze etching on the extracellular surface of erythrocyte ghosts to which they have been adsorbed (Tillack and Marchesi, 1970), and while erythrocyte actin can be readily polymerized into double helical filaments seen by negative staining after the addition of magnesium ions to extracts from the membrane, after trypsinization of the membrane, or after incubation of ghosts with HMM S-1, F-actin filaments have never been observed on or within unmolested erythrocyte ghost membranes by any electron microscopic technique.

Despite this, it has since been clearly established that in fact erythrocyte actin exists *in vivo* in a polymerized form, as very-short double-helical filaments termed protofilaments by Brenner and Korn (1980), in equilibrium with a small cytoplasmic G-actin monomer pool (Pinder et al., 1981). Cohen and Branton (1979) showed that the membrane skeleton contains elements that would nucleate the polymerization of G-actin, Sheetz (1979b) observed that DNase I, which is known to bind to and depolymerize F-actin, was effective in dissociating isolated erythrocyte membrane skeletons, and Lin and Lin (1979) obtained a direct estimate of the titre of F-actin filament ends from cytochalasin binding experiments. Knowing this and the number of actin molecules per cell, one can calculate that the F-actin protofilaments in the native erythrocyte membrane skeleton contain on average between 12 and 18 G-actin monomers (Pinder et al., 1981), each thus extending only about one half of one F-actin double helical turn, with an approximate length of between 30 and 40 nm. Since these protofilaments are considerably shorter than an individual spectrin molecule, it is perhaps not surprising that they should not have been recognized as F-actin filaments *in situ*, particularly since it is only recently, by the application of the tannic acid fixation procedure (Tsukita et al., 1980), that it has been possible to distinguish individual spectrin molecules within the intact membrane.

C. Band 4·1

Band 4·1 is an apparently globular protein with a molecular weight of approximately 82 000, which has not yet been given a specific name. On SDS polyacrylamide gels it runs as a single band in the continuous buffer system of Fairbanks et al. (1971), but can be resolved into two distinct bands, 4·1a and 4·1b, using the discontinuous buffer system of Laemmli (1970) (Edwards et al., 1979). However, these appear to be different forms of the same polypeptide, differing perhaps in the extent of their phosphorylation. Tyler et al. (1980b) have shown that "aged" band 4·1 preparations have a tendency to form dimers and higher aggregates which can also bind to spectrin, and Gratzer (1981) and Pinder et al. (1981) suggest that the dimer is the functional form of band 4·1, but without citing the evidence for this claim. It is clear, as discussed below, that band 4·1 is able both to bind to spectrin in the absence of actin, and to promote and stabilize the association between actin and spectrin by forming a ternary complex with them.

D. The Molecular Associations of Spectrin, Actin and Band 4·1 within the Membrane Skeleton

The principal components of the membrane skeleton, spectrin, actin and band 4·1 have an approximate molar stoichiometry *in vivo* of 1 spectrin tetramer: 5 actin monomers: 6 band 4·1 monomers (Pinder et al., 1981). Thus on average an actin protofilament of 15 G-actin molecules would be expected to have attached to it six spectrin tetramer ends, assuming all the possible spectrin-actin binding interactions are made. In addition band 4·9, a minor polypeptide with a molecular weight of about 50 000, appears to be an intrinsic component of the membrane skeleton, remaining associated with the principal components during the preparation of clean Triton shells from which all other proteins have been removed (Sheetz, 1979a), and in which the selective loss of actin and band 4·1 has reduced their molar ratios with respect to each spectrin tetramer to 4 actin monomers and two band 4·1 monomers (Pinder *et al.*, 1981). Band 4·9 is also found in the high molecular weight aggregates of spectrin, actin and band 4·1 isolated from low ionic strength extracts of erythrocyte ghosts (Calvert *et al.*, 1980a; Fowler and Taylor, 1980). Holdstock and Ralston (1980) have reported that *p*-mercuribenzene sulphonate (PMBS) disrupts Triton shells by specific interaction with band 4·9, but in other respects this protein is largely uncharacterized, and its role within the membrane skeleton is at present unknown.

The state of investigation into the interactions between the components of the membrane skeleton, currently a very active area of research, has recently been well reviewed by Lux (1979a), Gratzer (1981) and Pinder *et al.* (1981). Results from several laboratories have shown that *in vivo* spectrin is crosslinked into the extensive isotropic protein meshwork which constitutes the membrane skeleton by strong ternary complexes formed between the distal ends of bivalent spectrin tetramers, the multivalent junction points afforded by the F-actin protofilaments, and the stabilizing protein band 4·1. A binary complex can be formed between spectrin and F-actin alone, as first demonstrated by Brenner and Korn (1979), but its weakness in the absence of band 4·1 is illustrated by the failure of Ungewickell *et al.* (1979) to observe this rapidly sedimenting product in mixtures of F-actin and spectrin tetramers alone, and by its dissociation by shearing forces after being formed *in vitro* and detected by low-shear viscometry (Cohen and Korsgren, 1980). As might be expected, no complexes are formed with actin when it is in its monomeric G-form, but muscle G-actin can be induced to polymerize to long actin filaments when seeded *in vitro* with a complex of spectrin plus actin (Brenner and Korn, 1980), or with the intact ternary complex of spectrin, F-actin protofilaments and band 4·1 isolated from erythrocyte membranes and purified by gel chromatography (Pinder *et al.*, 1979). Figure 39 shows such muscle actin filaments, with the nucleating complex of erythrocyte proteins on the left at the non-growing end. Similarly, Cohen *et al.* (1978) and Cohen and Branton (1979) showed that inside-out erythrocyte membrane vesicles provided nucleation sites for the polymerization of muscle G-actin (but attached no particular biological significance to this beyond the demonstration of pre-existing F-actin protofilament seeds in the erythrocyte membrane skeleton, since the direction of the newly-polymerized F-actin filaments

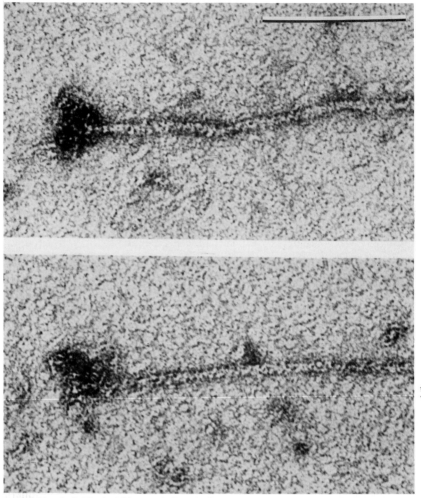

Fig. 39. Two individual rabbit muscle F-actin filaments polymerized by seeding a preparation of rabbit muscle G-actin by the addition of a purified oligomeric fraction containing spectrin, actin and band 4·1, which had been prepared from a low ionic strength extract of human erythrocyte ghosts. Erythrocyte membrane protein complexes can be seen clustered at the left end of each F-actin filament. Negatively stained with 1% uranyl acetate (from Pinder et al. (1979). *FEBS Lett.* **104**, 396–400, with permission). Magnification ×390 000; bar=100 nm.

Fig. 40. (a) The electron microscopic appearance of complexes formed between rabbit muscle F-actin and human erythrocyte spectrin tetramers and band 4·1, showing the F-actin filaments decorated and crosslinked by thin flexible fibres. (b) As (a) except with spectrin heterodimers in place of tetramers, showing less distinct decoration and no cross-bridges. (c) A decorated erythrocyte F-actin filament with projecting spectrin molecules, formed by incubating together spectrin, actin, band 4·1 and band 4·9 under re-associating conditions, after extracting these molecules from Triton shells with inositol hexaphosphate. ((a) And (b) from Ungewickell et al. (1979), reprinted by permission from *Nature* **280**, pp. 811–814. Copyright © 1979 Macmillan Journals Ltd. (c) Unpublished micrograph kindly supplied by Don L. Siegel, Harvard University.) All negatively stained with 1% uranyl acetate. Magnifications (a) and (b) ×120 000; (c) ×225 000; bars=100 nm.

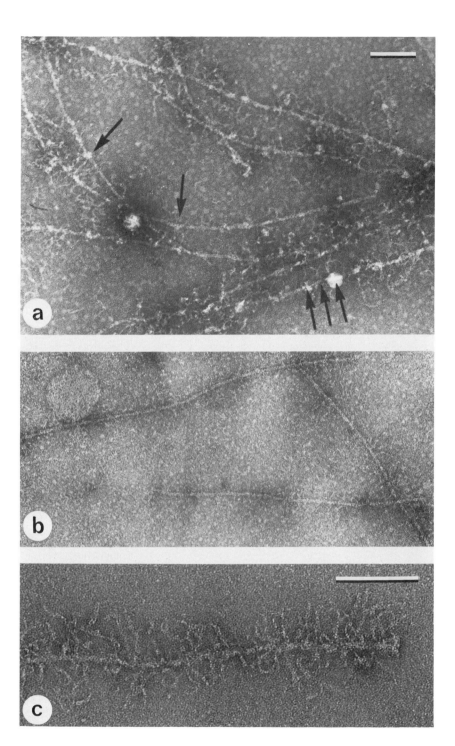

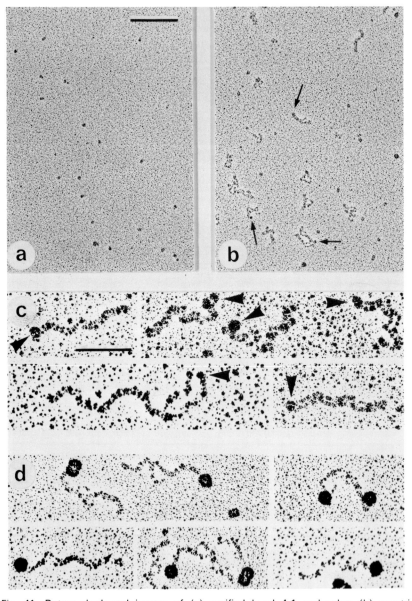

Fig. 41. Rotary shadowed images of (a) purified band 4·1 molecules; (b) spectrin heterodimers with band 4·1, showing indistinctly the binding of band 4·1 molecules to one end of some of the spectrin molecules (arrows); (c) spectrin tetramers with "aged" band 4·1, in which more prominent variable-sized aggregates of the band 4·1 monomers can be seen as small blobs bound to the ends of individual spectrin tetramers (arrows); and (d) spectrin tetramers with biotin-conjugated band 4·1 monomers subsequently labelled with an

in vitro, as determined by HMM S-1 binding, was the reverse of that normally found for membrane associated actin filaments *in vivo*), and Cohen and Foley (1980) and Fowler *et al.* (1981) studied the binding of F-actin to the cytoplasmic surface of the erythrocyte membrane, and have shown it to be mediated by spectrin dimers or tetramers.

The first *in vitro* formation of the complete ternary complex between purified spectrin tetramers, purified band 4·1 and erythrocyte or muscle F-actin was achieved by Ungewickell *et al.* (1979) and is shown in Fig. 40a. Spectrin tetramers can be seen attached terminally to and crosslinking the long F-actin filaments. Given the correct proportions of component proteins, these interactions are extensive enough to convert the solution into a firm gel. No macromolecular complexes were formed under the conditions of these experiments if one of the three components was omitted, if G-actin from either source was used intead of F-actin, or if the spectrin was present as the monovalent heterodimer, rather than the divalent tetramer. In this later instance, however, negative staining (Fig. 40b) revealed that some of the spectrin dimers had attached to F-actin, visible as fine threads extending laterally from the actin filaments, although they were unable to form the cross-bridges required for the assembly of a macromolecular complex. Similar images of spectrin/actin/4·1/4·9 complexes have been obtained by Don Siegel (personal communication, 1981) after extracting these polypeptides from Triton shells with inositol hexaphosphate and incubating them under re-associating conditions (Fig. 40c). Other reconstitution experiments have recently been undertaken by Fowler and Taylor (1980) and Cohen and Korsgren (1980), confirming and extending these results.

Clearer images of these interactions have been obtained by employing the low-angle shadowing technique developed for spectrin by Shotton *et al.* (1979). Tyler *et al.* (1979, 1980b) used this method to study the binding of purified 4·1 to isolated spectrin heterodimers and tetramers, with the results shown in Fig. 41. Individual band 4·1 molecules appeared as small roughly globular molecules with diameters between 5 and 7 nm (Fig. 41a). When mixed with spectrin dimers, stable complexes formed, but it was difficult to discern the small 4·1 molecules bound to one end of some of the spectrin heterodimers, since they were of approximately the same diameter (Fig. 41b). However, if the band 4·1 preparation was "aged" it tended to oligomerize, and after incubation of such an "aged" preparation with spectrin tetramers, more obvious labelling occurred at the extreme ends of the spectrin tetramers, i.e. at

avidin-ferritin preparation, in which the prominent ferritin molecules (black spheres) are clearly visible, attached to one or both ends of the spectrin tetramers ((a) and (b) from Tyler *et al.* (1979). *Proc. Natn. Acad. Sci. USA* **76**, 5192–5196, with permission; (c) and (d) from Tyler *et al.* (1980b). *J. Biol. Chem.* **255**, 7034–7039, with permission). Magnifications: (a) and (b) ×65 000; bar=200 nm; (c) and (d) ×150 000; bar=100 nm.

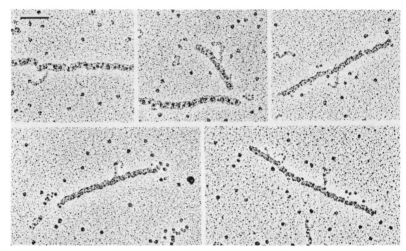

Fig. 42. Rotary shadowed images of short F-actin filaments polymerized from rabbit muscle G-actin in the presence of human erythrocyte spectrin heterodimer, at an actin:spectrin molar ratio of approximately 50:1. The dimers bind to the actin filaments by one end, and can be seen projecting from them in several places. The small round objects scattered over the background are unpolymerized G-actin molecules) (from Cohen *et al.* (1980). *Cell* **21**, 875–883, with permission). Magnification ×80 000; bar=100 nm.

those ends of the component spectrin dimers distal to the dimer-dimer interaction site (Fig. 41c). By ferritin-labelling band 4·1, employing a biotin:avidin conjugate (Fig. 41d), the band 4·1 binding sites on spectrin were made much more clearly visible, and their locations were confirmed to be at one or in some cases both ends of the spectrin tetramers. These experiments thus clearly defined the site of 4·1 binding, and in so doing confirmed the symmetrical divalent nature of the spectrin tetramer which results from its head-to-head mode of assembly from two heterodimers proposed by Shotton *et al.* (1979). Tyler *et al.* (1980b) also observed similar terminal band 4·1 binding on at least two arms of individual spectrin hexamers, confirming the molecular arrangement for these molecules shown in Fig. 35.

Cohen *et al.* (1980) extend this work by shadowing ternary complexes between spectrin, band 4·1 and rabbit G-actin, mixed in various proportions under well characterized conditions. Figure 42 shows the binding of spectrin dimers by one end to sites along short actin filaments formed by incubating

---

Fig. 43. As Fig. 42, except that G-actin is polymerized in the presence of an approximately 1:50 molar ratio of spectrin tetramers, (a) and (b) without and (c) and (d) with the additional presence of band 4·1. In both cases spectrin tetramers bind to actin filaments by one or both ends, and crosslink adjacent filaments (from Cohen *et al.* (1980). *Cell* **21**, 875–883, with permission). Magnifications all ×110 000; bar=200 nm.

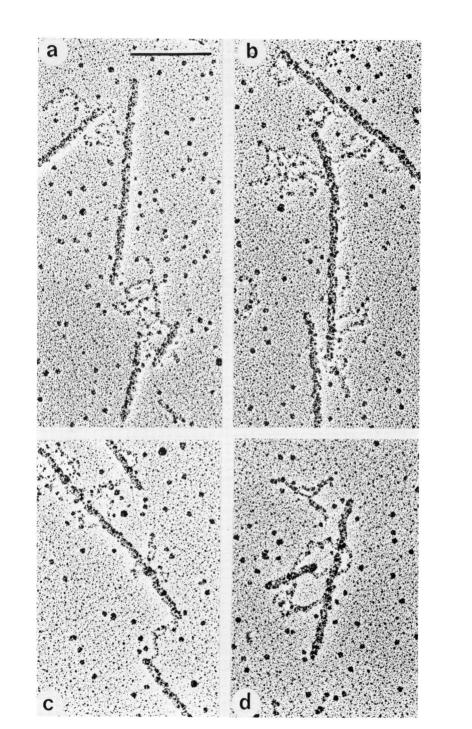

spectrin dimers with rabbit G-actin in a molar ratio of 1:50 under actin polymerizing conditions. When tetramers were similarly incubated with G-actin, the actin filaments became crosslinked by the divalent spectrin molecules (Fig. 43a and b). Other spectrin tetramers curved and bound at both ends to the same actin filament, while yet others bound at one end only. Very similar results were obtained if band 4·1 was also present in the polymerization mixture (Fig. 43c and d).

Many fewer binding interactions were observed when either spectrin dimers or tetramers were incubated with preformed F-actin in the absence of band 4·1. However, when this protein was added to mixtures of F-actin and spectrin dimers, nearly every actin filament observed had one or more spectrin dimers projecting from it, and the addition of band 4·1 to mixtures of F-actin and spectrin tetramers caused the formation of extensive aggregated networks of these proteins (micrographs not shown here).

When a more physiological molar ratio between spectrin and actin of 1:5 was employed, no aggregation occurred when spectrin dimers and G-actin were mixed under polymerizing conditions in the absence of band 4·1 (Fig. 44a). However, the additional inclusion of band 4·1 resulted in the formation not of F-actin filaments but of isolated dense nuclei with spectrin dimers radiating from them (Fig. 44b and 44c), while at this molar ratio mixtures of spectrin tetramers and G-actin generated large amorphous complexes under polymerizing conditions whether or not band 4·1 was present (Fig. 44d).

In summary, these striking micrographs clearly show that spectrin dimers and tetramers bind to actin filaments in an end-on orientation, the spectrin dimers being monovalent and the tetramers symmetrically divalent towards actin. In addition, they demonstrate that the actin binding site on the spectrin dimer lies very close to that for band 4·1 itself, explaining the further observation that band 4·1 promotes the binding of spectrin molecules to preformed F-actin filaments, although under most of the conditions tried it appeared to have very little effect upon the associations formed between spectrin and G-actin in polymerizing buffers. While the long actin filaments crosslinked by spectrin tetramers formed at spectrin:actin molar ratio of 1:50 (Fig. 43) clearly do not resemble the membrane skeleton *in vivo*, the irregular complexes resulting from mixing spectrin, actin and band 4·1 *in vitro* at the more physiological spectrin:actin molar ratio of 1:5 (Fig. 44b, 44c and 44d) seem to be formed by the attachment of radiating spectrin molecules to a short actin protofilament core, and thus more closely resemble the basic structural unit from which the membrane skeleton is constructed *in vivo* (compare Figs 21, 22 and 26).

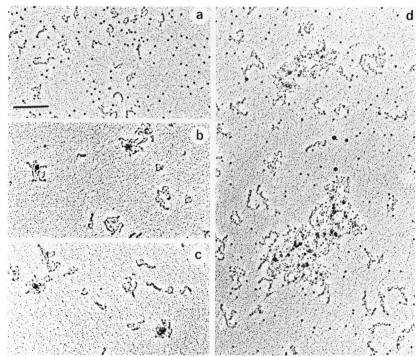

Fig. 44. Rotary shadowed images of the complexes formed after incubation of rabbit muscle G-actin and human erythrocyte spectrin at a physiological actin:spectrin molar ratio of approximately 5:1. (a) Actin and spectrin heterodimers—no complexes form. (b) And (c) actin, spectrin heterodimers and band 4·1—spectrin dimers can be seen radiating from the centres of small complexes, each presumably containing a short F-actin protofilament. (d) Actin, spectrin tetramers and band 4·1—the divalent spectrin tetramers crosslink the individual macromolecular complexes into large amorphous networks (from Cohen et al. (1980). Cell **21**, 875-883, with permission). Magnifications all ×48 000; bar=200 nm.

E. The Attachment of the Membrane Skeleton to Integral Proteins: Ankyrin and Glycoconnectin

The two-dimensional protein meshwork which constitutes the erythrocyte membrane skeleton can only influence the viscoelastic properties of the erythrocyte membrane and limit the translational mobility of its integral membrane proteins if it is intimately associated with them. That these associations are of a simple quantifiable nature was first demonstrated by Bennett and Branton (1977), who showed that the association between spectrin and the cytoplasmic surface of spectrin-depleted human erythrocyte membrane vesicles was mediated by a trypsin-sensitive site which was distinct from actin, glycophorin-A and band 3. It showed binding behaviour which was pH sensitive, saturable at about 200 μg of spectrin bound per mg of membrane protein with a dissociation constant at pH 7·6 of 45 μg spectrin per ml, and abolished by thermal denaturation of spectrin. More recently Goodman and Weidner (1980) have showed that spectrin tetramers and heterodimers bind with indistinguishable affinity.

Bennett (1978) demonstated that the rebinding of spectrin to the cytoplasmic surface of spectrin-depleted inside-out vesicles could be prevented competitively by a soluble 72 000 dalton proteolytic fragment derived from erythrocyte membranes and presumed to come from the then unidentified spectrin binding protein. Bennett and Stenbuck (1979a) identified this by raising antibodies to the 72 000 dalton fragment which they demonstrated were potent at inhibiting spectrin rebinding, and cross-reacted specifically with band 2·1, a reaction which was abolished in the presence of the 72 000 daton fragment. They concluded that band 2·1 was the membrane protein responsible for the binding of spectrin to the cytoplasmic surface of the erythrocyte membrane *in vivo*, and demonstrated that the 72 000 dalton polypeptide was a proteolytic fragment of it. Because of its role in anchoring the cytoskeleton to the phospholipid bilayer, they proposed the name *ankyrin* for band 2·1. Bennett and Stenbuck (1979b) went on to show that a specific 1:1 molecular association also existed between ankyrin and a proportion of the more abundant anion channel molecules (band 3) embedded within the lipid bilayer. These were shown to remain firmly associated with the membrane skeletal proteins, by virtue of their mutual association with ankyrin, in Triton shells from which the majority of band 3 had been extracted. This identification of band 2·1 as the spectrin binding protein was confirmed by peptide mapping by Luna *et al.* (1979) and independently by Yu and Goodman (1979), who also demonstrated that the minor bands 2·2 to 2·6 running just below band 2·1 on polyacrylamide gels (Fig. 2a) are proteolytic fragments of this molecule, forming a family of sequence-related proteins all having the capacity to bind spectrin, for which they proposed the name *syndeins*. Siegel *et al.* (1980) further demonstrated that these minor shorter versions of band 2·1 were present even after the most careful precautions had been taken to prevent proteolysis during the preparation of the erythrocyte membrane proteins, and thus concluded that they exist as such *in vivo*, although their significance is not understood.

The binding of ankyrin to spectrin molecules has been elegantly demonstrated by the low-angle shadowing studies of Tyler *et al.* (1979, 1980b) to occur at two sites both approximately 20 nm from the midpoint of the spectrin tetramer, as shown in Fig. 45, again confirming the symmetrical head-to-head association of spectrin dimer molecules to form such tetramers, and showing that the sites of attachment of spectrin molecules to the integral proteins of the membrane is separated by almost the whole length of the spectrin dimer from

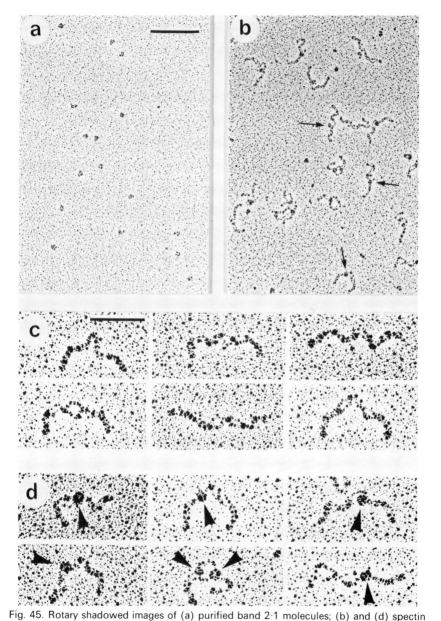

Fig. 45. Rotary shadowed images of (a) purified band 2·1 molecules; (b) and (d) spectin tetramers with band 2·1, showing the binding of individual band 2·1 molecules to one or both of two sites equidistant and approximately 20 nm from the midpoint of the spectrin tetramer (arrows); in comparison with (c) unlabelled spectrin tetramers ((a) and (b) from Tyler et al. (1979). *Proc. Natn. Acad. Sci. USA* **76**, 5192–5196, with permission; (c) and (d) from Tyler et al. (1980b). *J. Biol. Chem.* **255**, 7034–7039, with permission). Magnifications: (a) and (b) ×65 000; bar=200 nm; (c) and (d) ×150 000; bar=100 nm.

its site of interaction with actin and band 4·1. Calvert et al. (1980a) have shown this ankyrin site to be located on band 2 of the spectrin molecule.

Mueller and Morrison (1980) reported evidence that band 4·1 also has the ability to bind specifically to the sialoglycoprotein glycoconnectin (PAS 2), as discussed in Section III above. These additional binding interactions would serve to keep the distal ends of the long spectrin molecules intimately associated with the cytoplasmic surface of the lipid bilayer at the points of its tertiary association with band 4·1 and actin, constraining the skeleton to adopt a two-dimensional conformation closely apposed to the bilayer rather than extending any distance into the cytoplasm, as demonstrated *in vivo* by the frozen-section labelling studies with ferritin-conjugated anti-spectrin of Ziparo et al. (1978) (Fig. 23) discussed in Section IV above.

VI. OTHER MAJOR MEMBRANE PROTEINS

A. Polypeptides of Medium Molecular Weight

The erythrocyte membrane proteins which appear clearly on polyacrylamide gels (Figs 2 and 3, and Table I) but which have not been discussed above are few in number. Those with subunit molecular weights in the range 35 000 to 100 000 daltons are band 4·2 and band 6, plus a number of largely uncharacterized polypeptides running with band 3 and in the band 4·5 region. Band 4·2 and band 6 are both tetrameric peripheral proteins which bind in a saturable fashion to distinct sites on the cytoplasmic N-terminal domain of band 3 (Steck, 1972; Steck and Kant, 1973; Yu and Steck, 1974, 1975b; Kliman and Steck, 1980). The function of band 4·2 is unknown, but Tanner and Gray (1971) recognized from amino-acid sequence homologies that band 6 was glyceraldehyde-3-phosphate dehydrogenase, a glycolytic enzyme also present in solution in the erythrocyte cytoplasm. It has subsequently been shown that band 3 also binds the glycolytic enzymes aldolase and phosphofrucktokinase (Strapazon and Steck, 1977; Richards et al., 1979). While band 6 is present in considerable quantities on erythrocyte ghost membranes prepared by hypotonic lysis (Dodge et al., 1963), it is readily eluted if these membranes are resuspended in isotonic buffers, but Kliman and Steck (1980) have attributed this to excessive dilution and have provided data to suggest that two thirds of the cellular glyceraldehyde-3-phosphate dehydrogenese is bound to the membrane *in vivo*. The physiological significance of this binding remains to be elucidated. Since no electron microscopic studies of these proteins have been published, they will not be discussed further here.

B. Polypeptides of Low Molecular Weight: Torin and Cylindrin

1. Early Studies

In the late 1960s Harris (1968, 1969a) discovered that when bovine or human erythrocyte membranes were dialysed against distilled water or dilute sodium dodecyl sulphate solution and then disrupted by freezing and thawing, large multimeric protein molecules were released from them which could easily be seen by negative staining (Fig. 46). The basic structural unit of these proteins from both species was identical, being a hollow cylinder of diameter approximately 12·5 nm and length approximately 17·0 nm. Seen end-on, these hollow cylinders appeared as rings with the negative stain penetrating the hollow core. When seen on their side, it became clear that these molecules were each composed of four hollow rings stacked one upon another, the outer pair appearing more distinct than the central pair.

In these same studies, single rings or "torus structures" of similar diameter were also seen, and initially it was naturally assumed that these were the

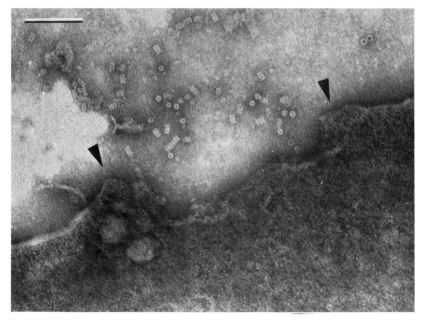

Fig. 46. Cylindrin molecules released onto the electron microscope grid support film from a bovine erythrocyte ghost (lower right of micrograph) in the vicinity of a membrane lesion (between arrowheads). Both side and end views of the molecules are visible. Negatively stained with 2% ammonium molybdate, pH 7·0 (from Harris (1971). *J. Ultrastr. Res.* **36**, 587–594, with permission). Magnification ×160 000; bar=100 nm.

dissociated component rings of the four-ringed hollow cylinder proteins. Spectrin was also extracted in large quantities under the conditions used to liberate these cylinder and torus proteins, and tended to dominate the biochemical and immunogenic properties of such extracts. However, its presence went largely unnoticed as it contributed only "amorphous background material" to the negatively-stained images. A measure of confusion and uncertainty thereby arose in the studies on these proteins by Harris and his colleagues (reviewed by Harris, 1980) and by Howe and Bachi (1978) as to the biochemical composition and location of the cylinders and torus structures. This was finally resolved by Harris and Naeem (1978) and White and Ralston (1979), who for the first time employed SDS polyacrylamide gel electrophoresis to allow specific identification of the extracted polypeptides and the purified proteins. These studies clearly showed not only that the cylinder and the torus proteins were chemically distinct from one another, but also that they were composed of low molecular weight polypeptides, neither being related to either of the two high molecular weight polypeptide subunits of spectrin.

2. Torin

In his review of the purification and properties of these molecules, Harris (1980) proposed the names *torin* and *cylindrin* for the torus structure protein and the hollow cylinder protein, respectively. Torin can be selectively extracted at pH 5·2 from erythrocyte membranes which have first been dialysed against distilled water and frozen and thawed, since at this pH spectrin and cylindrin both sediment with the membrane pellet. It can then be subsequently purified by gel filtration. Upon negatively staining (Fig. 47), the individual torin molecules appear either as faint rings when viewed down the torus axis or as double dots when viewed edge-on in deep stain (Harris, 1969b, 1971a; Harris and Naeem, 1978). The molecule exhibits a sedimentation constant of 9·0S in the analytical ultracentrifuge, runs as a single band on nondenaturing gels and dissociates on SDS polyacrylamide gels to subunits which run as a single band just below the band 8 region, with a molecular weight of approximately 20 000 (Fig. 48; Harris and Naeem, 1978; Harris,

---

Fig. 47. Negatively-stained torin molecules from human erythrocyte ghosts. In face view, the molecules appear as circles. When end on, they appear as bars if embedded in shallow regions of negative stain or as double dots if surrounded by deep stain (indicated by arrowheads in (b), where they are less frequent). Oblique orientations are also present but less easy to distinguish. (a) Stained with 2% sodium phosphotungstate, pH 7·0. (micrograph kindly provided by J. R. Harris). Magnification ×150 000; bar=100 nm. (b) Stained with 2% uranylacetate, pH 4·5, by the flotation technique of Horne and Pasquali Ronchetti (1975) (from Harris (1980). *Nouvelle Revue Française d'Hématologie* **22**, 411–448, with permission). Magnification ×240 000; bar=100 nm.

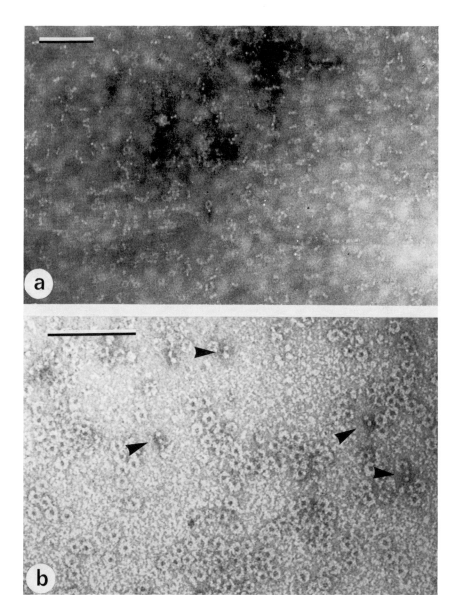

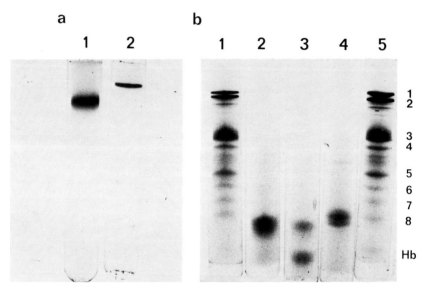

Fig. 48. 7% polyacrylamide gel electrophoretograms. (a) In a 50 mM Tris-HCl, pH 8·0, non-denaturing continuous buffer system, of purified torin (gel 1) and purified cylindrin (gel 2). (b) In a 0·1% SDS, 50 mM Tris-HCl, pH 8·0, denaturing continuous buffer system, of erythrocyte membranes (gels 1 and 5), purified torin (gel 2), torin plus haemoglobin (gel 3) and purified cylindrin (gel 4). The bands from the erythrocyte membranes have been labelled according to the scheme given in Fig. 2a (Fairbanks et al., 1971; Steck, 1974) (from Harris and Naeem (1978). Biochim. Biophys. Acta **537**, 495–500, with permission).

1980). Using a rotational superposition image enhancement technique (Markham et al., 1963), Harris (1969b) claimed that each torus has ten-fold symmetry, implying ten subunits and a molecular weight of about 200 000, but since the superposition patterns for five-, six- and eight-fold rotations were not published, and the more rigorous rotational power spectrum analysis methods of Crowther and Amos (1971) have not been subsequently undertaken (Harris, 1980), this claim must be regarded as unestablished. The protein has no known function.

3. Cylindrin

The various purification schemes developed for cylindrin are reviewed by Harris (1980). It is best prepared by elution, together with spectrin, at pH 8·0 from frozen and thawed erythrocyte membranes which have been dialysed against distilled water, following the prior elution of torin at pH 5·2. The bulk of the spectrin can then be precipitated by the addition of 30 mM $CaCl_2$ and the cylindrin molecules purified by sucrose density gradient centrifugation.

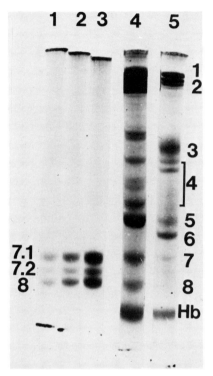

Fig. 49. 5·6% polyacrylamide gel electrophoretograms of increasing amounts of purified cylindrin (gels 1 to 3), together with samples of the crude water-soluble protein extract from ghosts (gel 4) and of erythrocyte membranes (gel 5), in the presence of 1% SDS as described by Fairbanks et al. (1971). The bands from the erythrocyte membranes have been labelled according to the scheme given in Fig. 2a (Fairbanks et al., 1971; Steck 1974), while the three component polypeptides of cylindrin have been labelled 7·1, 7·2 and 8 (from White and Ralston (1979). *Biochim. Biophys. Acta* **554**, 469–478, with permission).

The negatively-stained appearances of human and bovine cylindrin molecules are shown in Fig. 50 (Harris 1968; Harris and Kerr, 1976) in which it can be seen that the bovine molecules have a marked tendency to associate end-to-end. Harris (1971b, 1980) has presented micrographs of these proteins associated with disrupted erythrocyte membranes (Fig. 46) and within stromalytic projections from presumably "aged" bovine erythrocyte ghosts (Fig. 51; compare to Fig. 7), in support of his suggestion that cylindrin normally exists in its assembled form on the cytoplasmic surface of the native erythrocyte membrane. However, these characteristic cylindrical structures have not been seen by other workers after conventional negative staining, thin sectioning or freeze etching of normal erythrocyte ghost membranes which have not been subjected to disruption by dialysis against water or detergents

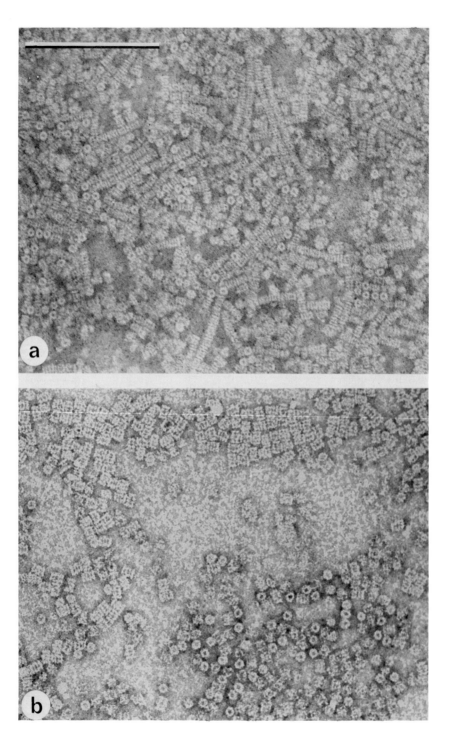

and by freezing and thawing. It is thus uncertain whether these proteins pre-exist as multimeric cylinders, or whether these structures are formed by the self-assembly of the component subunits in response to experimental manipulation. Recently it has been shown (Harris, 1983) that cylindrin isolated from erythrocytes of several mammals has a quaternary conformation very similar, if not identical, to the high molecular weight rabbit reticulocyte aminoacyl-tRNA synthetase complex. Commercially available samples of aminoacyl-tRNA synthetase from rabbit and bovine liver and yeast were also shown to contain a cylindrin-like particle. Preliminary enzymic analysis performed on human erythrocyte cylindrin has shown the presence of aminoacyl-tRNA synthetase activity together with Mg-ATPase and amino acid activated Mg-ATPase (Harris, personal communication). Exactly why this enzyme, involved in protein synthesis in the immature cell, should become associated with the membrane of the mature cell remains unclear. It is possible that some selective retention of high molecular weight cytosolic proteins occurs during the osmotic haemolysis and washing procedure used to prepare haemoglobin-free erythrocyte ghosts.

The exact subunit composition of cylindrin is still uncertain. Harris and Naeem (1978) and Harris (1980) report cylindrin as having a sedimentation coefficient of 22·5S at infinite dilution and a molecular weight of $896\,000 \pm 19\,000$ or $725\,000 \pm 23\,000$ (two analytical ultracentrifuge determinations), to run as a single band on non-denaturing gels and to be composed of two polypeptides running in the band 8 region, with molecular weights just greater than 20 000 (Fig. 48). White and Ralston (1979) found a sedimentation coefficient of 19·3S at approximately 0·5 mg/ml and a molecular weight of $747\,000 \pm 38\,000$ (one analytical ultracentrifuge determination), in general agreement with Harris' findings, and resolved the purified molecule on SDS polyacrylamide gels into three clearly separated polypeptides which they labelled band 7·1, band 7·2 and band 8 (Fig. 49), with apparent molecular weights of 28 000, 25 000 and 22 000, respectively. They further determined the stoichiometry of these bands to be approximately 2:1:2, and suggested the subunit composition of the intact cylindrin molecule to be twelve polypeptides each of band 7·1 and band 8 and six bands 7·2 polypeptides. This would make it highly likely that the entire molecule has

---

Fig. 50. Negatively-stained cylindrin molecules from (a) bovine and (b) human erythrocyte membranes, showing both side and end views. Note that the bovine molecules have a tendency to aggregate end-to-end, while those from human erythrocytes prefer to associate end-to-side, thus frequently forming tetrameric squares. (a) Stained with 2% sodium phosphotungstate, pH 7·0 (this micrograph, kindly supplied by J. R. Harris, is similar to Fig. 4 of Harris, 1968). (b) Stained with 2% uranyl acetate, pH 4·5, by the flotation method of Horne and Pasquali Ronchetti (1975) (from Harris and Kerr (1976). *J. Microscopy* **108**, 51–59. with permission). Magnifications ×183 000; bar=200 nm.

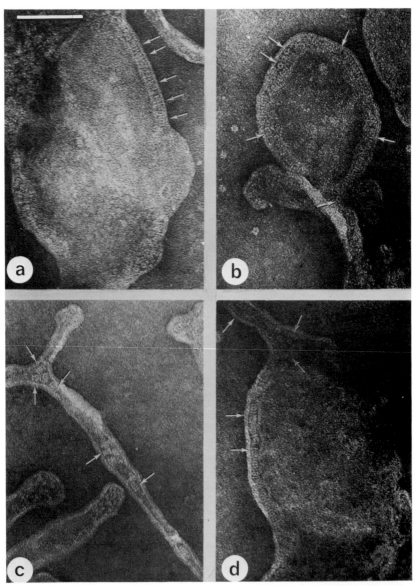

Fig. 51. (a–d) Selected regions of bovine erythrocyte ghosts which show stromalytic projections. Arrows indicate internal cylindrin molecules. In (b) several free molecules can also be seen. Negatively stained with 2% ammonium molybdate, pH 7·0 (from Harris (1971). *J. Ultrastr. Res.* **36,** 587–594, with permission). Magnification ×180 000; bar=100 nm.

six-fold rotational symmetry when viewed down the cylindrical axis. However, the gel electrophoresis analyses of Harris and Naeem (1981) and Malech and Marchesi (1981) suggest that cylindrin may contain five polypeptides in the molecular weight range 21 000 to 29 000. From the results of rotational superposition image enhancement analysis (Markham et al., 1963), of which the results for seven-, eight- and nine-fold rotations, but not for six-fold rotation, were published, Harris (1980) claimed that cylindrin has eight-fold symmetry. This analysis is open to the same reservations as that of torin, discussed above, and the symmetry of this protein molecule must thus be regarded as presently unestablished.

White and Ralston (1979) and Ralston (personal communication) have also identified several other polypeptides running in the region of the torin and cylindrin polypeptides on SDS polyacrylamide gels (i.e., in the band 7–band 8 region; see Table I footnotes) of which little is known and no electron microscopic studies undertaken.

VII. THE ERYTHROCYTE MEMBRANE IN DISEASE

A. Hereditary Spherocytosis, Hereditary Elliptocytosis and Hereditary Pyropoikilocytosis

Hereditary spherocytosis (HS) is an inherited disorder in which the circulating erythrocytes are osmotically fragile and exhibit characteristic partially spherical shapes which resemble those reversibly induced by metabolic depletion in healthy erythrocytes. The cells lack the ability to pass easily through the small splenic fenestrations which normally serve to selectively retain aged erythrocytes prior to their removal from the circulation, and consequently the spleen becomes greatly enlarged in HS patients who suffer severe anaemia and consequent jaundice (Dacie, 1960). Surgical removal of the spleen totally alleviates the anaemia, and splenectomized patients survive remarkably well with the spherocytotic erythrocytes in their circulation. Among persons of English or nothern European descent, HS is the most common of the hereditary haemolytic anaemias, affecting one out of every 5 000 individuals (Young, 1955). The disease is normally inherited as an autosomal dominant trait, although many patients appear to be spontaneous mutants since they come from otherwise unaffected families (Young, 1955). While the severity of the disease in the heterozygous individual varies enormously, there are no confirmed reports of homozygotes, suggesting that this condition may be lethal (Dacie, 1960).

Hereditary elliptocytosis (HE) is another autosomal dominant trait which resembles HS in many respects, except that the diseased erythrocytes tend to be elliptical rather than spherical in shape (Dacie, 1960). More recently a third type of disorder, hereditary pyropoikilocytosis (HPP), has also been recognized (Zarkowsky et al., 1975), manifesting itself both by the presence of many very small erythrocytes (microspherocytes) and of angular cells (poikilocytes) in the peripheral circulation and also by an abnormally low temperature (45 °C, as opposed to 49 °C for normal cells) at which the erythrocytes exhibit thermally-induced vesiculation.

The defects in all three of these haemolytic anaemis seem to involve abnormal interactions between spectrin and other protein components of the membrane skeleton, although the early literature on this subject (reviewed by Lux, 1979b; Lux and Glader, 1980) is incomplete, confused and often contradictory, and detailed biochemical characterizations of these changes has not yet been achieved. The fact that the diseases are apparently inherited as dominant traits suggests strongly that the lesions are located on structural proteins rather than enzymes, since in the latter case one would expect each disease to be recessive, with normal enzyme molecules produced by the normal parental gene masking the expression of the defective ones. If, however, the defect is in a membrane skeletal protein which could be incorporated into macromolecular assemblies but which is either unable to function normally or which results in incomplete assembly of the membrane skeleton, one would expect dominant inheritance, since abnormal skeletons would occur even in the presence of the normal protein. Furthermore, the fact that the membranes are abnormally fragile and prone to fragmentation suggests a loss of integrity in some structural element, particularly as it is well known that normal erythrocyte membranes undergo intense fragmentation and vesiculation following the extraction at low ionic strength of spectrin and actin (see above).

Direct evidence for membrane skeletal involvement in HS has been provided by Sheehy and Ralston (1978) who found that in two HS cases spectrin was completely unextractable, a condition which could be mimicked in normal cells by warming to 50 °C, at which temperature spectrin is heat denatured, by Agre et al. (1982) who found spectrin deficiency in a rare case of severe hereditary spherocytosis, and by Goodman et al. (1982) and Wolfe et al. (1982) who have demonstrated defective spectrin molecules, unable to bind band 4·1, in certain HS families. Spectrin abnormalities have also been detected in HE and HPP. Liu et al. (1981, 1982) and Palek et al. (1982) have found defects in the normal spectrin dimer-dimer asociation to form stable tetramers in both HE and HPP patients, while Dhermy et al. (1982) and Lawler et al. (1982) have demonstrated the presence of abnormal spectrin band 2 polypeptides in similar cases in which the ability of the spectrin molecules to form stable $\alpha_2\beta_2$ tetramers was impaired. Furthermore Lawler et al. (1982) were able to correlate this defective function in HPP with a decreased amount, in partial tryptic digests, of the 80 kDa band 2 peptide shown by Morrow *et al.* (1980) to be specifically involved in dimer-dimer association, indicating that the disease had caused a conformational change in this area of the polypeptide, while Dhermy *et al.* (1982) showed the spectrin from their HE cases to lack the normal band 2 phosphorylation. In separate studies, Mueller and Morrison (1981) have reported a patient completely lacking band 4·1 and glycoconnectin (PAS 2), and showing very low retention of spectrin in membrane skeletons prepared by Triton X-100 extraction of these erythrocyte membranes, and Tchernia *et al.* (1981) report a family in which three siblings homozygous for HE have erythrocytes which completely lack band 4·1 (although with normal levels of glycoconnectin), while heterozygotes in this family have 50% of the normal levels of band 4·1.

Little ultrastructural work has been done on HS erythrocytes. A limited study of one family of HS sufferers (the father and all three children) and of four other unrelated patients (Burke, 1981; Burke and Shotton, 1983) has given evidence, from spectrin characterization and particle aggregation studies, for a variety of defects of membrane skeletal organization in seven out of the eight cases. These included an unusually high proportion of spectrin dimers in a 0 °C extract from one patient's erythrocyte ghosts under conditions which yield spectrin tetramers from normal erythrocytes, suggesting defective spectrin dimer-dimer association, and an increased degree of particle aggregation seen after freeze-fracture of HS ghosts from two other unrelated patients which had been subjected to standard pretreatment and aggregation conditions (Elgsaeter and Branton, 1974; Elgsaeter *et al.*, 1976). The ghosts from one of these patients showed the additional unusual feature that, after aggregation, a large proportion of the intramembrane particles were *not* associated with the aggregates (Fig. 52). In addition the family of four HS patients all showed greatly *decreased* spectrin and actin extractability under mild pretreatment conditions and yet slightly *increased* particle aggregation when these pretreated ghosts were subjected to a standard aggregation assay (micrographs not shown), which was unexpected since Elgsaeter *et al.* (1976) showed that particle aggregation in normal erythrocytes depended upon the prior removal of these proteins.

In view of the extreme variability of these ultrastructural findings and of the biochemical results reported above, one must conclude that the case for a single molecular defect resulting in each of these diseases is no longer tenable. Changes in band 2 and band 4·1 have already been demonstrated to give rise to similar symptoms, and it is not difficult to imagine that malfunction of any one of the membrane skeletal proteins could produce similar overall effects. It is therefore likely that there may be several distinct molecular lesions of the erythrocyte membrane skeleton which give rise to this group of haemolytic anaemias, and that the titles "hereditary spherocytosis", "hereditary ellipto-

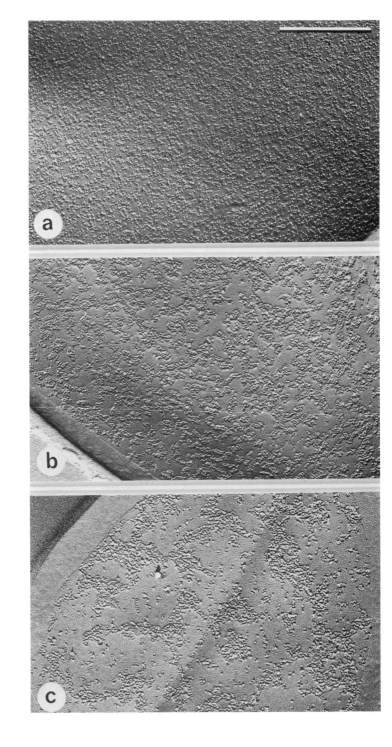

cytosis" and "hereditary pyropoikilocytosis" refer not to fundamentally distinct diseases but rather to different clinical expressions of this general class of erythrocyte membrane skeleton defects.

B. Spectrin-deficient Mutants in Mice

The best experimental demonstration of the importance of the membrane skeleton in determining the viscoelastic properties of the erythrocyte membrane comes from studies on spectrin-deficient recessive mutants of the common house mouse *Mus musculus*, which exhibit severe haemolytic anaemia. Greenquist *et al.* (1978) have studied one of these, "spherocytosis" (*sph/sph*), while Lux (1979b) reported on work that he and his colleagues have done on this and three similar mutants, "haemolytic anaemia" (*ha/ha*), "normoblastic" (*nb/nb*) and "jaundiced" (*ja/ja*). These rank in order of clinical severity $ja/ja > sph/sph > ha/ha > nb/nb$, with reticulocyte counts between 90% and 100% for the first three and between 50% and 60% for *nb/nb*. The erythrocytes show intense spherocytosis and budding *in vivo*, while ghosts freshly prepared from them undergo spontaneous vesiculation similar to that seen after spectrin depletion of normal erythrocyte ghosts, indicating that these cells have very unstable membranes. Analysis of their protein composition by SDS polyacrylamide gel electrophoresis shows that they lack spectrin to varying degrees in proportion to the clinical severity of the anaemia, *ja/ja* ghosts having no detectable spectrin, *sph/sph* ghosts lacking spectrin band 1 but having a trace of band 2, *ha/ha* ghosts having 25%–30% of the normal amount of spectrin and *nb/nb* ghosts having between 45% and 50% of the normal amount. Sheetz *et al.* (1980) and Koppel *et al.* (1981) also claim *sph/sph* ghosts to be deficient in the other components of the membrane skeleton, namely ankyrin, actin, band 4·1 and band 4·9, and conclude from fluorescence recovery measurements after photobleaching that the integral proteins diffuse about 50 times faster in the *sph/sph* than in the normal mouse erythrocytes. While the causes of these protein deficiencies are unknown, these data taken together highlight the central role of spectrin in the structural organization of the erythrocyte membrane skeleton, and of that skeleton in

Fig. 52. Comparison of the degree of IMP$_P$ aggregation induced by a standard pH 5·5 aggregation assay (Elgsaeter and Branton, 1974) between freeze-etched erythrocyte ghost membranes from a healthy donor and a patient with hereditary spherocytosis (HS). (a) Normal fresh ghosts. (b) Normal ghosts which have been pretreated (by incubation at 37 °C for 18 h in 20 mOsm sodium phosphate buffer, pH 7·6) and the incubated on ice for 30 min in 20 mOsm sodium phosphate buffer, pH 5·5, prior to freezing. (c) HS ghosts treated as in (b), showing increased IMP$_P$ aggregation and also the unusual feature of a large number of particles *not* associated with any aggregate (from Burke and Shotton (1983). *Br. J. Haematol.* **54**, 173–187, with permission). All magnifications ×50 000; bar=500 nm.

the restriction of translational mobility of the integral membrane proteins. This has been most clearly demonstrated by the reconstitution studies of Shohet (1979), who inserted normal spectrin by exchange haemolysis into spectrin-deficient erythrocytes from the *sph/sph* mouse, and found that this resulted in enhanced membrane stability and prevented the abnormal spontaneous membrane fusion to which the *sph/sph* erythrocytes are prone. Furthermore Williamson *et al.* (1982) have shown the importance of spectrin in maintaining phospholipid asymmetry in the normal membrane, by demonstrating that in *sph/sph* mice, in which spectrin was virtually absent, such lipid asymmetry was largely lost.

C. Glycophorin-deficient Mutants in Man

In certain rare cases, characterized by the absence of the En^a human erythrocyte antigen and denoted En(a-), glycophorin A is totally absent from the erythrocyte membrane (Gahmberg *et al.*, 1976; Tanner and Anstee, 1976; Furthmayer, 1978b). Surprisingly, the red cells of these homozygous individuals show no clinical abnormality, although associated changes are found in the degree of glycosylation of band 3 (Gahmberg 1976; Tanner *et al.*, 1976). Nigg *et al.* (1980a, b) have recently demonstrated an *in vivo* association between glycophorin A and band 3, but find no significant difference in the rotational diffusion of band 3 between En(a-) and normal erythrocyte membranes, and therefore conclude that glycophorin can contribute no more than 15% to the cross-sectional diameter of the band 3-glycophorin A complex in the plane of the normal membrane. This view is supported by results from freeze-fracture studies by Bächi *et al.* (1977) and Ghamberg *et al.* (1978) who showed that the shape and distribution of intramembrane particles are unaltered in En(a-) cells when compared to normal erythrocytes, and concluded that glycophorin A is not a quantitatively important constituent of the P-face intramembrane particles. This is entirely consistent with what is known of the conformations of the transmembrane hydrophobic domains of those two integral proteins (see Section III above) and with the finding of Edwards *et al.* (1979) that while band 3 partitions with the P half of the fractured bilayer, glycophorin A molecules may be covalently cleaved by the fracture process or partition as intact monomers with the E half.

D. Sickle Cell Anaemia

The disease of sickle cell anaemia is caused by a recessive single-site mutation in the structural gene for the haemoglobin β chain, resulting in the incorporation of a valine rather than a glutamic acid residue in position 6 of the polypeptide. The initial elongation ("sickling") of the erythrocytes is the passive result of the intracellular precipitation of the mutant haemoglobin S under conditions of low oxygen tension. While normally reversible, this shape change can become irreversible if the cells are allowed to remain for long in the sickled configuration. Palek (1977) has reviewed the evidence which suggests that the defect responsible for the permanent shape change of irreversibly sickled cells (ISC) is localized in the cell membrane. Lux *et al.* (1976) demonstrated that the sickled shape is retained when ghosts of the irreversibly sickled cells are converted to Triton shells by Triton X-100 extraction of the lipids and integral proteins (Fig. 53). However, the appearance of the negatively-stained membrane skeleton of reversibly and irreversibly sickled cells (Fig. 53a and b) appears essentially identical to that of normal healthy erythrocytes (Figs 25 and 26).

It has been suggested that such irreversible sickling, which is accompanied by a rise in internal calcium levels and a fall in ATP concentration, is due to the calcium-stimulated crosslinking of lysine and glutamine side chains of adjacent proteins by the membrane-bound enzyme transglutaminase, which can be brought about in normal cells by artificially increasing membrane permeability to extracellular calcium ions using the inophore A23187 (Lorand *et al.*, 1975, 1976, 1979; Siefring *et al.*, 1978). However, Palek and Liu (1979a, b) found that high molecular weight polymers of spectrin did *not* exist in the untreated membranes of irreversibly sickled cells, and were not formed when these membranes were subjected to catalytic oxidation in the presence of copper sulphate and *o*-phenanthroline or exposed to glutaraldehyde, while ATP-depleted normal cells formed polymers in both cases. They thus concluded that it was unlikely that spectrin crosslinking contributed to the permanent shape fixation of irreversibly sickled cells, presumably because the rise in calcium concentration and the ATP depletion in them did not reach the levels required for transglutaminase activity. The molecular basis of the membrane skeleton change which leads to irreversible sickling is thus at present unestablished.

E. Muscular Dystrophy

1. Introduction

There are, in man, several diseases of the skeletal musculature termed muscular dystrophies which are characterized by the progressive weakening and atrophy of the muscle tissue, with a severity and distribution in the body which is characteristic for each disease. The most severe is Duchenne muscular dystrophy, which is caused by a recessive mutation on the X chromosome and hence is usually expressed only in males, with an incidence of about 1 in every 3 000 live male births, heterozygous females acting as unaffected carriers. The

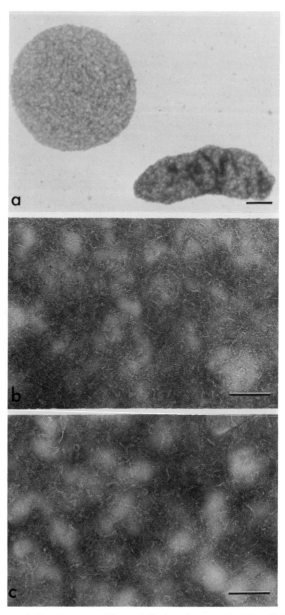

Fig. 53. Membrane skeletons prepared from sickled erythrocytes by Triton X-100 extraction, negatively stained with 1% ammonium molybdate, pH 7·4. A reversibly sickled cell (left) and an irreversibly sickled cell (right) are shown at low power in (a), the latter having retained its sickled shape through the procedures of hypotonic lysis and Triton X-100 extraction. Higher magnification views of the membrane skeleton of (b) a reversibly sickled cell and (c) an irreversibly sickled cell show them to be similar amorphous meshworks of filamentous material (compare with Figs 25 and 36) (from Lux et al. (1976). J. Clin. Invest. **58**, 955–963, with permission). (a) Magnification ×7 000; bar=1 μm. (b) And (c) magnifications ×110 000; bar=100 nm.

rare heterozygous female exceptions who do exhibit typical Duchenne symptoms all show X-autosomal chromosome translocations and selective inactivation of the normal X-chromosome, allowing expression of the recessive Duchenne condition. The first symptoms usually appear in the first five years of life, and the disease progresses rapidly, affecting primarily the proximal fast twitch muscle of the limbs (i.e. those nearest the torso), with later cardiac involvement and a small and variable degree of intellectual impairment, wheelchair immobilization in relatively few years and death, usually caused by respiratory or cardiac complications, in early adulthood. other common human dystrophic condition is myotonic muscular dystrophy, which is an autosomal dominant trait expressed in both male and female heterozygotes as a slow progressive dystrophy of the distal musculature, characterized by a prolonged muscle contraction (myotonia) upon stimulation, which is accompanied by a variety of other serious symptoms including cardiac abnormalities, cataracts, endocrine disturbances, testicular atrophy and mental retardation (Dubowitz, 1979). At present there is no treatment to cure or arrest the progress of these diseases, although several drug therapy trials on dystrophic chickens, which form a good animal model for Duchenne dystrophy, have recently proved encouraging (Barnard and Barnard, 1978, 1979, 1980; Stracher et al., 1978; Hudecki et al., 1979; Park et al., 1979).

While the molecular causes of the various human dystrophies are not known, it has been suggested by several workers that at least the Duchenne and myotonic muscular dystrophies may involve general membrane defects affecting many cell types, the phenotypic expression of which happens to be particularly severe in fast skeletal muscle fibres (see reviews by Lucy, 1980; Shotton, 1982). In support of this proposal, a host of chemical, biophysical and electron microscopic changes in the membranes of circulating dystrophic erythrocytes or their ghosts have been described. Unfortunately these data are incomplete and provide a far from coherent story, since the results of different workers using dissimilar techniques often do not permit strict comparisons.

2. Biochemical and Biophysical Differences

Despite early reports to the contrary, certain membrane characteristics appear not be affected. Thus Plishker and Appel (1980) have reviewed the many conflicting observations of lipid abnormalities in dystrophic erythrocyte membranes in the light of their own findings, and conclude that myotonic and Duchenne muscular dystrophies are not associated with any major change in lipids, a conclusion supported by the work of Godin et al. (1978) and McLaughlin and Engel (1979). Similarly, no important differences have been demonstrated in the polypeptide profiles of dystrophic erythrocyte membranes after polyacrylamide gel electrophoresis. In a separate more recent study, Mawatari et al. (1981) have reviewed and reinvestigated the many contradictory reports of changes in the Na^+ and K^+-ATPase activity and the effects of ouabain on this enzyme from the erythrocyte membranes of Duchenne patients, and have concluded that no significant difference exists between these and normal controls.

However, both Nagano et al. (1980) and Tsuchiya et al. (1981) report a significant decrease in spectrin extractability from ghosts of Duchenne patients, and in a series of papers, Roses and Appel and their collaborators reported clear changes in membrane protein kinase activity in erythrocyte ghosts from patients with myotonic and Duchenne dystrophy (Roses and Appel, 1973b, 1975, 1976; Roses et al., 1975, 1976), it being lower in myotonic muscular dystrophy, leading to a decreased level of phosphorylation of band 3, but higher in Duchenne dystrophy, resulting in the increased phosphorylation both of band 3 and also of band 2 of spectrin, in the latter case the increased phosphorylation being localized to one particular cyanogen bromide peptide (Roses et al., 1980). This latter effect was also noted in the mothers of affected sons and was proposed as a novel non-invasive method for carrier detection in Duchenne dystrophy (Roses et al., 1976). Unfortunately, others have obtained different results. Thus, in studying the phosphorylation of erythrocyte membranes from patients with myotonic dystrophy, Iyer et al. (1977) observed an increase in total phosphorylation, while Vickers et al. (1979) found no difference in the rate of phosphorylation of band 3 or spectrin at 37 °C, but observed that the protein kinase from these patients showed a decreased temperature response, probably caused by lipid changes in the myotonic membranes. In similar investigations of erythrocyte membranes from Duchenne patients, both Iyer et al. (1977) and Tortolero et al. (1979) found no differences in the endogenous phosphorylation of spectrin and band 3. Each group has attributed these discrepancies in findings between laboratories to different assay conditions.

Various observations have also been made which relate to the deformability and fluidity of the erythrocyte membrane in the dystrophies. Plishker et al. (1978) found that erythrocytes from patients with myotonic dystrophy accumulate calcium ions by passive influx at a significantly higher rate in vitro than did normal cells, but appear to have a normal calcium pump for active outwards transport of these ions. No observable differences were noted in Duchenne dystrophy (Plishker and Appel, 1980). Kim et al. (1980) found that the osmotic fragility of Duchenne erythrocytes was increased, even in young cells, suggesting a membrane abnormality in early stages of red cell maturation.

Using an assay of deformability based upon the aspiration of erythrocytes into very small (1 to 2 μm) micropipettes, Percy and Miller (1975) found those of Duchenne patients and carriers significantly less deformable than controls.

Butterfield et al. (1974, 1976) employed various electron spin resonance probes to measure membrane lipid fluidity, and found evidence for a greater fluidity in the erythrocyte membranes from patients with myotonic dystrophy and congenital myotonic (a rare disease characterized by myotonia without dystrophy of the muscle tissue), but could detect no differences between those from Duchenne patients and normal controls. In discussing this work, Butterfield (1977) suggests that the increased erythrocyte membrane lipid fluidity is related to the myotonic aspects of myotonic dystrophy, while the changes in membrane proteins noted above may be more specifically related to the dystrophy of the muscle. More recently Sato *et al.* (1978) have presented further esr data to suggest that while in Duchenne dystrophy the fluidity of the polar regions of the erythrocyte membrane bilayer is unchanged, esr probes in the non-polar regions do show increased fluidity and altered temperature and pH responses which distinguish them from normal controls.

Recent comprehensive reviews of the often contradictory reports of biochemical changes in the erythrocyte membranes of patients with Duchenne muscular dystrophy have been given by Rowland (1980) and Mawatari (1980), to which readers are referred for further information.

3. Scanning Electron Microscopic Studies

The first report of scanning electron micrographs of erythrocytes from Duchenne patients and carriers by Matheson and Howland, in 1974, excited great hopes that this technique could be used to supplement conventional serum creatine kinase enzyme assays for the non-invasive detection of carriers. They reported that Duchenne erythrocytes, prepared for scanning, after initial saline washes, by glutaraldehyde fixation, dehydration and critical point or air drying, appeared quite distinct from those of normal controls, have a far higher proportion of cells (between 21% and 98% in the four patients tested) which had adopted a crenated echinocytic morphology, compared with less than 8% in controls, while the three carriers tested also showed elevated proportions (34% to 40%) of echinocytes. Coming hard on the heels of an earlier report (Morse and Howland, 1973) that erythrocytes from dystrophic mice exhibited abnormal surface distortions, their paper prompted similar studies elsewhere. Lumb and Emery (1975) and Howells (1976) also reported increased proportions of echinocytes, although the differences in their hands were much less marked than those of Matheson and Howland. However Miale *et al.* (1975) found no significant differences in the appearance of erythrocytes from all the carriers and all but one of the Duchenne patients tested, and Roses and Appel (1974) and Miller (1976) found that erythrocytes from both Duchenne and myotonic patients showed increased numbers of stomatocytes rather than echinocytes, to a degree proportional to the pretreatment and with one false positive control, and so advised caution on reliance upon scanning electron microscopic results for diagnostic purposes. Then, in a thorough reinvestigation of the original findings using a larger number of subjects, Matheson *et al.* (1976) failed to confirm any distinct differences in appearance between normal and Duchenne erythrocytes, found that the morphology of the cells from any source was very sensitive to various washing and fixation regimes, and stated that morphology in the scanning electron microscope should therefore *not* be used as a diagnostic test. These investigations thus highlight both the delicate balance between the metabolic state of the membrane skeleton which determines cell morphology and the experimental manipulations to which the cell is subjected, and also the problems of maintaining cell shape during the unsupported drying and shrinkage experienced by scanning electron microscopic specimens, but leave open the possibility that appropriate experimental manipulations may in fact accentuate subtle differences between the membranes of normal and dystrophic erythrocytes (Iwasaki, 1980).

4. Freeze-fracture Morphology

In order to obtain a more intimate view of the possible membrane changes accompanying muscular dystrophy, the diseased erythrocytes have also been examined by the freeze-fracture technique, in parallel with similar studies in the affected muscles. Studies on animal models have shown striking decreases in the number of intramembrane particles on the fracture faces of dystrophic erythrocytes, both from mice (Shivers and Atkinson, 1979) and chickens (Shafiq *et al.*, 1976), in comparison with healthy controls. A similar analysis of fixed and glycerinated erythrocyte membranes from human Duchenne patients by Wakayama *et al.* (1978, 1979) also showed a significant reduction in intramembrane particle density, but no significant differences were noted for myotonic dystrophy nor other neuromuscular disorders.

While these results, summarized in Table II, are much more clear cut than those concerning the particle densities on the dystrophic muscle sarcolemmal fracture faces themselves, reviewed by Shotton (1982), it should be noted that the absolute particle densities recorded for the normal control erythrocytes in the above studies differ significantly from those reported by Weinstein *et al.* (1980) ($IMP_P = 2430 \pm 220$ per μm^2; see Section II above), highlighting the counting problems caused by variability in replica quality and subjective criteria of image interpretation and particle recognition to which such studies are prone.

5. Conclusion

While these erythrocyte results support suggestions of widespread basic membrane abnormalities in muscular dystrophy, it must be borne in mind that

TABLE II
The Effect of Muscular Dystrophy Upon Erythrocyte Intramembrane Particle Densities

Reference	Species	n	Normal Particle densities[a]		Dystrophic Particle densities[a]			
			PF	EF	n	PF	EF	% of control
Shaifiq et al. (1976)[b]	Chicken	2	3170±430	690±60	2	2140±610	400±80	58%
Shivers & Atkinson (1979)[b]	Mouse (Bar Harbor REJ 129)		3349±21	—		2228±12	—	68% 67%
Wakayama et al. (1979)[c]	Human: Duchenne	8	1813±17	98±2	8	1589±37	68±8	82%
	Myotonic	11	1793±17	99±1	10	1758±20	98±2	N.S. 69% N.S.
Weinstein et al. (1980)	Human		2430±220					

[a] All particle densities are per square micrometer.
[b] Values given are means ± standard deviation.
[c] Values given are means ± standard error of the mean.
n = Number of individuals studied (several replica fields from intact erythrocytes of each individual were counted).
N.S. = not significantly different.

such membrane abnormalities have not yet been unequivocably demonstrated in any other non-excitable tissue except fibroblasts (Jones and Witkowski, 1981, 1983). Furthermore it should be realized that the changes observed in the circulating erythrocyte membranes of dystrophic patients may be entirely secondary phenomena unrelated to the true molecular basis of the disease, resulting from their chronic exposure to dystrophic plasma which contains abnormally high levels of enzymes leaking from the damaged muscles. These are known to include creatine kinase and pyruvate kinase, which may be elevated even in apparently unaffected Duchenne carriers, but may also include calcium-activated neutral proteases and other enzymes of muscle origin for which specific assays have not been made.

However, while the suggestion of widespread membrane defects in the muscular dystrophies must at this stage still be treated as an unproven hypothesis, the evidence for changes in the plasma membrane of the dystrophic muscle fibres themselves is much firmer. Of particular interest and relevance to the present discussion is our recent finding (Appleyard *et al.*, 1983a, b) that monoclonal antibodies to band 2 of human erythrocyte spectrin cross-react with a spectrin-like molecule on the cytoplasmic surface of the sarcolemma of both normal and Duchenne dystrophic human muscle fibres, the immunochemical staining being stronger on the dystrophic fibres than the normal ones, indicating that the muscle spectrin molecules are either more abundant or more accessible to antibody labelling in the diseased state. The ultrastructural distribution and functional role of these molecules is under current investigation.

VIII. CONCLUSION: THE ERYTHROCYTE MEMBRANE AS A PARADIGM

Our present knowledge of the properties and molecular conformations of the major integral and peripheral proteins which comprise the erythrocyte membrane, and of the dynamic interactions which occur between them, exceeds that for any other multifunctional multicomponent membrane. As I have shown above, an enormous amount of this evidence has been derived from a wide variety of electron microscopic studies, both of the membrane itself and of its component proteins *in vitro*. In Fig. 54, Brian Burke and I have attempted to show the locations, shapes, sizes and interactions of these various major proteins of the erythrocyte membrane, in which the individual molecules are diagrammatically represented as nearly as possible to scale. As such it differs from many published diagrams of the erythrocyte membrane, in which the circa 50:1 ratio of size between the length of the spectrin tetramer

(194 nm) and the thickness of the lipid bilayer (approximately 4 nm) have usually been seriously under represented.

It is clear that the erythrocyte membrane is highly specialized, and that the fluid mosaic model of membrane structure, as usually presented, is an inadequate description of it. In particular, the membrane is a laminated structure, having an elaborate membrane skeleton of considerable thickness on the cytoplasmic surface of the lipid bilayer, which limits the translational mobility of the integral proteins, and which gives the membrane its extraordinary visco-elastic properties and mechanical strength. This membrane skeleton contains short actin protofilaments and the large fibrous tetrameric divalent actin-binding protein spectrin. There are undoubtedly features of the construction of the erythrocyte membrane which are of relevance to the structure of cell membranes in general, and as such the erythrocyte membrane can serve us as a paradigm or model.

In particular, a class of proteins immunologically similar to erythrocyte spectrin have recently been detected on the cytoplasmic surface of the plasma membrane in a wide variety of tissues including neurones, Schwann cells, lens cells, skeletal and cardiac muscle fibres and endothelial and epithelial cells and in many cultured cell types. Detailed studies of these non-erythroid spectrins (reviewed by Lazarides and Nelson, 1982) have shown them to be high molecular weight $\alpha_2\beta_2$ tetramers which resemble erythrocyte spectrin by exhibiting similar elongated molecular structures, being capable of crosslinking F-actin and inducing the polymerization of G-actin, and sharing antigenic sites and calmodulin binding properties on a 240 kDa polypeptide with band 1 of erythrocyte spectrin. In contrast, the muscle spectrin present in human striated muscle fibres is recognized only by monoclonal antibodies specific for band 2 of human erythrocyte spectrin (Appleyard *et al.*, 1983a, b).

Since this field of research is still in its infancy, such comparisons between erythrocyte spectrin and these non-erythroid homologues are at present far from complete. They have, however, already clearly shown that non-erythroid spectrin molecules exist in many (but not all) tissues, where they are associated with the plasma membrane and may be expected to play similar roles as erythrocyte spectrin. This likelihood has been strengthened by the reports of Bennett (1979) and Bennett and Davis (1981) and by Cohen *et al.* (1982) of proteins in non-erythroid cells which cross-react with antibodies to erythrocyte ankyrin and to band 4·1, respectively.

Thus we may conclude that the interactions between the proteins of the erythrocyte membrane skeleton are not unique, but rather are a highly specialized version of the protein-protein interactions to be found in many other cell types, where they play a key role in linking cytoskeletal elements to the plasma membrane and in limiting translational mobility of components within that membrane. In addition, Tyler *et al.* (1980a) have demonstrated

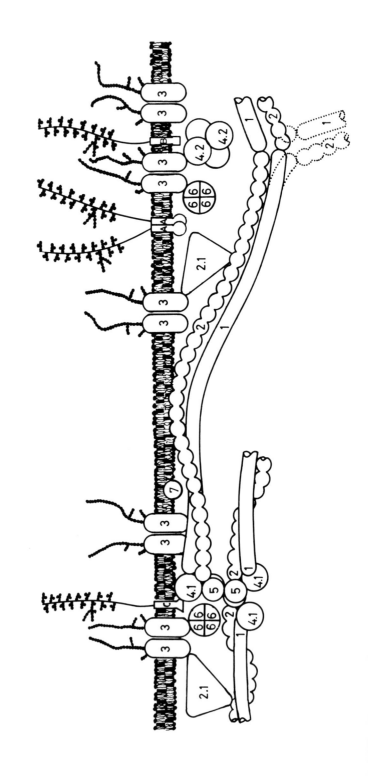

Fig. 54. A model of the erythrocyte membrane in transverse section, giving a highly simplified diagrammatic representation of the molecular interactions known to occur with it. The individual lipid, oligosaccharide and protein components are drawn approximately to scale, their relative sizes being calculated on the assumption that the G-actin monomer is a 5 nm diameter sphere with a molecular weight of 43 000 daltons. The scale approximately 1·2 mm=1 nm. The lipid bilayer is represented by double-tailed black circles for the phospholipids and black rectangles for the cholesterol molecules, in approximately the correct molar ratio. Similarly, the extracellular oligosaccharide chains on the glycoproteins are represented by strings of black circles, but no attempt has been made to differentiate between the individual lipid or sugar molecules nor to represent the glycolipids. The arrangement of the oligosaccharides on the glycophorins is according to the data of Thomas and Winzler (1969) and Yoshima et al. (1980), while the more complex oligosaccharides on band 3 (Tsuji et al., 1981b) are represented schematically. The protein nomenclature used follows that of Steck (1974), as in Fig. 2 and Table I. The various polypeptides have been given distinctive shapes solely for clarity of representation. The overall shape of band 2·1 is that described by Tyler et al. (1979), but its linear dimensions have been increased by approximately 20% over their published values in order to preserve the scale of this molecule in the model with respect to its mass. 7 represents the integral protein that electrophoreses in the band 7 region. The diagram is highly simplified, with the peripheral proteins band 4·9, torin and cylindrin, and also the polypeptides 2·2–2·6 which are related to band 2·1, being omitted, since their association with the membrane are not yet clear. A, B, and C denote glycophorin A, glycophorin B and glycophorin C (glycoconnectin), respectively. The diameter of their transmembrane α-helical segments has been exaggerated about four times in the model, while their extracellular N-terminal domains have been drawn as extended segments of polypeptide chain, as aids to clarity. Band 3 is represented as a dimer, despite recent evidence that it may be tetrameric (Weinstein et al., 1980). Every protein-lipid and protein-protein contact in this model represents a specific molecular association for which evidence has been discussed or referenced above. In order to correctly represent the size of the very large spectrin molecules in relationship to the phospholipid bilayer, it has been possible to include only one complete spectrin heterodimer in the diagram. At the point of its interaction with another heterodimer to form a spectrin tetramer (lower right of diagram), an alternative possible bonding arrangement involving a third spectrin heterodimer to form a spectrin hexamer is shown dotted. While band 1 and band 2 have been drawn as distinct from one another, no attempt has been made to represent the structural domains into which each of the polypeptides is known to fold. Band 2 is shown as interacting with the lipid bilayer as a reminder that spectrin does contain hydrophobic sequences and associates specifically with the anionic phospholipids, as discussed in Section V. The number of band 5 subunits shown in the protofilament is the minimum required to demonstrate its role as a polyvalent attachment site for spectrin molecules, although it is known that in vivo such protofilaments are longer (Pinder et al., 1981). It should be noted that the majority of integral membrane proteins are not assigned any specific association with the membrane skeletal proteins, the relative numbers of these two protein classes being approximately that found in vivo. For clarity, the association known to occur between glycophorin A and band 3 (Nigg et al., 1980a) is not shown.

that spectrin has certain structural similarities with mammalian pulmonary macrophage actin-binding protein (ABP) and chicken gizzard filamin, both of which are large fibrous divalent homodimers capable of crosslinking actin microfilaments by binding end-on to F-actin at the distal ends of their monomer strands in manner very reminiscent of spectrin binding, but which do not involve binding of actin to membranes (Hartwig and Stossel, 1981). It seems likely that the different structural organization of the homodimeric ABP and filamin molecules on the one hand and of the more complex spectrin tetramers on the other reflects the additional membrane-associated binding functions which are specific to spectrin.

Lessons can also be drawn from comparisons of the principal erythrocyte integral membrane proteins. The sialoglycoproteins and the anion protein are transmembrane integral proteins with radically different molecular structures. Each of the former has an extracellular N-terminus and its functional groups are situated in the hydrophilic compartments, its polypeptide spanning the hydrophobic domain of the lipid bilayer in the most economical fashion possible by means of a single short α-helix. In contrast, the latter has a cytoplasmic N-terminus and is specifically involved in transmembrane transport, for which purpose it has a large globular domain embedded within the bilayer, within which the polypeptide chain traverses the bilayer several times. It is probable that they are representative of two entirely different classes of integral membrane proteins, possibly inserted into the lipid bilayer by distinct biosynthetic routes, which are likely to be widely represented in the membranes of other cells.

IX. ACKNOWLEDGEMENTS

I am greatly indebted to my colleagues who have kindly provided original prints of their published and unpublished electron micrographs for this review, and in some cases also prepublication copies of their papers, which have been most helpful. My own recent contributions to this field have been made possible by project grants from the Medical Research Council and the Wellcome Trust, whose generous support is gratefully acknowledged.

X. REFERENCES

Agre, D., Orringer, E. P. and Bennett, V. (1982). *New Engl. J. Med.* **306**, 1155–1161.
Anderson, J. M. (1979). *J. Biol. Chem.* **254**, 939–944.
Anderson, J. M. and Tyler, J. M. (1980). *J. Biol. Chem.* **255**, 1259–1265.
Appleyard, S., Dunn, M., Dubowitz, V., Scott, M., Pittman, S. and Shotton, D. M. (1983a). *Eur J. Cell Biol.* Suppl. **1**, 7.

Appleyard, S., Dunn, M., Dubowitz, V., Scott, M., Pittman, S. and Shotton, D. M. (1983b). *Proc. Natn. Acad. Sci. USA* (in press).
Asano, A. and Sekiguchi, K. (1980). *J. Supramolec. Struct.* **9**, 441–452.
Avissar, N., DeVries, A., Ben-Shaul, Y. and Cohen, I. (1975). *Biochim. Biophys. Acta* **375**, 35–43.
Avruch, J. and Fairbanks, G. (1974). *Biochemistry* **13**, 5507–5514.
Bächi, T., Aguet, M. and Howe, C. (1973). *J. Virology* **11**, 1004–1012.
Bächi, T., Whiting, K., Tanner, M. J. A., Metaxas, M. N. and Anstee, D. J. (1977). *Biochim. Biophys. Acta* **464**, 635–639.
Barnard, P. J. and Barnard, E. A. (1978). *In* "The Biochemistry of Myasthenia Gravis and Muscular Dystrophy" (G. G. Lunt and R. M. Marchbanks, eds) pp. 319–329. Academic Press, New York.
Barnard, E. A. and Barnard, P. J. (1979). *Annals NY Acad. Sci.* **317**, 374–399.
Barnard, P. J. and Barnard, E. A. (1980). *In* "Muscular Dystrophy Research: Advances and New Trends" (C. Angelini, G. A. Danieli and D. Fontanari, eds) pp. 242–249, Excerpta Medica, Amsterdam. International Congress Series No. 527.
Bennett, V. (1978). *J. Biol. Chem.* **253**, 2292–2299.
Bennett, V. (1979). *Nature* **281**, 597–599.
Bennett, V. and Branton, D. (1977). *J. Biol. Chem.* **252**, 2753–2763.
Bennett, V. and Davis, J. (1981). *Proc. Natn. Acad. Sci. USA* **78**, 7550–7554.
Bennett, V. and Stenbuck, P. J. (1979a). *J. Biol. Chem.* **254**, 2533–2541.
Bennett, V. and Stenbuck, P. J. (1979b). *Nature* **280**, 468–473.
Bessis, M. (1973a). "Living Blood Cells and their Ultrastructure". Springer-Verlag, Berlin, Heidelberg and New York.
Bessis, M. (1973b). *In* "Red Cell Shape: Physiology, Pathology and Ultrastructure" (M. Bessis, R. I. Weed and P. F. LeBlond, eds) pp. 1–25. Springer-Verlag, Berlin, Heidelberg and New York.
Bessis, M. (1974). "Corpuscles: An Atlas of Red Blood Cell Shapes". Springer-Verlag, Berlin, Heidelberg and New York.
Birchmeier, W. and Singer, S. J. (1977). *J. Cell Biol.* **73**, 657–659.
Branton, D. (1971). *Phil. Trans. R. Soc. Ser. B.* **261**, 133–138.
Branton, D., Bullivant, S., Gilula, N. B., Karnovsky, M. J., Moor, H., Mühlethaler, K., Northcote, D. H., Packer, L., Satir, B., Satir, P., Speth, V., Staehlin, L. A., Steere, R. L. and Weinstein, R. S. (1975). *Science* **190**, 54–56.
Branton, D. and Kirchanski, S. (1977). *J. Microscopy* **111**, 117–124.
Brenner, S.-L. and Korn, E. D. (1979). *J. Biol. Chem.* **254**, 8620–8627.
Brenner, S.-L. and Korn, E. D. (1980). *J. Biol. Chem.* **255**, 1670–1676.
Bretscher, M. S. and Raff, M. C. (1975). *Nature* **258**, 43–49.
Burke, B. E. (1981). PhD Thesis, University of London.
Burke, B. E. and Shotton, D. M. (1982). *EMBO J.* **1**, 505–508.
Burke, B. E. and Shotton, D. M. (1983). *Br. J. Haematol.* **54**, 173–187.
Butterfield, D. A. (1977). *Accounts of Chemical Research* **10**, 111–116.
Butterfield, D. A. Chesnut, D. B., Roses, A. D. and Appel, S. H. (1974). *Proc. Natn. Acad. Sci. USA* **71**, 909–913.
Butterfield, D. A., Chesnut, D. B., Appel, S. H. and Roses, A. D. (1976). *Nature* **263**, 159–161.
Cabantchik, Z. I., Knauf, P. A. and Rothstein, A. (1978). *Biochim. Biophys. Acta* **515**, 239–302.
Calvert, R., Bennett, P. and Gratzer, W. (1980a). *Eur. J. Biochem.* **107**, 355–361.
Calvert, R., Bennett, P. and Gratzer, W. (1980b). *Eur. J. Biochem.* **107**, 363–367.

Casaly, H. and Sheetz, M. P. (1980). *J. Cell Biol.* **87,** 216a.
Caspar, D. L. D. and Klug, A. (1962). *Cold Spring Harb. Symp. Quant. Biol.* **27,** 1–24.
Cassoly, R., Daveloose, D. and Leterrier, F. (1980). *Biochim. Biophys. Acta* **601,** 478–489.
Cherry, R. J., Burkli, A., Busslinger, M., Schneider, G. and Parish, G. R. (1976). *Nature* **263,** 389–393.
Cohen, C. and Branton, D. (1979). *Nature* **279,** 163–165.
Cohen, C. M. and Foley, S. F. (1980). *J. Cell Biol.* **86,** 694–698.
Cohen, C. M., Foley, S. F. and Korsgren, C. (1982). *Nature* **299,** 648–650.
Cohen, C. M. and Korsgren, C. (1980). *Biochem. Biophys. Res. Commun.* **97,** 1429–1425.
Cohen, C. M., Jackson, P. L. and Branton, D. (1978). *J. Supramolec. Struct.* **9,** 113–124.
Cohen, C. M., Tyler, J. M. and Branton, D. (1980). *Cell* **21,** 875–883.
Cooper, R. A., (1978). *J. Supramolec. Struct.* **8,** 413–430.
Cooper, R. A. and Jandl, J. H. (1969). *J. Clin. Invest.* **48,** 736–744.
Cotmore, S. F., Furthmayr, H. and Marchesi, V. T. (1977). *J. Molec. Biol.* **113,** 539–553.
Crowther, R. A. and Amos, L. A. (1971). *J. Molec. Biol.* **60,** 123–130
Dacie, J. V. (1960). "The Haemolytic Anaemias, Congenital and Acquired". J. and A. Churchill Ltd, London.
Danon, D., Goldstein, L. Marikovsky, Y. and Skutelsky, E. (1972). *J. Ultrastruct. Res.* **38,** 500–510.
De Kruijff, B., van Zoelen, E. J. J. and van Deenen, L. L. M. (1978). *Biochim. Biophys. Acta* **509,** 537–542.
Dhermy, D., Lecomte, M. C., Garbarz, M., Bournier, O., Galand, C., Gautero, H., Feo, C., Alloisio, N., Delaunay, J. and Boivin, P. (1982). *J. Clin. Invest.* **70,** 707–715.
Dodge, J. T., Mitchell, C. and Hanahan, D. J. (1963). *Arch. Biochem. Biophys.* **100,** 119–130.
Dubowitz, V. (1979). *Ann NY Acad. Sci.* **317,** 431–439.
Edwards, H. H., Mueller, T. J. and Morrison, M. (1979). *Science* **203,** 1343–1346.
Elgsaeter, A. and Branton, D. (1974). *J. Cell Biol.* **63,** 1018–1030.
Elgsaeter, A., Espevik, T. and Kopstad, G. (1980). *38th Ann. Proc. Electron Microscopy Soc. Amer.* (San Francisco, California, 1980). (G. W. Bailey, ed), pp. 752–755.
Elgsaeter, A., Shotton, D. M. and Branton, D. (1976). *Biochim. Biophys. Acta* **426,** 101–122.
Elliott, A., Offer, G. and Burridge, K. (1976). *Proc. R. Soc. Ser. B* **193,** 45–53.
England, B. J., Gunn, R. B. and Steck, T. L. (1980). *Biochim. Biophys. Acta* **623,** 171–182.
Engstrom, L. H. (1970). PhD dissertation, University of California, Berkeley.
Espevik, T. and Elgsaeter, A. (1981). *J. Microscopy* **122,** 159–163.
Evans, E. A. (1973). *Biophys. J.* **13,** 926–940.
Evans, E. A. and Hochmuth, R. M. (1977). *J. Membrane Biol.* **30,** 351–362.
Evans, E. A. and Hochmuth, R. M. (1978). *Curr. Top. Membr. Transp.* **10,** 1–64.
Evans, E. A. and Lacelle, P. L. (1975). *Blood* **45,** 29–43.
Fairbanks, G., Steck, T. L. and Wallach, D. F. H. (1971). *Biochemistry NY* **10,** 2607–2617.
Fairbanks, G., Avruch, J., Dino, J. E. and Patel, V. P. (1978). *J. Supramolec. Struct.* **9,** 97–112.

Fisher, K. A. (1975). *Science* **190**, 983-984.
Fisher, K. A. (1976). *Proc. Natn. Acad. Sci. USA* **73**, 173-177.
Fisher, K. A. (1978). *Proc. 9th Congr. Electron Microscopy*, Toronto. Vol. III, p. 521.
Fisher, K. and Branton, D. (1974). *Methods Enzymol.* **32**, 35-44.
Fowler, V. and Bennett, V. (1978). *J. Supramolec. Struct* **8**, 215-221.
Fowler, V. and Branton, D. (1977). *Nature* **268**, 23-26.
Fowler, V. and Taylor, D. L. (1980). *J. Cell Biol.* **85**, 361-376.
Fowler, V., Luna, E. J., Hargreaves, W. R., Taylor, D. L. and Branton, D. (1981). *J. Cell Biol.* **88**, 388-395.
Fukuda, M., Eshdat, Y., Tarone, G. and Marchesi, V. T. (1978). *J. Biol. Chem.* **253**, 2419-2428.
Fuduka, M. and Fuduka, M. N. (1981). *J. Supramolec. Struct. Cell Biochem.* **17**, 313-324.
Furthmayr, H. (1978a). *J. Supramolec. Struct.* **9**, 79-95.
Furthmayr, H. (1978b). *Nature* **271**, 519-524.
Gahmberg, C. G., Myllyla, G., Leikola, J., Pirkola, A. and Nordling, S. (1976). *J. Biol. Chem.* **251**, 6108-6116.
Gahmberg, C. G., Taurén, G., Virtanen, I. and Wartiovaara, J. (1978). *J. Supramolec. Struct.* **8**, 337-347.
Gerritsen, W. J. Verkleij, A. J., Zwall, R. F. A. and van Deenen, L. L. M. (1978). *Eur. J. Biochem.* **85**, 255-261.
Gerritsen, W. J., Henricks, P. A. J., de Kruijff, B. and van Deenen, L. L. M. (1980). *Biochim. Biophys. Acta* **600**, 607-619.
Godin, D. V., Bridges, M. A. and MacLeod, P. J. M. (1978). *Res. Commun. Chem. Pathol. Pharmacol.* **20**, 331-349.
Golan, D. E. and Veatch, W. (1980). *Proc. Natn. Acad. Sci. USA* **77**, 2537-2541.
Goldman, R., Pollard, T. and Rosenbaum, J. (eds) (1976). "Cell Motility". Cold Spring Harbor Laboratory, New York.
Goodman, S. R., Shiffer, K. A., Casoria, L. A. and Eyster, M. E. (1982). *Blood* **60**, 772-784.
Goodman, S. R. and Weidner, S. A. (1980). *J. Biol. Chem.* **255**, 8082-8086.
Grant, C. W. M. and McConnell, H. M. (1974). *Proc. Natn. Acad. Sci. USA* **71**, 4653-4657.
Gratzer, W. B. (1981). *Biochem. J.* **198**, 1-8.
Greengard, P. (1978). *Science* **199**, 146-152.
Greenquist, A. C., Shohet, S. B. and Bernstein, S. E. (1978). *Bood* **51**, 1149-1155.
Guidotti, G. (1972). *Ann. Rev. Biochem.* **41**, 731-752.
Haest, C. W. M., Plasa, G., Kamp, D. and Deuticke, B. (1978). *Biochim. Biophys. Acta* **509**, 21-32.
Hainfeld, J. F. and Steck, T. L. (1977). *J. Supramolec. Struct.* **6**, 301-311.
Hall, C. E. and Doty, P. (1958). *J. Amer. Chem. Soc.* **80**, 1269-1274.
Hall, C. E. and Slayter, H. S. (1959). *J. Biophys. Biochem. Cytol.* **5**, 11-16.
Harris, H. W. and Lux, S. E. (1980). *J. Biol. Chem.* **255**, 11512-11520.
Harris, H. W., Levin, N. and Lux, S. E. (1980). *J. Biol. Chem.* **255**, 11521-11525.
Harris, J. R. (1968). *Biochim. Biophys. Acta* **150**, 534-537.
Harris, J. R. (1969a). *Biochim. Biophys. Acta* **188**, 31-42.
Harris, J. R. (1969b). *J. Mol. Biol.* **46**, 329-335.
Harris, J. R. (1971a). *Biochim. Biophys. Acta* **229**, 761-770.
Harris, J. R. (1971b). *J. Ultrastruct. Res.* **36**, 587-594.
Harris, J. R. (1980). *Nouv. Rev. Fr. Hématol.* **22**, 411-448.

Harris, J. R. and Kerr, J. (1976). *J. Microscopy* **108**, 51–59.
Harris, J. R. and Naeem, I. (1978). *Biochim. Biophys. Acta* **537**, 495–500.
Harris, J. R. (1983). *Micron* (in press).
Hartwig, J. H. and Stossel, T. P. (1981). *J. Mol. Biol.* **145**, 563–568.
Holdstock, S. J. and Ralston, G. B. (1980). *I.R.C.S. Med. Sci. Biochem. Cell Membrane Biol. Haematol.* **8**, 723–724.
Horne, R. W. and Pasquali Ronchetti, I. (1974). *J. Ultastruct. Res.* **47**, 361–383.
Hosey, M. M. and Tao, M. (1977). *J. Biol. Chem.* **252**, 102–109.
Howe, C. and Bachi, T. (1973). *Exp. Cell Res.* **76**, 321–332.
Howells, K . F. (1976). *Res. Exp. Med. (Berlin)* **168**, 213–217.
Hudecki, M. S., Pollina, C. M., Bhargava, A. K., Hudecki, R. S. and Heffner, R. R. (1979). *Muscle Nerve* **2**, 57–67.
Huxley, H. E. (1963) *J. Molec. Biol.* **7**, 281–308.
Imhof, B. A., Acha-Orbea, H. J., Libermann, T. A., Reber, B. F. X., Lanz, J. H., Winterhalter, K. H. and Birchmeier, W. (1980). *Proc. Natn. Acad. Sci. USA* **77**, 3264–3268.
Iwasaki, Y. (1980). *Adv. Neurol. Sci.* **24**, 772–779.
Iyer, S. L., Hoening, B. A., Sherblom, A. P. and Howland, J. L. (1977). *Biochem. Med.* **18**, 384–391.
Ji, T. H., Kiehm, D. J. and Middaugh, C. R. (1980). *J. Biol. Chem.* **255**, 2990–2993.
Johnson, R. M. and Kirkwood, O. H. (1978). *Biochim. Biophys. Acta* **509**, 58–66.
Jones, G. E. and Wilkowski, J. A. (1981). *J. Cell Sci.* **48**, 291–300.
Jones, G. E. and Wilkowski, J. A. (1983). *J. Neurol. Sci.* **58**, 159–174.
Juliano, R. L. (1973). *Biochim. Biophys. Acta* **300**, 341–378.
Juliano, R. L. Kimelberg, H. K. and Papahadjopoulos, D. (1971). *Biochim. Biophys. Acta* **241**, 894–905.
Kiehm, D. J. and Ji, T. H. (1977). *J. Biol. Chem.* **252**, 8524–8531.
Kim, H. D., Luthra, M. G., Watts, R. P. and Stern, L. Z. (1980). *Neurology* **30**, 726–731.
Kirkpatrick, F. H. (1976). *Life Sci.* **19**, 1–18.
Kliman, H. J. and Steck, T. L. (1980). *J. Biol. Chem.* **255**, 6314–6321.
Knutton, S. (1977). *J. Cell Sci.* **28**, 89–210.
Koppel, D. E. and Sheetz, M. P. (1981). *Nature* **293**, 159–161.
Koppel, D. E., Sheetz, M. P. and Schindler, M. (1981). *Proc. Natn. Acad. Sci. USA* **78**, 3576–3580.
Laemmli, U. K. (1970). *Nature* **227**, 680–687.
Lalazar, A. Reichler, Y. and Loyter, A. (1976). *Proc. 6th Europ. Cong. Electron Microscopy (Jerusalem)*, pp. 321–323.
Lawler, J., Liu, S.-C., Palek, J. and Prchal, J. (1982). *J. Clin. Invest.* **70**, 1019–1030.
Lazarides, E. and Nelson, W. J. (1982). *Cell* **31**, 505–508.
Lin, D. C. and Lin S. (1979). *Proc. Natn. Acad. Sci. USA* **76**, 2345–2349.
Liu, S.-C., Palek, J. and Prchal, J. T. (1982). *Proc. Natn. Acad. Sci. USA* **79**, 2072–2076.
Liu, S.-C., Palek, J., Prchal, J. T. and Castleberry, R. P. (1981). *J. Clin. Invest.* **68**, 597–605.
Lorand, L., Shishido, R., Parameswaren, K. N. and Steck, T. L. (1975). *Biochem. Biophys. Res. Commun.* **67**, 1158–1166.
Lorand, L., Weissmann, L. B., Epel, D. L. and Bruner-Lorand, J. (1976). *Proc. Natn. Acad. Sci. USA* **73**, 4479–4481.

Lorand, L., Siefring, G. E. Jr. and Lowe-Krentz, L. (1979). *Seminars in Haematology* **16**, 65–74.
Lovrein, R. E. and Anderson, R. A. (1980). *J. Cell Biol.* **85**, 534–548.
Lowey, S., Slayter, H. S., Weeds, A. G. and Baker, H. (1969). *J. Mol. Biol.* **42**, 1–29.
Lucy, J. A. (1980). *Br. Med. Bull.* **36**, 187–192.
Lumb, E. M. and Emery, A. E. H. (1975). *Br. Med. J.* **3**, 467–468.
Luna, E. J., Kidd, G. H. and Branton, D. (1979). *J. Biol. Chem.* **254**, 2526–2532.
Lutz, H. U., Lomant, A. J., McMillan, P. and Wehrli, E. (1977). *J. Cell Biol.* **74**, 389–398.
Lutz, H. U., von Däniken, A., Semenza, G. and Bächi, T. (1979). *Biochim. Biophys. Acta* **552**, 262–280.
Lux, S. E. (1979a). *Nature* **281**, 426–429.
Lux, S. E. (1979b). *Seminars in Haematology* **16**, 21–51.
Lux, D. E. and Glader, B. E. (1980). In "Haematology of Infancy and Childhood" (D. G. Nathan and F. A. Oski, eds), W. B. Saunders, Philadelphia.
Lux, S. E. and John, K. M. (1977). In "Cell Shape and Surface Architecture" (C. F. Fox, J. P. Ravel and U. Henning, eds) pp. 481–491. A. R. Liss, New York.
Lux, S. E., John, K. M. and Karnovsky, M. J. (1976). *J. Clin. Invest.* **58**, 955–963.
Malech, H. L. and Marchesi, V. T. (1981). *Biochim. Biophys. Acta* **670**, 385–392.
Marchesi, M. T. (1979a). *J. Membane Biol.* **51**, 101–131.
Marchesi, V. T. (1979b). *Seminars in Haematology* **16**, 3–20.
Marchesi, V. T. and Palade, G. E. (1967a). *J. Cell Biol.* **35**, 385–404.
Marchesi, V. T. and Palade, G. E. (1967b). *Proc. Natn. Acad. Sci. USA* **58**, 991–995.
Marchesi, V. T. and Steers, E., Jr (1968). *Science* **159**, 203–204.
Marchesi, V. T., Furthmayr, H. and Tomita, M. (1976). *Ann. Rev. Biochem.* **45**, 667–697.
Margaritis, L. H., Miller, K. R. and Branton, D. (1976). *Proc. 6th Europ. Cong. Electron Microscopy (Jerusalem)*, pp. 128–130.
Margaritis, L. H., Elgsaeter, A. and Branton, D. (1977). *J. Cell Biol.* **72**, 47–56.
Markham, R., Frey, S. and Hills, G. J. (1963). *Virology* **20**, 88–102.
Markowitz, S. and Marchesi, V. T. (1981). *J. Biol. Chem.* **256**, 6463–6468.
Marton, L. S. G. and Garvin, J. E. (1973). *Biochem. Biophys. Res. Commun.* **52**, 1457–1462.
Matheson D.W. and Howland, J. E. (1974) *Science* **184**, 165–166.
Matheson, D. W., Engel, W. K. and Derrer, E. C. (1976). *Neurology* **26**, 1182–1183.
Mawatari, S. (1980). *Adv. Neurol. Sci.* **24**, 790–796.
Mawatari, S., Igisu, H., Kuroiwa, Y. and Miyoshino, S. (1981). *Neurology* **31**, 293–297.
McLaughlin, J. and Engel, W. K. (1979). *Archs. Neurol. Chicago* **36**, 351–354.
Miale, T. D., Frias, J. L. and Lawson, D. L. (1975). *Science* **187**, 453–454.
Middaugh, C. R. and Ji, T. H. (1980). *Eur. J. Biochem.* **110**, 587–592.
Miller, S. E., Roses, A. D. and Appel. S. A. (1976). *Arch. Neurol.* **33**, 172–174.
Mohandas, N., Greenquist, A. C. and Shohet, S. B. (1978). *J. Supramolec. Struct.* **9**, 453–458.
Mombers, C., de Gier, J., Demel, R. A. and van Deenen, L. L. M. (1980). *Biochim. Biophys. Acta* **603**, 52–62.
Mombers, C., van Dijck, P. W. M., van Denen, L. L. M., de Gier, J. and Verkleij, A. J. (1977). *Biochim. Biophys. Acta* **470**, 152–160.
Mombers, C., Verkleij, A. J., de Gier, J. and van Deenen, L. L. M. (1979). *Biochim. Biophys. Acta* **551**, 271–281.

Morrow, J. S. and Marchesi, V. T. (1981). *J. Cell Biol.* **88**, 463–468.
Morrow, J. S., Speicher, D. W., Knowles, W. J., Hsu, C. J. and Marchesi, V. T. (1980). *Proc. Natn. Acad. Sci. USA* **77**, 6592–6596.
Morse, P. F. and Howland, J. L. (1973). *Nature* **245**, 156–157.
Mueller, T. J., Dow, A. W. and Morrison, M. (1976). *Biochem. Biophys. Res. Commun.* **72**, 94–99.
Mueller, T. J., Li, Y.-T. and Morrison, M. (1979). *J. Biol. Chem.* **254**, 8103–8106.
Mueller, T. J. and Morrison, M. (1974). *J. Biol. Chem.* **249**, 7568–7573.
Mueller, T. J. and Morrison, M. (1977). *J. Biol. Chem.* **252**, 6573–6576.
Mueller, T. J. and Morrison, M. (1980). *J. Cell Biol.* **87**, 202a.
Mueller, T. J. and Morrison, M. (1981). *J. Supramolec. Struct.* (Suppl. 5), 131 (Abstract).
Nagano, Y., Wong, P. and Roses, A. D. (1980). *Clin. Chim. Acta* **108**, 469–474.
Nakashima, K. and Beutler, E. (1979). *Proc. Natn. Acad. Sci. USA* **76**, 935–938.
Nakashima, H. and Makino, S. (1980). *J. Biochem. (Tokyo)* **88**, 933–947.
Nermut, M. V. and Williams, L. D. (1980). *J. Microscopy* **118**, 453–461.
Nicolson, G. L. and Painter, R. G. (1973). *J. Cell Biol.* **59**, 395–406.
Nicolson, G. L., Marchesi, V. T. and Singer, S. J. (1971). *J. Cell Biol.* **51**, 265–272.
Nigg, E. A. and Cherry, R. J. (1979a). *Biochemistry NY* **18**, 3457–3465.
Nigg, E. A. and Cherry, R. J. (1979b). *Nature* **277**, 493–494.
Nigg, E. A. and Cherry, R. J. (1980). *Proc. Natn. Acad. Sci. USA* **77**, 4702–4706.
Nigg, E. A., Bron, C., Girardet, M. and Cherry, R. J. (1980a). *Biochemistry NY* **19**, 1887–1893.
Nigg, E. A., Gahmberg, C. G. and Cherry, R. J. (1980b). *Biochim. Biophys. Acta* **600**, 636–642.
Ohnishi, T. (1962). *J. Biochem. (Tokyo)* **52**, 307–308.
Ohnishi, T. (1977). *Br. J. Haemat.* **35**, 453–458.
Owens, J. W., Mueller, T. J. and Morrison, M. (1980). *Archs. Biochem. Biophys.* **204**, 247–254.
Palek, J. (1977). *Br. J. Haemat.* **35**, 1–9.
Palek, J. and Liu, S.-C. (1979a). *J. Supramolec. Struct.* **10**, 79–96.
Palek, J. and Liu, S.-C. (1979b). *Seminars in Haematology* **16**, 75–93.
Palek, J., Liu, S.-C., Liu, P.-Y., Prchal, J. T. and Castleberry, R. P. (1981). *Blood* **57**, 130–139.
Park, J. H., Hill, E. J., Chou, T.-H., LeQuire, V., Roeloffs, R. and Park, C. R. (1979). *Ann. NY Acad. Sci.* **317**, 356–369.
Patel, V. P. and Fairbanks, G. (1981). *J. Cell Biol.* **88**, 430–440.
Percy, A. K. and Miller, M. E. (1975). *Nature* **258**, 147–148.
Pinder, J. C., Bray, D. and Gratzer, W. B. (1975). *Nature* **258**, 765–766.
Pinder, J. C., Bray, D. and Gratzer, W. B. (1977). *Nature* **270**, 752–754.
Pinder, J. C., Ungewickell, E., Bray, D. and Gratzer, W. B. (1978). *J. Supramolec. Struct.* **8**, 439–445.
Pinder, J. C., Ungewickell, E., Calvert, R., Morris, E. and Gratzer, W. B. (1979). *FEBS Letters* **104**, 396–400.
Pinder, J. C., Clark, S. E., Baines, A. J., Morris, E. and Gratzer, W. B. (1981). *In* "The Red Cell" (G. J. Brewer, ed) *Prog. Clin. Biol. Res.* **55**, 343–354. A. R. Liss, New York.
Pinto da Silva, P. (1972). *J. Cell Biol.* **53**, 777–787.
Pinto de Silva, P. (1973). *Proc. Natn. Acad. Sci. USA* **70**, 1339–1343.
Pinto da Silva, P. and Branton, D. (1970). *J. Cell Biol.* **45**, 598–605.

Pinto da Silva, P. and Nicolson, G. L. (1974). *Biochim. Biophys. Acta* **363,** 311–319.
Pinto da Silva, P. and Torrisi, M. R. (1981). *J. Cell Biol.* **91,** 263a.
Pinto da Silva, P., Moss, P. S. and Fudenberg, H. H. (1973). *Exp. Cell Res.* **81,** 127–138.
Plishker, G. A., Gitelman, H. J. and Appel, S. H. (1978). *Science* **200,** 323–325.
Plishker, G. A. and Appel, S. H. (1980). *Muscle Nerve* **3,** 70–81.
Plut, D. A., Hosey, M. M. and Tao, M. (1978). *Eur. J. Biochem.* **82,** 333–337.
Poo, M. and Cone, R. A. (1974). *Nature* **247,** 438–441.
Potempa, L. A. and Garvin, J. E. (1976). *Biochem. Biophys. Res. Commun.* **72,** 1049–1055.
Ralston, G. B. (1975). *Aust. J. Biol. Sci.* **28,** 259–266.
Ralston, G. B. (1976). *Biochim. Biophys. Acta* **443,** 387–393.
Ralston, G. B., Dunbar, J. and White, M. (1977). *Biochim. Biophys. Acta* **491,** 345–348.
Richards, S., Higashi, T. and Uyeda, K. (1979). *Fedn Proc. Fedn Amer. Soc. Exp. Biol.* **38,** 798.
Romans, A. Y., Yeagle, P. L., O'Connor, S. E. and Grisham, C. M. (1979). *J. Supramolec. Struct.* **10,** 241–251.
Rosenmann, E. *et al.* (1982). *Nature* **298,** 563–565.
Roses, A. D. and Appel, S. H. (1973a). *J. Biol. Chem.* **248,** 1408–1411.
Roses, A. D. and Appel, S. H. (1973b). *Proc. Natn. Acad. Sci. USA* **70,** 1855–1859.
Roses, A. D. and Appel, S. H. (1974). *Lancet* **2,** 1400.
Roses, A. D. and Appel, S. H. (1975). *J. Membrane Biol.* **20,** 51–58.
Roses, A. D. and Appel, S. H. (1976). *J. Neurol. Sci.* **29,** 185–193.
Roses, A. D., Herbstreith, M. H. and Appel, S. H. (1975). *Nature* **254,** 350–351.
Roses, A. D., Roses, M. J., Miller, S. E., Hull, K. L. and Appel, S. H. (1976). *New Engl. J. Med.* **294,** 193–198.
Roses, A. D., Herbstreith, M. H. and Shile, P. (1980). *Neurology* **30,** 423.
Rowland, L. P. (1980). *Muscle Nerve* **3,** 3–20.
Rubin, C. S. and Rosen, O. M. (1975). *Ann. Rev. Biochem.* **44,** 831–877.
Sato, B., Nishikida, K., Samuels, L. T. and Tyler, F. H. (1978). *J. Clin. Invest.* **61,** 251–259.
Schekman, R. and Singer, S. J. (1976). *Proc. Natn. Acad. Sci. USA* **73,** 4075–4079.
Schindler, M., Koppel, D. E. and Sheetz, M. P. (1980). *Proc. Natn. Acad. Sci. USA* **77,** 1457–1461.
Segrest, J. P. and Kohn, L. D. (1973). *In* "Protides of the Biological Fluids 21st Colloquium, A.2. Conformation and Structure of Membranes" (H. Peeters, ed), pp. 183–189. Pergamon Press, Oxford.
Segrest, J. P., Gulik-Krzywicki, T. and Sardet, C. (1974). *Proc. Natn. Acad. Sci. USA* **71,** 3294–3298.
Shafiq, S. A., Leung, B. and Schutta, H. S. (1976). *J. Neurol. Sci.* **30,** 299–302.
Sheehy, R. and Ralston, G. B. (1978). *Blut* **36,** 145–148.
Sheetz, M. P. (1979a). *Biochim. Biophys. Acta* **557,** 122–134.
Sheetz, M. P. (1979b). *J. Cell Biol.* **81,** 266–270.
Sheetz, M. P. and Casaly, J. (1980). *J. Biol. Chem.* **255,** 9955–9960.
Sheetz, M. P. and Sawyer, D. (1978). *J. Supramolec. Struct.* **8,** 399–412.
Sheetz, M. P. and Singer, S. J. (1974). *Proc. Natn. Acad. Sci. USA* **71,** 4457–4461.
Sheetz, M. P. and Singer, S. J. (1976). *J. Cell Biol.* **70,** 247–251.
Sheetz, M. P. and Singer, S. J. (1977). *J. Cell Biol.* **73,** 638–646.
Sheetz, M. P., Painter, R. G. and Singer, S. J. (1976a). *Biochemistry NY* **15,** 4486–4492.
Sheetz, M. P., Painter, R. G. and Singer, S. J. (1976b). *In* "Cell Motility" (R. Goldman,

J. L. Rosenbaum and T. D. Pollard, eds) Vol. B, pp. 651–664. Cold Spring Harbor Laboratory, New York.
Sheetz, M. P., Schindler, M. and Koppel, D. E. (1980). *Nature* **285**, 510–512.
Shivers, R. R. and Atkinson, B. G. (1979). *Amer. J. Path.* **94**, 97–102.
Shohet, S. B. (1979). *J. Clin. Invest.* **64**, 483–494.
Shotton, D. M. (1982). *J. Neurol. Sci.* **57**, 161–190.
Shotton, D., Thompson, K., Wofsy, L. and Branton, D. (1978). *J. Cell Biol.* **76**, 512–531.
Shotton, D., Burke, B. and Branton, D. (1978). *Biochim. Biophys. Acta* **536**, 313–317.
Shotton, D. M., Burke, B. E. and Branton, D. (1979). *J. Mol. Biol.* **131**, 303–329.
Shulte, T. H. and Marchesi, V. T. (1979). *Biochemistry NY* **18**, 275–279.
Siefring, G. E., Apostol, A. B., Velasco, P. T. and Lorand, L. (1978). *Biochemistry NY* **17**, 2598–2604.
Siegel, D. L., Goodman, S. R. and Branton, D. (1980). *Biochim. Biophys. Acta* **598**, 517–527.
Singer, S. J. (1974). *Ann. Rev. Biochem.* **43**, 805–833.
Singer, J. and Nicolson, G. (1972). *Science* **175**, 720–731.
Sjöstrand, F. S. (1979). *J. Ultrastruct. Res.* **69**, 378–420.
Slayter, H. S. (1976). *Ultramicroscopy* **1**, 341–357.
Slayter, H. S. and Lowey, S. (1967). *Proc. Natn. Acad. Sci. USA* **58**, 1611–1618.
Sleytr, U. B. and Robards, A. W. (1977a). *J. Microscopy* **110**, 1–25.
Sleytr, U. B. and Robards, A. W. (1977b). *J. Microscopy* **111**, 77–100.
Sobue, K., Kanda, J., Inui, M., Morimoto, K. and Kakiuchi, S. (1982). *FEBS Lett.* **148**, 221–225.
Southworth, D., Fisher, K. and Branton, D. (1975). *In* "Techniques of Biochemical and Biohysical Morphology" (D. Glick and R. Rosenbaum, eds) Vol. 2, pp. 247–282. John Wiley and Sons, New York.
Speicher, D. W., Morrow, J. S., Knowles, W. J. and Marchesi, V. T. (1980). *Proc. Natn. Acad. Sci. USA* **77**, 5673–5677.
Steck, T. L. (1972). *J. Mol. Biol.* **66**, 295–305.
Steck, T. L. (1974). *J. Cell Biol.* **62**, 1–19.
Steck, T. L. (1978). *J. Supramolec. Struct.* **8**, 311–324.
Steck, T. L. and Kant, J. A. (1973). *J. Biol. Chem.* **248**, 8457–8464.
Steck, T. L. and Kant, J. A. (1975). *Methods Enzym.* **31**, 172–180.
Steck, T. L. and Yu, J. (1973). *J. Supramolec. Struct.* **1**, 220–232.
Stracher, A., McGowan, E. B. and Shafiq, S. A. (1978). *Science* **200**, 50–51.
Strapazon, E. and Steck, T.L. (1977). *Biochemistry NY* **16**, 2966–2871.
Sweet, C. and Zull, J. E. (1970). *Biochem. Biophys. Res. Commun.* **41**, 135–141.
Takahshi, K. (1978). *J. Biochem. (Tokyo)* **83**, 905–908.
Tanner, M. J. A. and Anstee, D. J. (1976). *Biochem. J.* **153**, 271–277.
Tanner, J. J. A. and Gray, W. R. (1971). *Biochem. J.* **125**, 1109–1117.
Tanner, J. J. A., Jenkins, R. E., Anstee, D. J. and Clamp, J. R. (1976). *Biochem. J.* **155**, 701–703.
Tchernia, G., Mohandes, N. and Shohet, S. B. (1981). *J. Clin. Invest.* **68**, 454–460.
Thomas, D. B. and Winzler, R. J. (1969). *J. Biol. Chem.* **244**, 5943–5946.
Thompson, S., Rennie, C. M. and Maddy, A. H. (1980). *Biochim. Biophys. Acta* **600**, 756–768.
Tillack, T. W. and Marchesi, V. T. (1970). *J. Cell Biol.* **45**, 649–653.
Tillack, T. W., Scott, R. E. and Marchesi, V. T. (1972). *J. Exp. Med.* **135**, 1209–1227.
Tilney, L. G. (1974). *J. Cell Biol.* **63**, 349a.

Tilney, L. G. (1975). *In* "Molecules and Cell Movement" (S. Inoué and R. E. Stephens, eds). Raven Press, New York.
Tilney, L. G. and Detmers, P. (1975). *J. Cell Biol.* **66**, 508–520.
Tokuyasu, K. T., Schekman, R. and Singer, S. J. (1979). *J. Cell Biol.* **80**, 481–486.
Tomita, N. and Marchesi, V. T. (1975). *Proc. Natn. Acad. Sci. USA* **72**, 2964–2968.
Tomita, M., Furthmayr, H. and Marchesi, V. T. (1978). *Biochemistry NY* **17**, 4756–4770.
Tortolero, M., Fischer, S., Piau, J. P., Delaunay, J. and Schapira, G. (1979). *Biomedicine* **30**, 47–52.
Tsuchiya, Y., Sugita, H., Ishiura, S. and Imahori, K. (1981). *Clin. Chim. Acta* **109**, 285–293.
Tsuji, T., Irimura, T. and Osawa, T. (1980). *Biochem. J.* **187**, 677–686.
Tsuji, T., Irimura, T. and Osawa, T. (1981a). *Carbohydrate Res.* **92**, 328–332.
Tsuji, T., Irimura, T. and Osawa, T. (1981b). *J. Biol. Chem.* **256**, 10 497–10 502.
Tsukita, S., Tsukita, S. and Isaikawa, H. (1980). *J. Cell Biol.* **85**, 567–576.
Tuech, J. K. and Morrison, M. (1974). *Biochem. Biophys. Res. Commun.* **59**, 352–360.
Tyler, J. M., Hargreaves, W. R. and Branton, D. (1979). *Proc. Natn. Acad. Sci. USA* **76**, 5192–5196.
Tyler, J. M., Anderson, J. M. and Branton, D. (1980a). *J. Cell Biol.* **85**, 489–495.
Tyler, J. M., Reinhardt, B. N. and Branton, D. (1980b). *J. Biol. Chem.* **255**, 7034–7039.
Ungewickell, E. and Gratzer, W. (1978). *Eur. J. Biochem.* **88**, 379–385.
Ungewickell, E., Bennett, P. M., Calvert, R., Ohanian, V. and Gratzer, W. B. (1979). *Nature* **280**, 811–814.
van Zoelen, E. J. J., Verkleij, A. J., Zwall, R. F. A. and van Deenen, L. L. M. (1978a). *Eur. J. Biochem.* **86**, 539–546.
van Zoelen, E. J. J., van Dijck, P. W. M., de Kruijff, B., Verkleij, A. J. and van Deenen, L. L. M. (1978b). *Biochim. Biophys. Acta* **514**, 9–24.
Vickers, J. D., McComas, A. J. and Rathbone, M. P. (1979). *Neurology* **29**, 791–796.
Wakayama, Y., Hodson, A., Pleasure, D., Bonilla, E. and Schotland, D. L. (1978). *Ann. Neurol.* **4**, 253–256.
Wakayama, Y., Hodson, A. Bonilla, E., Pleasure, D. and Schotland, D. L. (1979). *Neurology* **29**, 670–675.
Wang, K. and Richards, F. M. (1974). *J. Biol. Chem.* **249**, 8005–8018.
Weed, R. I. and Lacelle, P. L. (1969). *In* "Red Cell Membrane: Structure and Function" (G. A. Jamieson and T. J. Greenwalt, eds) pp. 318–338. J. B. Lippincott Co., Philadelphia and Toronto.
Weed, R. I., Lacelle, P. L. and Merrill, E. W. (1969). *J. Clin. Invest.* **48**, 795–809.
Weinstein, R. S. (1969). *In* "Red Cell Membrane: Structure and Function" (G. A. Jamieson and T. J. Greenwalt, eds) pp. 36–76. J. B. Lippincott Co., Philadelphia and Toronto.
Weinstein, R. S. (1974). *In* "The Red Blood Cell" (D. M. Surgenor, ed.) Vol. 1. 2nd edn, pp. 213–268. Academic Press, New York.
Weinstein, R. S., Khodadad, J. K. and Steck, T. L. (1978). *J. Supramolec. Struct.* **8**, 325–335.
Weinstein, R. S., Khodadad, J. K. and Steck, T. L. (1980). *J. Cell Biol.* **87**, 209a.
White, M. D. and Ralston, G. B. (1979). *Biochim. Biophys. Acta* **554**, 469–478.
Williamson, P., Bateman, J., Kozarsky, K., Mattocks, K., Hermanowicz, N., Choe, H.-R. and Schlegel, R. A. (1982). *Cell* **30**, 725–733.
Wolfe, L. C., John, K. M., Falcone, J. C., Byrne, A. M. and Lux, S. E. (1982). *New Engl. J. Med.* **307**, 1367–1374.

Wolfe, L. C., Lux, S. E. and Ohanian, V. (1980). *J. Cell Biol.* **87**, 203a.
Yoshima, A., Furthmayr, H. and Kobata, A. (1980). *J. Biol. Chem.* **255**, 9713–9718.
Young, L. E. (1955). *Amer. J. Med.* **18**, 486–497.
Yu, J. and Branton, D. (1976). *Proc. Natn. Acad. Sci. USA* **73**, 3891–3895.
Yu, J. and Branton, D. (1977). *In* "Cell Shape and Surface Architecture", pp. 453–458. A. R. Liss, Inc., New York.
Yu, J. and Goodman, S. R. (1979). *Proc. Natn. Acad. Sci. USA* **76**, 2340–2344.
Yu, J. and Steck, T. L. (1974). *Fedn Proc. Fedn Amer. Soc. Exp. Biol.* **33**, 1532.
Yu, J. and Steck, T. L. (1975a). *J. Biol. Chem.* **250**, 9170–9175.
Yu, J. and Steck, T. L. (1975b). *J. Biol. Chem.* **250**, 9176–9184.
Yu, J., Fischman, D. A. and Steck, T. L. (1973). *J. Supramolec. Struct.* **1**, 233–248.
Zacharius, R. M., Zell, T. E., Morrison, J. H. and Woodlock, J. J. (1969). *Anal. Biochem.* **30**, 148–152.
Zarkowsky, H. S., Mohandas, N., Speaker,C. S. and Shohet, S. B. (1975). *Br. J. Haematol.* **29**, 537–545.
Ziparo, E., Lemay, A. and Marchesi, V. T. (1978). *J. Cell Sci.* **34**, 91–101.

4. Plasma Membrane Intercellular Junctions. Morphology and Protein Composition

C. A. L. S. COLACO AND W. H. EVANS

Imperial Cancer Research Fund, Lincoln's Inn Fields, London, England and National Institute for Medical Research, Mill Hill, London, England

I.	Introduction	332
II.	Cell-Cell Adhesion Junctions	333
	A. The Macula Adherens or Spot Desmosome	333
	1. Morphology	333
	2. Protein composition	335
	B. Hemidesmosomes	337
	C. The Zonula Adherens or Belt Desmosome	337
	1. Morphology	338
	2. Protein composition	339
	D. The Fascia Adherens	339
	1. Morphology	339
	2. Protein composition	339
III.	Cell-Cell Communication Junctions	342
	A. Synaptic Junctions	342
	1. Morphology	342
	2. Protein composition	343
	B. Gap Junctions	348
	1. Morphology	348
	2. Protein composition	354
IV.	The Tight Junction or Zonula Occludens	356
V.	Septate Junctions	357
VI.	Future Prospects	358
VII.	Acknowledgements	360
VIII.	References	360

I. INTRODUCTION

In multicellular organisms, tissues and organs, the spatial and functional assembly of the constituent cell types is made possible by a variety of specific cell-cell and cell-substratum interactions. The nature of these interactions, which account for the highly coordinated summation of function in multicellular assemblies, is illustrated by the developmental events observed during embryogenesis and organogenesis, the metabolic changes in hormonally and neuronally stimulated secretory-absorptive epithelia, and the contrasting levels of control seen in synchronously beating heart cells or the slow spread of contraction in smooth muscle. Such cellular interactions can be studied directly at the plasma membrane level, now widened to include its intracellular and extracellular appendages, and more specifically at those membrane regions where the intercellular junctions mediating these interactions are located.

Morphology has long pointed to the presence in animal cells of a bewildering variety of intercellular junctions (Farquhar and Palade, 1963; Staehelin, 1974), and it has always been assumed that different morphologies mediate different cellular functions. In this chapter, morphological data relating to the three major categories of intercellular junctions are briefly reviewed and correlated with currently available information on the protein composition and molecular architecture of the junctions. While it is still true that knowledge of junctional morphology, extensively extended in the last ten years by application of the freeze-fracture technique, is heavily outweighed by the paucity of information on biochemical composition, advances in cell membrane and membrane protein fractionation, coupled with the use of monoclonal and polyclonal antibodies at the cell and molecular level, are now rapidly contributing to our knowledge of the structure and function of intercellular junctions. The account of the morphological features of junctional structure and assembly will therefore be highly abbreviated to allow a more extensive reviewing of data on the composition and arrangement of the protein constituents. A number of general reviews have been published focussing mainly on the structure and function of intercellular junctions, and the reader is referred to those by Gilula (1978), McNutt and Weinstein (1973), Overton (1975) and Staehelin (1974). Also, several reviews that concentrate on specific types of intercellular junctions are available, for example: gap junctions (Peracchia, 1980), chemical synapses (Matus, 1978; Gurd, 1982), tight junctions (Van Deurs, 1980) and septate junctions (Noirot-Timothée and Noirot, 1980). Neuromuscular junctions, and their associated nicotinic receptors, form a separate subject and are not covered in this chapter.

II. CELL-CELL ADHESION JUNCTIONS

Three major categories of adherens junctions of similar morphological appearance have been described: *maculae*, *fasciae* and *zonulae adherentes*. These contribute to intercellular adhesion and thus help tissues resist mechanical stress.

A. The Macula Adherens or Spot Desmosome

1. Morphology

This junction consists of two straight and parallel cell membranes with the junctional area ranging from 0·1 to 1·4 μm in diameter. The membranes are separated by an extracellular gap (25–35) nm containing electron dense material bisected by a central lamina (Fig. 1A). In thin section this intercellular cement, in tissues impregnated with lanthanum, appears to consist of a series of fibrils linked at the central lamina. Adjacent to the cytoplasmic faces of the apposed membranes are found densely-staining circular plaques where 10–12 nm tonofilaments converge (Kelly, 1966; Rayns *et al.*, 1969). Protofilaments (3–5 nm in diameter) attach the tonofilaments to the plaques, and these may be identical to those fibrils present in the intercellular cleft (LeLoup *et al.*, 1979). In freeze-fractured preparations, the plasma membrane zones where desmosomes are located contain clusters of 12–20 nm elongated particles regularly arranged on the P-fracture face. Cross-fracture reveals a 5 nm midline of particles corresponding to the central lamina and associated with the cross-bridging 3 nm filamentous profiles (Fig. 1C). These filamentous profiles are thought to correspond with the protofilaments seen in thin sections (LeLoup *et al.*, 1979; Shimoro and Clementi, 1976). Models of desmosome structure based on morphological studies have been advanced (Fig. 1D).

Desmosome formation has been studied *in vitro* using reassociating embryonic cells (Overton, 1974) and *in vivo* using thin sections of bovine nasal epidermis, a tissue in which there is continual synthesis and turnover of these junctions (LeLoup *et al.*, 1979). The following major stages in desmosome formation are identified: (a) condensation of the intercellular cement; (b) thickening and straightening of the apposed plasma membranes and (c) formation of the central lamina and circular plaques. A role for lectin receptors in the formation of the *macula adherens* in embryonic corneal epithelial cells has been advanced (Overton and DeSalle, 1980).

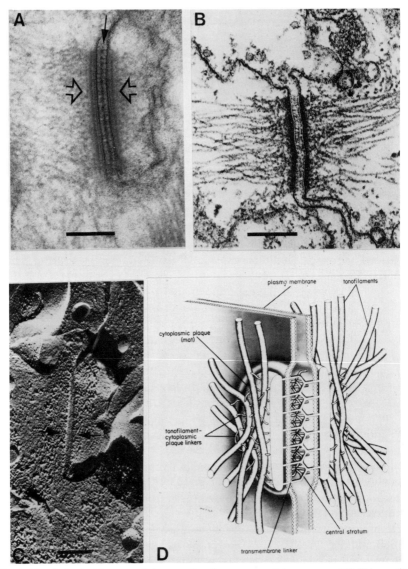

Fig. 1. Morphological features of the *macula adherens* or spot desmosome. (A) Thin section of *macula adherens* of frog epidermis showing central lamina (arrowed) and cytoplasmic plaques (arrowheads). (B) Thin section of ruthenium red stained *macula adherens* of newt epidermis showing 10 nm tonofilaments attached to cytoplasmic plaques. (C) Freeze fracture (cross-fracture) of *macula adherens* of mouse palate epithelium showing central lamina (between arrows). (D) Model of *macula adherens* structure. (B) And (C) by kind permission of D. E. Kelly and (D) Dr L. A. Staehelin. Scale bars (A and B) = 200 nm; (C) = 400 nm.

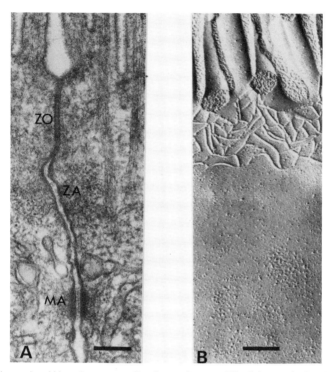

Fig. 2. Thin section (A) and corresponding freeze-fracture (B) of the terminal web region of intestinal epithelium showing *zonula occludens* (Z.O.), *zonula adherens* (Z.A.) and *macula adherens* (M.A.) intercellular junctions. By kind permission of Dr L. A. Staehelin. Scale bars = 200 nm.

2. Protein Composition

The intercellular cement in desmosomes stains with ruthenium red (Fig. 1B) and colloidal iron, and is probably enriched in glycoproteins (Benedetti and Emmelot, 1968; Kelly, 1966; Overton, 1974). A number of glycoproteins are identified in subcellular fractions highly enriched in *maculae adherentes* junctions, especially two major components of approximate molecular weight 130 000 and 115 000 (Drochmans et al., 1978; Mueller and Franke, 1983; Skerrow and Matolsty, 1974). These preparative methods exploit the relative stability of desmosomes of bovine nose epidermis to acidic buffers (Skerrow, 1979; Skerrow and Matolsty, 1974) and neutral detergents (Drochmans et al., 1978). The major polypeptides in bovine nose junctions were of molecular weight 250 000, 215 000, 200 000, 175 000 and 164 000 (Fig. 3) (Franke et al., 1981b). Two of these polypeptides (desmoplakin I and II, molecular weights 250 000 and 215 000) have been identified as major constituents of the

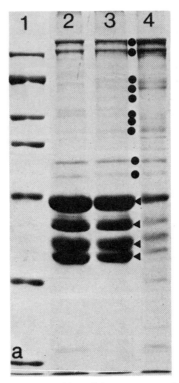

Fig. 3. Polyacrylamide gel of polypeptides of desmosome enriched fractions. Channel 1, reference proteins (from top to bottom: myosin heavy chain, clathrin, β-galactosidase, phosphorylase a, bovine serum albumin, actin, chymotrypsin); Channel 2, desmosome-tonofilament fraction isolated in pH 9·0 buffer; Channel 3, desmosome-tonofilament after additional extraction in high salt buffer containing Triton X-100; Channel 4, desmosome fraction isolated in pH 2·3 citric acid buffer. Dots denote major polypeptides of desmosome rich fractions; arrowheads indicate major prekeratin bands. Reproduced from Franke et al. (1981b), with permission.

desmosome plaques in epithelial and myocardial cells of diverse species using specific antibodies (Franke et al., 1981a, b). Peptide mapping shows the desmoplakins are closely related proteins (Mueller and Franke, 1983). Staining with periodic acid-dansyl hydrazine or fluorescent lectins of polyacrylamide gels in which cow nasal desmosome fractions were electrophoresed also identifies a number of glycoproteins that may be associated with the intercellular cleft material (Gorbsky and Steinberg, 1981; Shrida et al., 1982).

The preparation from rat intestine of residual desmosome plaques has led to the identification of three polypeptides of molecular weight 55 000, 48 000 and 40 000 that probably correspond to tonofilament components (Franke et al., 1981a). Antibodies raised against various desmosome components are expected to feature increasingly in immunocytochemical studies for specifically locating these polypeptides to the component parts of this intercellular junction (Franke et al., 1981b).

B. Hemidesmosomes

These junctions (Fig. 4) resemble half a spot desmosome, but the single cytoplasmic plaque is ellipsoidal, and tonofilament arrangement may differ (Kelly, 1966; Staehelin, 1974). Hemidesmosomes are found mainly at the basal surface of epithelial cells, and are thought to attach cells to the extracellular matrix. No information on their biochemical composition has yet appeared.

C. Zonula Adherens or Belt Desmosome

1. Morphology

This intercellular junction is found as a component of the terminal web encircling and attaching epithelial cells. The parallel plasma membranes are separated by a 25–35 nm gap filled with electron-dense material (Fig. 2). Filamentous mats of poorly defined structure are visible on the cytoplasmic surfaces of the apposed membranes, and fialments of 6–7 nm converge on these (Farquhar and Palade, 1963; Hull and Staehelin, 1979). In freeze fracture, the intramembranous particles are of irregular size and arrangement

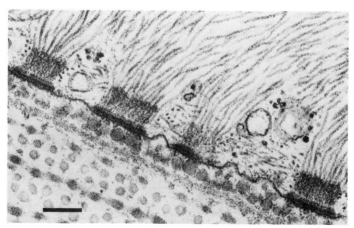

Fig. 4. Thin section of hemidesmosomes from newt epidermis showing 10 nm tonofilaments attached to a single cytoplasmic plaque. By kind permission of Dr D. E. Kelly. Scale bar = 200 nm.

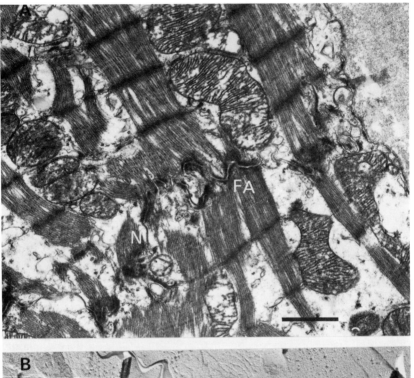

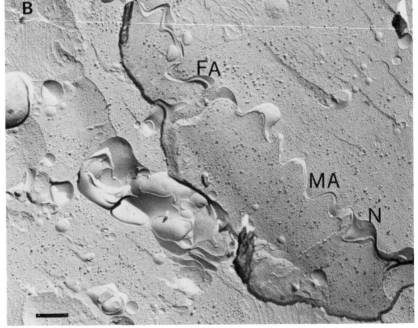

(Mukherjee and Staehelin, 1971). Studies on developing epithelia (Mercer, 1965; Deane and Wurzelmann, 1965), reassociating embryonic cells (Overton, 1962) and hepatocytes (Miettinen et al., 1978) show that zonula-adherens-like structures appear before the other junctions.

2. Protein Composition

Subcellular fractions enriched in zonula adherens junctions have yet to be isolated. When isolated brush borders are stained with heavy meromyosin, the 6–7 nm filaments that converge towards the *zonulae adherentes* junctions are identified as actin-containing filaments (Mooseker, 1976; Rodewald et al., 1976). Detergent extracted brush-border preparations, enriched in *zonulae adherentes* junctions and a core of microfilament bundles contract on addition of ATP and Ca^{2+} or Mg^{2+}, suggesting a role for these junctions in the contraction of microvilli in brush borders (Rodewald et al., 1976). Analysis of detergent-extracted brush-border microvilli shows the presence of myosin, actin, tropomyosin and α-actinin. Two actin-binding polypeptides of molecular weight 68 000 (fimbrin) and 95 000 (villin), are also present in these preparations and one of these (fimbrin) may effect the actin-membrane linkage (Bretscher and Weber, 1979; 1980).

D. The Fascia Adherens

1. Morphology

This is the major junction of the cardiac intercalated disc (Fig. 5). In thin section, the *fascia adherens* resembles the *zonula adherens*, but it does not encircle the cardiocyte (Fawcett and McNutt, 1969; McNutt, 1975). The apposed plasma membranes are separated by a fairly wide intercellular gap (20–40 nm), and no central lamina is present (Fig. 6). Thin filaments converge to form an opulescent cytoplasmic filamentous mat underneath the junctional membrane, and it is this structure that gives the intercalated disc its refractile image, and cardiac muscle its characteristic cross-striations (McNutt, 1975).

2. Protein Composition

Extraction of cardiac plasma membrane fractions containing intact intercalated discs with detergents, followed by centrifugation in sucrose density

Fig. 5. Thin section (A) and freeze-fracture (B) of intercalated disc region of mouse heart showing *macula adherens* (M.A.), *fascia adherens* (F.A.) and the gap junction or nexus (N). (B) Is by kind permission of Dr D. Gros. Scale bars (A) = 1·0 μm; (B) = 200 nm.

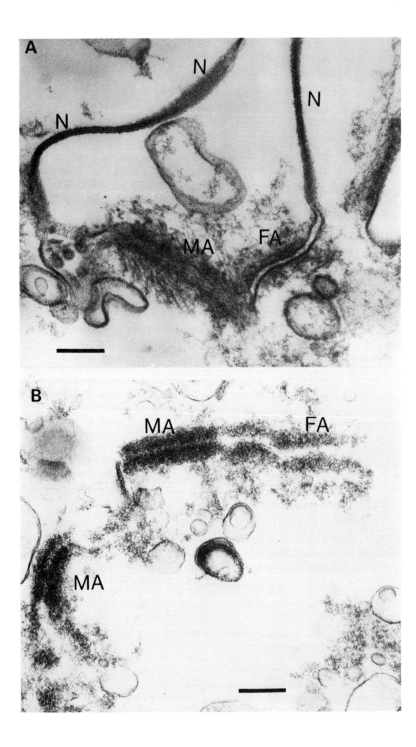

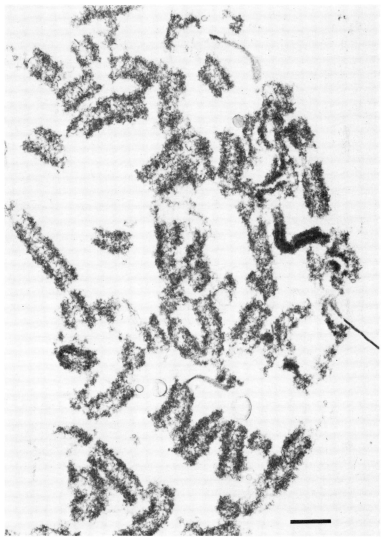

Fig. 7. Field view of *fascia adherens* fraction prepared from cardiac intercalated discs by extraction with detergents and centrifugation in sucrose density gradients. Scale bar=200 nm.

Fig. 6. Isolated cardiac intercalated disc before (A) and after (B) extraction with 1% Triton X-100, showing *macula adherens* (M.A.), *fascia adherens* (F.A.) and gap junction (N). Scale bars=200 nm.

gradients of the detergent-insoluble extract yields fractions containing large numbers of *fasciae adherentes* (Fig. 7) (Colaco and Evans, 1980, 1981, 1982). The major polypeptides in the *fascia adherens* fraction are of average molecular weights 140 000, 64 000 and 60 000 (Colaco and Evans, 1981). Since no residual membrane structure is observed in thin sections of *fascia adherens* fractions (Figs 6b, 7), these polypeptides probably represent components of the cytoplasmic filamentous mats and their linkages across the intercellular gap.

It is perhaps more than coincidental that these major polypeptides of the *fasciae adherentes* correspond closely in molecular weight to the immunochemically identified constituents in the intercalated disc, viz desmin, molecular weight 58 000 (Lazarides, 1980), fimbrin, molecular weight 68 000 (Bretscher and Weber, 1980) and vinculin, molecular weight 130 000 (Geiger et al., 1980). Desmin is a component of intermediate filaments, whereas fimbrin and vinculin are thought to be membrane microfilament bundle linkage proteins. It is notable that the polypeptide components of *fascia adherens* are different to those of macula adherens junctions (Colaco and Evans, 1981; Mueller and Franke, 1983).

III. CELL-CELL COMMUNICATION JUNCTIONS

Communication between cells comprising tissues and organs can occur either indirectly across the intercellular space, as in chemical synapses, or directly between cells as in the gap junction or nexus. Cytoplasmic bridges can also facilitate cell-cell communication (Woodruff and Telfer, 1980), but these are not ordinarily considered as intercellular junctions.

A. Synaptic Junctions

1. Morphology

There is considerable heterogeneity in the morphology of synaptic junctions of the central and peripheral nervous systems (Pfenninger and Rees, 1976; Manina, 1979), but some common features can be described briefly. The intercellular cleft (25–35 nm) separating the pre- and postsynaptic membranes contains electron-dense material which appears as a system of fibres after osmium fixation (Gray, 1969). The postsynaptic density or thickening, and the presynaptic dense projections represent cytoplasmic elaborations of their respective junction-forming synaptic plasma membranes, and the extent of their differentiation can vary from synapse to synapse (Fig. 8). Microtubules can be visualized in the pre- and postsynaptic regions, often linked to components in the cytoplasm (Westrum and Gray, 1976; Gray, 1976). In

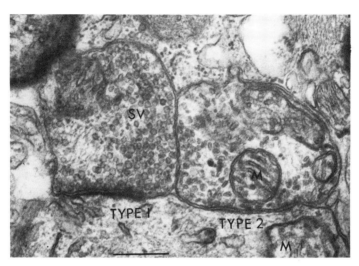

Fig. 8. Two synaptic bulbs in contact with a dendrite (d) in the spinal cord. On the left, an excitatory bulb containing many round synaptic vesicles (sv) and a Type 1 thickening at the synaptic cleft. On the right, an inhibitory bulb with flattened synaptic vesicles and a Type 2 cleft (m, mitochondrion). By kind permission of Dr G. Gray. Magnification ×38 000; scale bar=400 nm.

freeze fracture, distinct 8–9 nm particles occur in the postsynaptic membrane (Sandri et al., 1972). The presynaptic membrane contains fewer intramembranous particles than the postsynaptic membrane. Reports of pores, "synaptopores", in this membrane, that may correspond to points of vesicle fusion with the presynaptic membrane and release of transmitters, have also appeared (Pfenninger and Rees, 1976; Heuser et al., 1975; Matus, 1978). Studies of synapse formation in vitro indicate that during the transformation of the filapodia of the axonal tip into a presynaptic terminal, growth activity ceases when the axonal filapodia approach within about 6–8 nm of the postsynaptic neuron (Pfenninger and Rees, 1976). About 6–12 h later, thickening of the postsynaptic membrane occurs, followed by the appearance of synaptic vesicles and the interneuronal cleft increases in width to 15–20 nm (Slater, 1978). Observations in vivo also show that the formation of the postsynaptic density preceeds the appearance of synaptic vesicles (Nelson, 1975; Rakic, 1975).

2. Protein Composition

Defining the molecular components of the synaptic junctions has undoubtedly been delayed by the cellular and functional heterogeneity of the various regions of the central nervous system. Synaptic junctions are isolated normally from brain synaptosome preparations, but the extent of contamina-

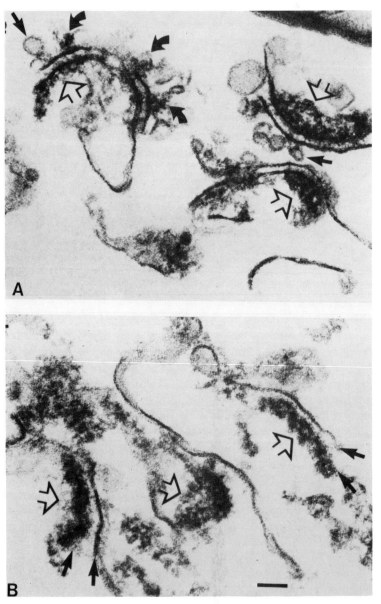

Fig. 9. Examples of four types of synaptic junctional extracted found in *n*-octyl glucoside fractions of rat brain synaptic membranes. (A) Synaptic junctional complexes. These junctions retain the presynaptic dense projections (curved arrows), many of which are associated with synaptic vesicles (arrows), and have intact pre- and postsynaptic membranes and a postsynaptic density (open arrows). (B) Synaptic junctions. Synaptic

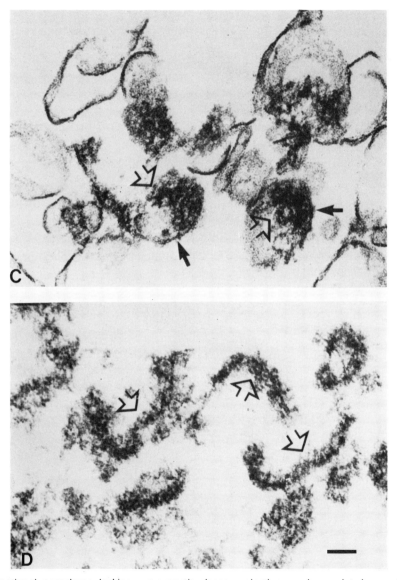

junctional complexes lacking pre-synaptic dense projections and associated synaptic vesicles. Arrows point to pre- and postsynaptic membranes. (C) Postsynaptic elements. Postsynaptic densities with overlying postsynaptic membranes (arrows). (D) Postsynaptic densities. The open arrows point to the cytoplasmic side. Photograph by Dr P. Gordon-Weeks. Scale bar=100 nm.

tion by non-synaptosomal membranes, especially those of glial origin is a major problem (Evans, 1978; Gurd, 1982; Matus et al., 1980). Many investigators have directed their efforts towards identifying constituents of the pre- and postsynaptic density complex after detergents are used to solubilize and remove non-junctional components (Fig. 9) (Kelly and Cotman, 1977; Matus and Jones, 1978; Carlin et al., 1980). Judicious use of low concentrations of mild detergents can leave the pre- and postsynaptic membranes and their elaborations largely intact (Fig. 9).

The major proteins identified at the synaptic junction are summarized in Table I. Figure 10 shows that as postsynaptic structures are purified, there is a progressive increase in the concentration of proteins of apparent molecular weight 180 000, 130 000, 110 000, 94 000, 65 000, 60 000 and 51 000 (Gurd et al., 1982). The 51 000 molecular weight protein probably corresponds to the postsynaptic density specific protein described by Kelly and Cotman (1978). This protein may be absent in cerebellar junctions (Carlin et al., 1980; Flanagan et al., 1982). A number of glycoproteins appear to be located in the postsynaptic density (Gurd, 1980, 1982), and this location has now been confirmed by using ferritin-labelled concanavalin A as an ultra-structural probe to show that these glycoproteins are present at the inner aspect of the

TABLE I

Proteins Identified in Synaptic Junctions and Postsynaptic Densities

Protein	Methods of identification	Reference
Tubulin	Immunochemistry	Matus et al. (1975)
	Peptide mapping	Kelly and Cotman (1978)
Actin	2-D gels	Kelly and Cotman (1978)
	Immunochemistry	Toh et al. (1976)
	Peptide mapping	Mushinsky et al. (1978)
Neurofilament	Immunochemistry	Berzins et al. (1977)
Calmodulin	Amino-acid analysis	Grab et al. (1979)
	Immunochemistry	Wood et al. (1980)
Proteins 1a* and 1b	Phosphorylation by cAMP-dependent protein kinase	Ueda et al. (1979) Kelly et al. (1979)
Thy-1	Immunochemistry	Acton et al. (1978)
Myosin	Peptide mapping Immunochemistry	Kelly (1981)
Fodrin (brain spectrin)	Peptide mapping Immunochemistry	Carlin et al. (1983)

* Protein 1 (synapsin) has now been clearly shown to be located in synaptic vesicles (De Camilli et al., 1983).

postsynaptic membrane (Gurd et al., 1982). Minor differences are seen in the polypeptide components of postsynaptic density complexes prepared from different species (Carlin et al., 1980; Gurd et al., 1982).

A number of enzymic activities have been measured in fractions containing synaptic junctions. These include cAMP-stimulated protein kinases (de Blas et al., 1979; Weller and Morgan, 1976), Ca^{2+}-activated protein kinases (Grab et al., 1979), protein phosphatases (Kelly et al., 1979), cyclic nucleotide phosphodiesterase (Therien and Mushynski, 1979) and glycoprotein sialidases (Gurd, 1980, 1982).

Undoubtedly, highly purified synaptic junctions (Fig. 9) are still heterogeneous, and at least 4 major types of junctional structures can be recognized. The use of monoclonal antibodies is expected to feature increasingly in distinguishing functionally between the various morphological entities remaining in fractions prepared using detergents to remove all components except the junctional complexes (Nieto-Sampedro et al., 1981; Gurd et al., 1983).

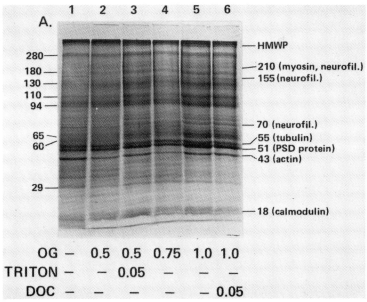

Fig. 10. Protein composition of rat brain synaptic junctional fractions extracted with varying concentrations (% w/v) of OG, n-octyl glucoside, Triton X-100, and DOC, deoxycholate. Extractions were carried out simultaneously with phase partitioning of a synaptosomal fraction in polyethylene glycol and dextran two phase polymer system. Proteins were analysed by polyacrylamide gel electrophoresis in sodium dodecyl sulphate-containing buffers. The molecular weights ($\times 10^{-3}$) and the possible identity of polypeptides are indicated. HMWP, high molecular weight tubulin associated proteins; neurofil, neurofilament. Increasing OG concentrations was paralled by an increasing purity of the postsynaptic densities. Reproduced from Gurd et al. (1982a), with permission.

B. Gap Junctions

These have now emerged as ubiquitous plasma membrane specializations providing private pathways of communication between cells with prohibition of access to the extracellular space. Gap junctions are permeable to intracellular molecules with a nominal diameter of 1·2 nm, i.e. a molecular weight of about 1000. Thus, they are permeable to small sugars, ions and nucleotides, but not to proteins, nucleic acids and other macromolecules. Gap junctions have the potential for mediating in the electrical and chemical (or metabolic) coupling of cells, thereby accounting for their functional definition as electrotonic synapses or junctions or as communicating junctions. Communication between cells across gap junctions can be detected using electrical methods, techniques that exploit the phenomenon of metabolic cooperation, and dye transfer (e.g. lucifer yellow, 6-carboxy-fluorescein) (Lowenstein, 1979).

1. Morphology

In thin sections, the gap junction is an intimate apposition of the junction-forming regions of the plasma membrane (Fig. 12B), but the cells are separated by a 2–4 nm gap of extracellular space permeable to heavy metal salts, e.g. lanthanum hydroxide (Revel and Karnovsky, 1967). Oblique sections frequently reveal a lattice of subunits that bridge the intercellular gap (Fig. 12B) (McNutt and Weinstein, 1973). However, the trademark of the gap junction is now provided by freeze fracture (Fig. 13), for a hexagonal lattice of 4–6 nm intramembranous particles with a centre-to-centre spacing of 8–9 nm on the P face, with a corresponding lattice of pits on the E-face is seen (Chalcroft and Bullivant, 1970; McNutt and Weinstein, 1973). A central depression of about 2·5 nm is often observed on the particles, and this is often equated with the intercellular channel seen as an electron-dense dot in negatively-stained preparations of isolated hepatic gap junctions (Fig. 14) (Peracchia, 1980; Unwin and Zampighi, 1980). Rapid-freeze techniques suggest that the hexagonal lattice seen in freeze-fractured junctions is an indication that the junction is in a closed or uncoupled state (Peracchia, 1977, 1978; Raviola et al., 1980).

Gap junctions have also been analysed using replicas of quick-frozen and deep-etched samples. Crystallized junctions were shown to be in the closed high-resistance state, with the channel-closing apparatus most likely located at or near the inside of the junctional membrane (Hirokawa and Heuser, 1982). Using fractured hepatocytes, the deep-etching procedure indicated that there were no prominent protrusions and no undercoatings of cytoskeletal elements on the inner (cytoplasmic) side of the junction. The smooth

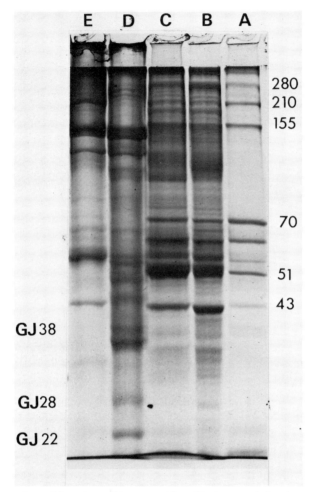

Fig. 11. Comparison of the polypeptide composition of brain synaptic fractions with liver gap junction and cardiac *fascia adherentes* fractions by polyacrylamide gel electrophoresis in sodium dodecyl sulphate-containing buffers. Columns represent: (A) neurofilament standard; (B) synaptic junctions prepared by the method of Cotman and Taylor (1974); (C) synaptic junctions prepared by two-phase partitioning in the presence of *n*-octyl glucoside; (D) a liver fraction enriched in gap junctions, and (E) a cardiac *fascia adherens* fraction. Molecular weights ($\times 10^{-3}$) are indicated. Major gap junction (GJ) polypeptides are indicated. Reproduced from Gurd *et al.* (1982a), with permission.

undersurface of gap junctions contrasts with the corrugated outer surface with uniform 8–9 nm protrusions with central pores seen in junctions split open by perfusion with hypertonic saline. These results will require current models (Fig. 15; Unwin and Zampighi, 1980) to be updated.

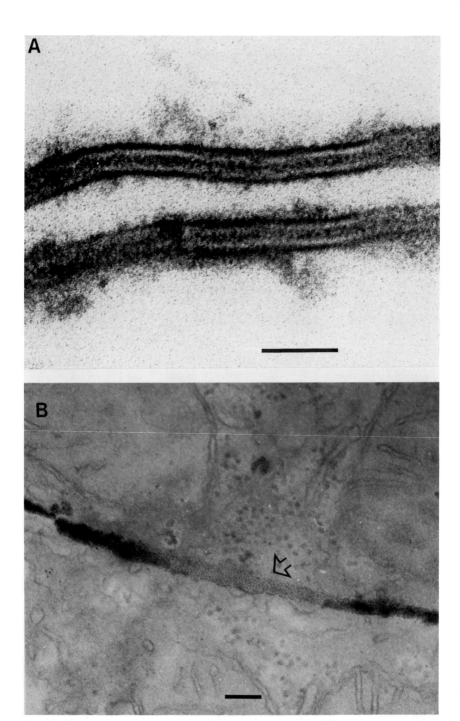

Fig. 12. Thin section of an isolated rat liver gap junction (A). (B) Section through liver impregnated with lanthanum showing how it outlines the hexagonal lattice of subunits (arrow). By kind permission of J. P. Revel. Scale bars (A)=50 nm, (B)=400 nm.

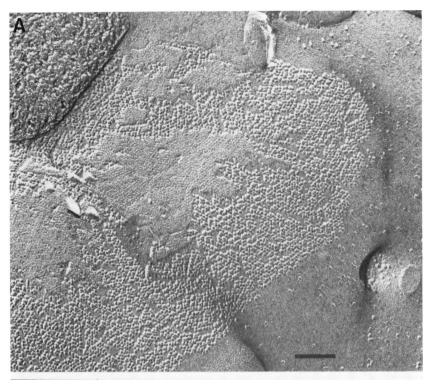

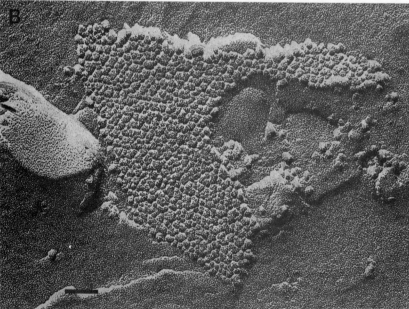

Fig. 13. Freeze fracture of rat liver (A) and an isolated rat heart (B) gap junction showing hexagonal lattice of subunits on the P-fracture face and corresponding pits on the E-fracture face in (A). (A) is by kind permission of Dr J. P. Revel.

Although gap junctions are morphologically similar in all phylla examined (e.g. hydra; Filshea and Flower, 1977, jellyfish; King and Spencer, 1979), a major exception are the arthropoda, where in thin section, the gap is 3–4·5 nm and the subunits are often seen to bridge the intercellular gap (Satir and Gilula, 1973; Zampighi et al., 1978). In freeze fracture, arthropod gap junctions contain larger subunits with greater lattice spacing (12–15 nm). Furthermore, the intramembranous particles are associated with the E-face

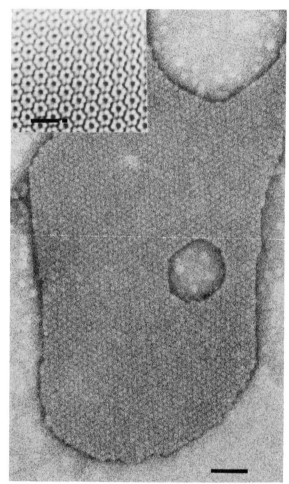

Fig. 14. Isolated rat liver gap junction negatively stained with 1% sodium silicotungstate, showing hexagonal lattice of subunits. *Inset.* Similar gap junction fragment but completely unstained and subjected to digitally filtered image analysis. The six-subunit construction of the lattice is clearly shown. Photograph kindly given by N. G. Wrigley and E. Brown. Scale bars=50 nm, inset=20 nm (Wrigley et al., 1984).

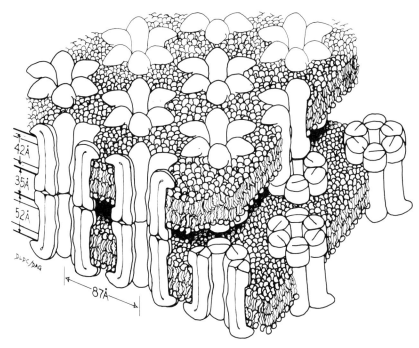

Fig. 15. Diagrammatic representation of the organization of a hepatic gap junction as inferred from X-ray diffraction and electron microscopical studies. This model proposes hexagonal symmetry with the hydrophilic channel allowing direct intercellular communication located between the subunits. Reproduced from Makowski et al. (1977), with permission.

(Staehelin, 1974; Graf, 1978). Intercellular junctions containing orthogonal arrays are present in freeze-fractured hypothalamus, but the function of such arrangements and their relationship to other junctions is obscure (Hatton and Ellisman, 1981).

The lattice regularity of gap junctions makes them attractive candidates for application of physical techniques to determine their three-dimensional structure. Models have been proposed on the basis of X-ray diffraction data (Fig. 15) (Makowski et al., 1977), low-irradiation microscopy (Baker et al., 1983) and image process analysis of negatively-stained isolated liver gap junctions (Unwin and Zampighi, 1980; Zampighi and Unwin, 1979). Models postulate that the transmembrane channel is formed from six hexagonally-arranged subunits containing the polypeptides studied biochemically (see below). Figure 14 shows an isolated negatively-stained gap junction and the image reconstruction obtained using unstained junctions.

Studies on gap junction development have been carried out *in vitro* using reaggregating cells (Johnson et al., 1974; Sheridan, 1978; Preus et al., 1981) and *in vivo* in embryonic systems (Ginzberg and Gilula, 1979; Shibata *et al.*, 1980) and regenerating liver (Meyer *et al.*, 1981). The initial step involves the appearance of "formation plaques" occurring when the paired membranes are separated by a 20–30 nm intercellular gap (Sheridan, 1978). In freeze fracture, these plaques appear as raised areas containing clusters of large intramembranous particles (Ginzberg and Gilula, 1979). The intercellular space then narrows to 2–4 nm, and this process is accompanied by the appearance of intramembranous particles similar in size to those of mature junctions (Staehelin, 1974), which slowly aggregate leaving particle-free aisles (Shibata *et al.*, 1980). Gap junctions, especially those in cardiac intercalated discs, are extremely susceptible to ischaemia (Ashraf and Halverson, 1978), leading to rapid transfer of junctions to the cell interior (Colaco, 1980). Whorls of internal gap junctions, as observed, for example, in eye lens epithelia (Fentiman *et al.*, 1982) may also indicate internalization routes that lead to junction degradation. Cytoplasmic gap junctions, frequently with annular profiles, are observed in several cell types and these are often associated with acid phosphatase activity suggesting that they may represent a stage in their degradation (Larsen and Tung, 1978). Gap junction formation in many tissues is hormonally regulated (Decker, 1976) and one of the most intensively investigated tissues from this viewpoint is the myometrium, especially with respect to their control by oestrogens (Garfield *et al.*, 1980). During liver regeneration, hepatocyte gap junctional area is reduced approximately 100-fold suggesting a role in cell proliferation (Meyer *et al.*, 1981).

2. Protein Composition

Important strides have been made in studying the proteins of gap junctions, this, undoubtedly, has been aided by the development of various methods for their isolation and their relative simple polypeptide composition. Gap junctions can account for 0·2–2% of the total plasma membrane area, but there can be much variation in different regions of the same tissue or organ. For example, gap junctions are about ten times more numerous in leading pacemaker cells of sinus node than in working myocardium; overall, macular gap junctions account for 0·2% of plasma membrane surface area in rabbit heart (Masson-Pevet *et al.*, 1979).

Extraction of plasma membrane fractions with detergents removes most of the nonjunctional membranes and leaves mainly a residue containing, in liver, gap junctions, collagen and undefined amorphous material (Culvenor and Evans, 1977). A variety of procedures have been described for purifying further gap junctions, including the use of collagenases of varying degrees of purity, sonication, various detergents, urea and alkaline washes (Culvenor and Evans, 1977; Duguid and Revel, 1975; Ehrart, 1981; Henderson *et al.*,

1979; Hertzberg and Gilula, 1979). These approaches have yielded varying results, but a major polypeptide of molecular weight 26 000–28 000, that rapidly proteolyses to yield fragments of molecular weight 21 000 and 10 000, has emerged as the major polypeptide of hepatic gap junctions (Finbow *et al.*, 1980; Nicholson *et al.*, 1981). Figure 16 shows the amino-acid sequence of the fifty-two amino acids at the NH_2-terminus of the gap junction polypeptide.

Subcellular fractions from heart enriched in gap junctions show a more heterogeneous polypeptide composition than those prepared from liver, probably reflecting fraction impurity, but polypeptides of approximate molecular weight 26 000 and 29 000 are present (Colaco and Evans, 1981; Kensler and Goodenough, 1980; Manjanath *et al.*, 1982). A great deal of attention has also been focused on eye lens as a potential source of gap junctions that can be prepared using simple subcellular fractionation routines (Benedetti *et al.*, 1976; Broekhuyse *et al.*, 1976, 1979; Bloemendal, 1977). Analysis of thin sections in the electron microscope shows that eye lens fibre junctions are identical to those of liver and other tissues, although major differences are sometimes observed by freeze-fracture analysis (Kistler and Bullivant, 1980; Zampighi *et al.*, 1982). Both liver and eye lens gap junction fractions contain a major polypeptide of molecular weight 26 000–28 000. However, despite superficial ultrastructural and compositional similarities, it has been clearly shown that these polypeptides are not homologous, using partial peptide mapping and by immunochemical criteria (Hertzberg *et al.*, 1982), although the immunochemical criteria are now questioned (Traub *et al.*, 1982; Traub and Willecke, 1982). The eye lens fibre junction may contain a unique gene product (Bok *et al.*, 1982) that is functionally different, a fact supported by poor evidence for electrotonic coupling between the compressed cells in the lens cortex (Rae, 1978). Eye lens epithelial cells are connected by conventional gap junctions (Fentiman *et al.*, 1982). The presence of these junctions, in addition to the predominent fibre cell junctions, thus complicates interpretations when whole eye lens junctions are examined (Colaco, 1981).

The identification of the major polypeptide component of liver gap junctions, and its degradation products has encouraged the application of methods for measuring their turn-over rates. Although analysis of gross gap

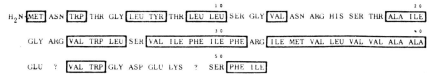

Fig. 16. Sequence of the fifty-two amino terminal residues of the major 28 000 molecular weight polypeptide of hepatic gap junctions. Hydrophobic residues are enclosed in boxes. Question marks indicate sequence regions where no amino acid could be identified. From Nicholson *et al.* (1981), with permission.

junction fractions indicated a slow rate of metabolic turnover (Evans and Gurd, 1973), studies on the incorporation of ^{3}H-leucine, ^{35}S-methionine and ^{14}C-bicarbonate into the junctional polypeptides indicate a more rapid turnover rate in liver with a half-life of between 5 and 19 h (Fallon and Goodenough, 1981; Yancey et al., 1981). Turnover of junctions from regenerating and cholestatic liver has been examined using an antibody against the 28 000 mol. wt polypeptide (Traub et al., 1983). It has been suggested that this rapid metabolic turnover is a correlate of the responsiveness of gap junctions and thus intercellular communication to physiological demands (Yancey et al., 1981).

The use of the freeze-fracture technique and filipin, a polyene antibiotic, has provided new insights into the distribution of cholesterol in non-junctional and junctional regions of the plasma membrane. Filipin-cholesterol complexes, recognized as 25 nm protuberances in the electron microscope, are extensive in non-junctional surface areas of hepatocytes, but are nearly absent in tight and gap junction formation zones (Robenek et al., 1982). It is hypothesized that the generation of plasma membrane zones deficient in cholesterol facilitates adhesion between cell surfaces and modulates lipid-protein interactions which control the assembly of junctions.

IV. TIGHT JUNCTIONS OR ZONULAE OCCLUDENTES

In the *zonulae occludentes* (Fig. 2), the plasma membranes are in close apposition along the length of the junction with points of total occlusion of the extracellular space (Farquhar and Palade, 1963) making the junction impermeable to heavy metal salts (Revel and Karnovsky, 1967). Freeze-fracture analysis shows a meshwork of 6–8 nm wide fibrils on the P-face with complementary grooves on the E-face (Fig. 2B) (Chalcroft and Bullivant, 1970; Staehelin, 1974). The complexity of the meshwork may be a measure of junction "leakiness", for the most physiologically "tight" junctions, e.g. Sertoli and rat choroid plexus junctions possess extremely complex anastomizing fibril networks (Bouvier and Bouchard, 1978).

The initial stage in the formation of tight junctions is the appearance of paired particle-free zones in the apposed membranes, followed by the assembly of linear arrays of particles in these zones. Finally, the particles fuse to form the intramembranous ridges (Porvanzik et al., 1976, 1978; Sheridan, 1978). Gap and tight junctions are often found in close proximity in epithelial tissues, and the same intramembranous components have also been implicated in the formation of gap junctions (Ginsberg and Gilula, 1979; Montesano et al., 1975). Studies using excised rat prostate tissue, which undergoes massive proliferative assembly of new tight junction strands along the lateral side of the columnar epithelial cells, have emphasized the speed of assembly (within 5 min) and the lack of effect of inhibitors of protein synthesis (e.g. cycloheximide) and metabolic uncouplers (e.g. dinitrophenol). These results suggest, as in the case of gap junctions, that the proliferative assembly of tight junctions involves their molecular reorganization from a pool of pre-existing components (Kachar and Pinto da Silva, 1981).

Despite their abundance in epithelia, no information on their isolation and therefore protein composition is available. In epithelial monolayers tight junctions not only control the passage of solutes across the monolayer, but also segregate the plasma membrane into biochemically distinctive basolateral (serosal) and apical (mucosal) zones. This functional segregation of the plasma membrane, which breaks down when cells are isolated (Pisam and Ripoche, 1976), has been shown experimentally using membrane-bound lectins that fail to pass through the tight junctional region. However, lipid soluble fluorescent probes diffused freely through the tight junction provided that they could "flip-flop" to the inner monolayer of the plasma membrane bilayer (Dragsten *et al.*, 1981).

The proposition that intrinsic membrane proteins are the principal structural elements of the strands comprising tight junctions is now questioned. Direct rapid freezing of newly-formed junctions between rat prostrate epithelial cells suggest that the individual tight junction strands are pairs of inverted cylindrical lipid micelles sandwiched between the fused membranes of adjacent cells (Kachar and Reese, 1982).

V. SEPTATE JUNCTIONS

Since their initial characterization, three main types of septate junctions have been described (Graf, 1978; Noirot-Timothée and Noirot, 1980). All these categories of junctions consist of paired membranes with electron-dense bars or "septae" of regular width and periodicity bridging the intercellular gap. The categories of septate junctions differ in the width of the septae, their periodicity and the width of the intercellular gap. Tangential sections display a honeycomb appearance, and in lanthanum-impregnated tissues, pleated or undulating bands of regular periodicity are seen. In freeze-fracture, parallel rows of intramembranous particles are observed on the P-face with corresponding rows of pits on the E-face. The categories of septate junctions also vary in intramembranous particle size and arrangement (Graf, 1978).

The physiological role of septate junctions is as yet unclear. An earlier hypothesis advocating a role in intercellular communication (Satir and Gilula, 1973) has been superseded by a suggestion of a role in the regulation of paracellular transepithelial flow (Lord and Dibona, 1976; Bilbaut, 1980).

Indeed, it may be that septate junctions in invertebrate epithelia and tight junctions in vertebrates are functionally analogous, both representing different membrane differentiations for control of transepithelial gradients and transport.

VI. FUTURE PROSPECTS

During the last twenty-five years, knowledge of intercellular junctions has advanced almost entirely through morphological observation in the electron microscope. The ultrastructural features seen predominantly in thin, stained sections have allowed the major classes of junctions to be defined (Farquhar and Palade, 1963), and important functional distinctions to be made, as, for example, between tight and gap junctions (Revel and Karnovsky, 1967). The introduction of the freeze-fracture technique in the mid-to-late sixties identified some further morphological properties of intercellular junctions, especially relating to the arrangement of intramembranous particles in the junctional domains of the plasma membrane lipid bilayers. Important clues about the events preceding and surrounding the development and assembly of junctions were obtained using the morphological techniques. The morphological approach will undoubtedly continue to make important contributions to our knowledge of the structure of intercellular junctions, with the adoption of new techniques, e.g. high-voltage electron microscopy, computer-image processing of stained and unstained specimens, electron microscopy of frozen sections and X-ray diffraction, low temperature embedding, freeze-fracture immunochemistry, etc.

However, there is no question that the final details of the molecular architecture of the various categories of intercellular junctions, their function, synthesis and turnover, will emerge from the application of new techniques in membrane biochemistry, immunology and molecular biology. New developments in the methods for isolating and sequentially disassembling junctions and for characterizing their constituent proteins, glycoproteins and lipids will result in the chemical structure of their constituents complementing and possibly displacing morphology as the absolute criterion of junctional type. The availability of antibodies prepared against discrete junctional proteins will allow them to be used as tools to analyse junctions at the fluorescence microscopy, immunoelectronmicroscopy and molecular level. Determination of the amino-acid sequence of junctional polypeptides is seen increasingly as a prelude to the identification of the genes that regulate their appearance and functioning.

Already, as illustrated in Section III.B, immunological and molecular approaches are being used to advance our knowledge of junctional structure.

The morphological regularity and apparent molecular parsimony in the construction of gap junctions has made them attractive candidates for analytical and functional dissection using modern methods and it has been shown, for example, that although similarities are to be found in the intramembranous particle arrangement and major polypeptide components of the hepatic and lens gap junctions, they are now seen to be the products of different genes (Hertzberg *et al.*, 1982). Morphological similarities between gap junctions in various tissues will soon be explored at the immunochemical, peptide mapping, and amino-acid sequence level. Information on amino-acid sequences will allow synthetic oligonucleotides, complementary to unique sequences in the messenger RNAs coding for the gap-junctional polypeptides, to be used as hybridization probes for the identification of cDNA clones and genes coding for the gap-junctional polypeptides, thereby allowing the study of the genetic control of intercellular communication across gap junctions and its modulation by hormones, etc.

Progress in unravelling the morphologically more complex desmosome (Franke *et al.*, 1981a, b) and synaptic junctions (Gurd, 1982) (Sections II.A, III.A) is being aided by the availability of a catalogue of antisera, soon, undoubtedly, to be widened by a variety of monoclonal antibodies against the major and minor junctional polypeptides and attached components of the cytoskeleton and extracellular matrix. Ferritin-labelled antibodies are being gradually displaced by those labelled with colloidal gold particles of various diameters that can be attached also to lectins, and other proteins, e.g. horseradish peroxidase. Thus, these electron microscopic immunolocalization studies can be used to study, in various tissues, the nature of the linkage of cytoskeletal elements to adhesion junctions. Furthermore, biochemical and immunochemical approaches will feature in unravelling the details of the assembly, dispersal, internalization and recycling of junctional polypeptides in embryogenesis, organogenesis, etc. The molecular nature of intramembranous particles concentrated at junctional areas of the plasma membrane will become clearer using, for example immunolabelled replicas.

Finally, morphology has shown that intercellular junctions are modified in malignancy, e.g. in carcinomas and metastases, and it is anticipated that the nature of these changes will become clearer when monoclonal antibody and gene cloning techniques are used to probe the changes leading to uncontrolled growth and division. The product of the oncogene of *Rous sarcoma* virus (pp 60) is concentrated in the region of adhesion plaques and cell-cell contact areas of cells, indicating that intercellular junctions may be open to modification in virally-induced cancers (Bishop, 1982).

VII. ACKNOWLEDGEMENTS

We thank Drs J. W. Gurd (Scarborough College, University of Toronto) and J. Hope for reading and commenting on the manuscript.

VIII. REFERENCES

Acton, R. T., Addis, J., Carl, G. F., McClain, L. D. and Bridges, W. (1978). *Proc. Natn. Acad. Sci. USA* **75**, 3283–3287.
Ashraf, M. and Halverson, C. (1978). *J. Mol. Cell Cardiol.* **10**, 263–269.
Baker, T. S., Casper, D. L. D., Hollingshead, C. J. and Goodenough, D. A. (1983). *J. Cell Biol.* **96**, 204–216.
Benedetti, E. and Emmelot, P. (1968). *J. Cell Biol.* **38**, 15–24.
Benedetti, E. L., Dunia, I., Bentzel, C. J., Vermorken, A. J. M., Kibbelaar, M. and Bloemendal, H. (1976). *Biochim. Biophys. Acta* **457**, 353–384.
Berzins, K., Cohen, R. S., Grab, D. and Siekevitz, P. (1977). *Soc. Neurosci. Abstr.* **3**, 331.
Bilbaut, A. (1980). *J. Cell Sci.* **41**, 341–348.
Bishop, J. M. (1982). *Sci. Amer.* **246**, 68–79.
Bloemendal, H. (1977). *Science* **197**, 127–138.
Bok, D., Dockstader, J. and Horwitz, J. (1982). *J. Cell Biol.* **92**, 213–220.
Bouvier, D. and Bouchaud, C. (1978). *Biol. Cellulaire* **31**, 109–112.
Bretscher, A. and Weber, K. (1979). *Proc. Natn. Acad. Sci. USA* **76**, 2321–2325.
Bretscher, A. and Weber, K. (1980). *J. Cell Biol.* **86**, 335–340.
Broekhuyse, R. M., Kuhlmann, E. D. and Stols, A. L. H. (1976). *Exp. Eye Res.* **23**, 365–371.
Broekhuyse, R. M., Kuhlmann, E. D. and Winkers, H. J. (1979). *Exp. Eye Res.* **29**, 303–313.
Carlin, R. K., Bartelt, D. C. and Siekevitz, P. (1983). *J. Cell Biol.* **96**, 443–448.
Carlin, R. K., Grab, D. J., Cohen, R. S. and Siekevitz, P. (1980). *J. Cell Biol.* **86**, 831–843.
Chalcroft, J. P. and Bullivant, S. (1970). *J. Cell Biol.* **47**, 49–60.
Colaco, C. A. L. S. (1980). PhD thesis. Council for National Academic Awards, UK.
Colaco, C. A. L. S. and Evans, W. H. (1980). *Biochem. Soc. Trans.* **8**, 328.
Colaco, C. A. L. S. and Evans, W. H. (1981). *J. Cell Sci.* **52**, 313–325.
Colaco, C. A. L. S. and Evans, W. H. (1982) *Biochim. Biophys. Acta* **684**, 40–46.
Culvenor, J. G. and Evans, W. H. (1977). *Biochem. J.* **168**, 475–481.
Deane, H. W. and Wurzelmann, S. (1965). *Amer. J. Anatomy* **117**, 91–133.
de Blas, A. L., Wang, Y. J., Sorensen, R. and Mahler, H. R. (1979). *J. Neurochem.* **33**, 647–659.
De Camilli, P., Harris, S. M., Huttner, W. B. and Greengard, P. (1983). *J. Cell Biol.* **96**, 1358–1373.
Decker, R. S. (1976). *J. Cell Biol.* **69**, 669–685.
Dragsten, P. R., Blumenthal, R. and Handler, J. S. (1981). *Nature* **294**, 718–722.
Drochmans, P., Freudenstein, C., Wanson, J.-C., Laurent, L., Keenan, T. W., Stadler, J., LeLoup, R. and Franke, W. (1978). *J. Cell Biol.* **79**, 427–443.
Duguid, J. R. and Revel, J. P. (1975). *Cold Spring Harb. Symp. Quant. Biol.* **40**, 45–47.

Ehrart, J. C. (1981). *Cell Biol. Int. Rep.* **5,** 1055–1061.
Evans, W. H. (1978). "Preparation and Characterisation of Mammalian Plasma Membranes". North Holland-Elsevier, Amsterdam.
Evans, W. H. and Gurd, J. W. (1973). *Eur. J. Biochem.* **36,** 273–279.
Fallon, R. F. and Goodenough, D. A. (1981). *J. Cell Biol.* **90,** 521–526.
Farquhar, M. G. and Palade, G. E. (1963). *J. Cell Biol.* **17,** 375–409.
Fawcett, D. W. and McNutt, N. S. (1969). *J. Cell Biol.* **42,** 1–67.
Fentiman, I., Hodges, G., Newman, R. and Stoker, M. (1982). *Exp. Cell Res.* **139,** 95–100.
Filshea, B. K. and Flower, N. E. (1977). *J. Cell Sci.* **23,** 151–172.
Finbow, M., Yancey, S. B., Johnson, R. and Revel, J.-P. (1980). *Proc. Natn. Acad. Sci. USA* **77,** 970–974.
Flanagan, S. D., Yost, B. and Crawford, G. (1982). *J. Cell. Biol.* **94,** 743–748.
Franke, W. W., Winter, S., Grund, C., Schmid, E., Schiller, D. L. and Jarasch, E. D. (1981a). *J. Cell Biol.* **90,** 116–127.
Franke, W. W., Schmid, E., Grund, C., Muller, H., Engelbrecht, R., Moll, R., Stadler, J. and Jarasch, E. D. (1981b). *Differentiation* **20,** 217–241.
Garfield, R. E., Kannan, M. S. and Daniel, E. E. (1980). *Amer. J. Physiol.* **238,** C81–C89.
Geiger, G., Tokuyasu, K. T., Dutton, A. H. and Singer, S. J. (1980). *Proc. Natn. Acad. Sci. USA* **77,** 4127–4131.
Gilula, N. B. (1978). In "Intercellular Junctions and Synapses" (J. Feldman, N. B. Gilula and J. D. Pitts, eds) pp. 3–22. Chapman and Hall, London.
Ginsberg, R. D. and Gilula, N. B. (1979). *Dev. Biol.* **68,** 110–129.
Gorbsky, G. and Steinberg, M. S. (1981). *J. Cell Biol.* **90,** 243–248.
Grab, D. J., Berzins, K., Chen, R. S. and Siekevitz, P. (1979). *J. Biol. Chem.* **254,** 8690–8696.
Graf, F. (1978). *Biol. Cellulaire* **33,** 55–62.
Gray, E. G. (1969) *Progr. Brain Res.* **31,** 141–155.
Gray, E. G. (1976). "The Synapse". Carolina Biology Readers. Burlington, N. Carolina.
Gurd, J. W. (1980). *Can. J. Biochem.* **58,** 941–1003.
Gurd, J. W. (1982). In "Molecular Approaches to Neurobiology". pp. 99–130. Academic Press, New York.
Gurd, J. W., Gordon-Weeks, P. and Evans, W. H. (1982b). *J. Neurochem.* **39,** 1117.
Gurd, J. W., Gordon-Weeks, P. and Evans, W. H. (1983). *Brain Res.* (in press).
Hatton, J. D. and Ellisman, M. H. (1981). *Cell Tissue Res.* **215,** 309–323.
Henderson, D., Eibl, H. and Weber, K. (1979). *J. Mol. Biol.* **132,** 193–218.
Hertzberg, E. L. and Gilula, N. B. (1979). *J. Biol. Chem.* **254,** 2138–2147.
Hertzberg, E. L., Anderson, D. J., Friedlander, M. and Gilula, N. B. (1982). *J. Cell Biol.* **92,** 53–59.
Heuser, J. E., Reese, T. S. and Landis, D. M. D. (1975). *Cold Spring Harb. Symp. Quant. Biol.* **40,** 17–24.
Hirokawa, N. and Heuser, J. (1982). *Cell* **30,** 395–406.
Hull, B. E. and Staehelin, L. A. (1979). *J. Cell Biol.* **81,** 67–82.
Johnson, R., Hammer, M., Sheridan, J. and Revel, J.-P. (1974). *Proc. Natn. Acad. Sci. USA* **71,** 4536–4540.
Kachar, B. and Reese, T. S. (1982). *Nature* **296,** 464–466.
Kachar, B. and Pintoa da Silva, P. (1981). *Science* **213,** 541–544.
Kelly, D. E. (1966). *J. Cell Biol.* **28,** 51–72.

Kelly, P. T. (1981). Quoted in Gurd, 1982.
Kelly, P. T. and Cotman, C. W. (1977). *J. Biol. Chem.* **252**, 786–793.
Kelly, P. T. and Cotman, C. W. (1978). *J. Cell Biol.* **79**, 173–183.
Kelly, P. T., Cotman, C. W. and Largen, M. (1979). *J. Biol. Chem.* **254**, 1564–1575.
Kensler, R. W. and Goodenough, D. A. (1980). *J. Cell Biol.* **86**, 755–764.
King, M. G. and Spencer, A. N. (1979). *J. Cell Sci.* **36**, 391–400.
Kistler, J. and Bullivant, S. (1980). *FEBS Lett.* **111**, 73–78.
Larsen, W. J. and Tung, H. H. (1978). *Tissue and Cell* **10**, 585–598.
Lazarides, E. (1980). *Nature* **283**, 249–256.
LeLoup, R., Laurent, L., Ronveaux, M.-F., Drochmans, P. and Wanson, J.-C. (1979). *Biol. Cellulaire* **34**, 137–152.
Lord, B. A. P. and Dibona, D. R. (1976). *J. Cell Biol.* **71**, 967–972.
Lowenstein, W. R. (1979). *Biochem. Biophys. Acta* **560**, 1–65.
Makowski, L., Caspar, D. L. D., Phillips, W. C. and Goodenough, D. A. (1977). *J. Cell Biol.* **74**, 629–645.
Manina, A. A. (1979). *Int. Rev. Cytol.* **57**, 345–383.
Manjunath, C. K., Goings, G. E. and Page, E. (1982). *Biochem. J.* **205**, 189–194.
Masson-Pevet, M., Bleeker, W. K. and Gros, D. (1979). *Circulation Res.* **45**, 621–629.
Matus, A. (1978). *In* "Intercellular Junctions and Synapses" (J. Feldman, N. B. Gilula and J. D. Pitts, eds) pp. 99–139. Chapman and Hall, London.
Matus, A. and Jones, D. H. (1978). *Proc. R. Soc. (Lond.)* **203**, 135–151.
Matus, A. I., Walters, B. B. and Mughal, S. (1975). *J. Neurocytol.* **4**, 733–744.
Matus, A., Pehling, G., Ackermann M. and Maeder, J. (1980). *J. Cell Biol.* **87**, 346–359.
McNutt, N. S. (1975). *Circulation Res.* **37**, 1–13.
McNutt, N. S. and Weinstein, R. S. (1973). *Progr. Biophys. Mol. Biol.* **23**, 45–101.
Mercer, E. H. (1965). *In* "Organogenesis" (R. L. De Haan and H. Ursprung, eds) pp. 29–54, Holt, Rinehart and Winston, New York.
Meyer, D. J., Yancey, S. B. and Revel, J.-P. (1981). *J. Cell Biol.* **91**, 505–523.
Miettinen, A., Virtanen, I. and Linder, E. (1978). *J. Cell Sci.* **31**, 341–353.
Montesano, R., Friend, D. S., Perrelet, A. and Orci, L. (1975). *J. Cell Biol.* **67**, 310–319.
Mooseker, M. S. (1976). *J. Cell Biol.* **71**, 417–433.
Mueller, H. and Franke, W. (1983). *J. Mol. Biol.* **163**, 647–671.
Mukherjee, T. M. and Staehelin, L. A. (1971). *J. Cell Sci.* **8**, 573–599.
Mushynski, W. E., Glen, S. and Therien, H. M. (1978). *Can. J. Biochem.* **56**, 820–830.
Nelson, P. G. (1975). *Cold Spring Harb. Symp. Quant. Biol.* **40**, 359–371.
Nicholson, B. J., Hunkapiller, M. W., Grim, L. B., Hood, L. E. and Revel, J.-P. (1981). *Proc. Natn. Acad. Sci. USA* **78**, 7594–7598.
Nieto-Sampedro, M., Bussineau, C. M. and Cotman, C. W. (1981). *J. Cell Biol.* **90**, 675–686.
Noirot-Timothée, C. and Noirot, C. (1980). *Int. Rev. Cytol.* **63**, 97–140.
Overton, J. (1962). *Dev. Biol.* **4**, 532–548.
Overton, J. (1974). *Dev. Biol.* **39**, 226–246.
Overton, J. (1975). *Curr. Topics Dev. Biol.* **10**, 1–34.
Overton, J. and DeSalle, R. (1980). *Dev. Biol.* **75**, 168–176.
Peracchia, C. (1977). *J. Cell Biol.* **72**, 628–641.
Peracchia, C. (1978). *Nature* **271**, 669–671.
Peracchia, C. (1980). *Int. Rev. Cytol.* **66**, 81–146.
Pfenninger, K. H. and Rees, R. F. (1976). *In* "Neuronal Recognition" (S. H. Barondes, ed.) pp. 131–178. Chapman and Hall, London.

Pisam, M. and Ripoche, P. (1976). *J. Cell Biol.* **71**, 907–920.
Porvaznik, K. M., Johnson, R. G. and Sheridan, J. D. (1976). *J. Ultrastruct. Res.* **55**, 343–359.
Porvaznik, K. M., Johnson, R. G. and Sheridan, J. D. (1979). *J. Supramolec. Struct.* **10**, 13–30.
Preus, D., Johnson, R. and Sheridan, J. (1981). *J. Ultrastruct. Res.* **77**, 248–262.
Rae, J. L. (1978). *Curr. Topics Eye Res.* **1**, 37–90.
Rakic, P. (1975). *Cold Spring Harb. Symp. Quant. Biol.* **40**, 333–346.
Raviola, E., Goodenough, D. A. and Raviola, G. (1980). *J. Cell Biol.* **87**, 273–279.
Rayns, D. G., Simpson, F. O. and Ledingham, J. M. (1969). *J. Cell Biol.* **42**, 322–325.
Revel, J.-P. and Karnovsky, M. J. (1967). *J. Cell Biol.* **33**, C7–C12.
Robenek, H., Jung, W. and Gebhardt, R. (1982). *J. Ultrastruct. Res.* **78**, 95–106.
Rodewald, R., Newman, S. B. and Karnovsky, M. J. (1976). *J. Cell Biol.* **70**, 541–554.
Sandri, C., Akert, K., Livingston, R. B. and Moore, H. (1972). *Brain Res.* **41**, 1–16.
Satir, P. and Gilula, N. B. (1973). *Ann. Rev. Entomol.* **18**, 143–166.
Sheridan, J. D. (1978). *In* "Intercellular Junctions and Synapses" (J. Feldman, N. B. Gilula and J. D. Pitts, eds) pp. 37–59. Chapman and Hall, London.
Shibata, Y., Nakata, K. and Page, E. (1980). *J. Ultrastruct. Res.* **71**, 258–271.
Shida, H., Gorbsky, G., Shida, M. and Steinberg, M. S. (1982). *J. Cell. Biochem.* **20**, 113–126.
Shimoro, M. and Clementi, F. (1976). *J. Ultrastruct. Res.* **56**, 121–136.
Skerrow, C. J. (1979). *Biochim. Biophys. Acta* **579**, 241–245.
Skerrow, C. J. and Matolsty, A. G. (1974). *J. Cell Biol.* **63**, 515–530.
Slater, C. R. (1978). *In* "Intercellular Junctions and Synapses" (J. Feldman, N. B. Gilula and J. D. Pitts, eds) pp. 215–239. Chapman and Hall, London.
Staehelin, L. A. (1974). *Int. Rev. Cytol.* **39**, 191–283.
Therien, H. M. and Mushynski, W. E. (1979). *Biochim. Biophys. Acta* **585**, 201–209.
Toh, B. H., Gallichio, H. A., Jeffrey, P. L., Livett, B. G., Muller, H. K., Cauchi, M. N. and Clarke, F. M. (1976). *Nature* **264**, 648–650.
Traub, O. and Willecke, K. (1982). *Biochim. Biophys. Res. Commun.* **109**, 895–901.
Traub, O., Druge, P. M. and Willecke, K. (1983). *Proc. Natn. Acad. Sci. USA* **80**, 755–759.
Traub, O., Jansen-Timmen, U., Druge, P. M., Demietzel, R. and Willecke, K. (1982). *J. Cell Biochem.* **19**, 27–44.
Ueda, T., Greengard, P., Berzins, K., Cohen, R. S., Blomberg, F., Grab, D. J. and Siekevitz, P. (1979). *J. Cell Biol.* **83**, 308–319.
Unwin, P. N. T. and Zampighi, G. (1980). *Nature* **283**, 547–549.
Van Deurs, B. (1980). *Int. Rev. Cytol.* **65**, 117–191.
Weller, M. and Morgan, I. G. (1976). *Biochim. Biophys. Acta* **433**, 223–228.
Westrum, L. E. and Gray, E. G. (1976). *Brain Res.* **105**, 547–550.
Wood, J. G., Wallace, R. W., Whitaker, J. N. and Cheung, W. Y. (1980). *J. Cell Biol.* **84**, 66–76.
Woodruff, R. I. and Telfer, W. H. (1980). *Nature* **286**, 84–86.
Wrigley, N. F., Brown, E. and Chillingworth, R. K. (1984). *Biophys. J.* **45**, 167–172.
Yancey, S. B., Nicholson, B. J. and Revel, J.-P. (1981). *J. Supramolec. Struct.* **16**, 221–242.
Zampighi, G., Ramon, F. and Duran, W. (1978). *Tissue Cell* **10**, 413–426.
Zampighi, G. and Unwin, P. N. T. (1979). *J. Mol. Biol.* **135**, 451–464.
Zampighi, G., Simon, S. A., Robertson, J. D., Mcintosh, T. J. and Costello, M. J. (1982). *J. Cell Biol.* **93**, 175–189.

Subject Index

A

Acanthamoeba
  myosin-like proteins of, 115
Actomyosin
  ATPase, 82, 98
Actin, 3–19, 75–81
  ATP binding site, 79
  blood platelet paracrystals, 65
  cross-linked by spectrin, 73
  decorated filaments, 65
  decoration with myosin, HMM and S-1, 165–174
  double-helical structure, 2, 3
  in erythrocyte membrane skeleton, 243, 279–282
  F-actin, 3–5
  F-actin binding to erythrocyte membrane cytoplasmic surface, 287
  F-actin paracrystals, 5–10
  F-actin subunit structure, 5–9
  in freeze-dried cytoskeletons, 5
  G-actin microcrystals, 16–18
  interaction with tropomyosin, 52–56
  in *Limulus* sperm, 69
  metal shadowed, 3
  microcrystals, 13–18
  negatively stained, 3–5
  in non-muscle cells, 64, 65
  optical filtering of F-actin, 5, 7
  partial decoration of filaments, 173, 174
  of *Pecten maximus* adductor muscle, 5
  polarity reversal at Z-line, 166
  polymerization by polyamines, 9, 10
  in sperm acrosomal process, 69
  structure of actin: DNase I, 18–19
  X-ray diffraction of F-actin, 10–13
Actin and thin filaments, 1–88
  actin-tropomyosin interaction, 52–56
  influence of Ca^{2+} on regulation, 46–52
  influence of troponin on regulation, 46–52
  muscle actin, 3–19
  in non-muscle cells, 63–75, 87, 88
  non-muscle cell actin-binding proteins, 73, 75
  non-muscle thin filament assemblies, 67–73
  non-muscle troponin-like proteins, 67
  3-D reconstructions, 45–52
  role of tropomyosin and troponin in regulation, 45–57
  tropomyosin, 19–34
  troponin, 34–44
  the Z-disc, 57–63
Actin-myosin interaction, 161–194
  actin-linked regulation, 176–178
  Ca^{2+} release from sarcoplasmic reticulum, 164
  in contracting muscle, 193–194
  decorated actin (arrowhead complex), 165–174
  difference between insect and vertebrate muscle, 189
  in filament lattice, 181–183
  geometric constraints in filament lattice, 183–192
  geometry in cross-bridge cycle, 193
  in intact muscle, 181–194
  in vitro studies, 165–181
  location of tropomyosin, 177, 178

Subject Index

Actin-myosin interaction (cont.)
  myosin-linked regulation, 178–181
  non-rigor cross-bridge attachment, 174–176
  partial decoration of actin, 173, 174
  rapid-freezing technique, 170, 171
  removal of regulatory light chains, 178–181
  steric-blocking mechanism, 176–178
  structural assay for F-actin, 171–173
  3-D reconstruction of arrowhead complex, 167–171
α-Actinin, 63
  antibody labelling of frozen sections, 63
Ankyrin, 253
  attachment of membrane skeleton, 291–294
  in erythrocyte membrane skeleton, 243
Antibody
  against desmosome proteins, 336, 337
  effect on protein rotational diffusion, 253, 255
  to erythrocyte membrane 72 000 dalton to proteolytic fragment, 292
  Fab fragment, 114
  against gap junction polypeptide, 356
  labelling of muscle M-line, 129, 131
  location of C-protein, 126
  to myosin regulatory light chain, 116
  to scallop myosin, 114
  against spectrin, 223
  to troponin C, 41
  to troponin components, 42
Antigens, MN blood group, 237
ATP
  action on contractile proteins, 64
  action on terminal web, 67
  filament formation by non-muscle myosin, 150
  interaction with actomyosin, 98
  membrane protein diffusion, 251
  muscular contraction, 161, 162, 164
  synthesis in erythrocyte, 208
  use of analogues of, 175
ATPase
  of fast and slow muscle, 115
  of HMM and S-1, 110
  of myosin head, 114

B

Blood platelets
  contractile proteins of, 64
  tropomyosin of, 65–67

C

Chemical cross-linking
  of myosin dimer, 134
α-Chymotrypsin
  digestion of myosin, 109, 110
Concanavalin A
  ferritin-labelled, 346
Cylindrin, 298–303
  function of, 301
  isolation and purification, 298
  negative staining of, 300, 302
  SDS-PAGE of, 299, 301, 303
  subunit composition, 301, 303
Cytokinesis
  induction by spermine, 79
Cytoskeleton, 242, 243

D

Desomosome
  antibodies to specific proteins, 336
  formation, 333
DNase I
  actin complex, 19
Duchenne muscular dystrophy, 309–316

E

Echinoderm
  sperm actin filaments, 69
Erythrocyte
  ATP concentration, 209, 210, 309
  ATP-depleted, 251
  ATP synthesis, 208
  Ca^{2+} level, 309
  cell shape, 208–210
  echinocytic form, 209, 210
  ferritin-antibody labelling, 257
  flexibility, 208
  haemolysis and ghost production, 209

Erythrocyte (cont.)
  membrane carbohydrates, 210
  membrane composition, 210, 211
  membrane lipids, 210, 211
  membrane properties, 208–210
  origin and lifespan, 207, 208
  physiological role, 208–210
Erythrocyte membrane
  action of drugs on shape, 246
  asymmetric distribution of phospholipids, 264
  binding of muscle F-actin to cytoplasmic surface, 287
  blebbing, 247
  cell shape changes, 244–247
  in disease, 303–316
  elasticity of, 244
  freeze etching of cytoplasmic surface, 233, 234
  freeze etching of, 215–234
  freeze etching of extracellular surface, 227–233
  glycophorin-deficient mutants in man, 308
  irreversible shape changes, 247
  model of, 316–320
  in muscular dystrophy, 309–316
  physical properties, 244
  reversible shape changes, 245, 246
  sickle cell anaemia, 308, 309
  skeletal control of protein diffusion, 249–255
  spectrin-deficient mutants in mice, 307, 308
  transglutaminase, 309
Erythrocyte membrane band 3 protein (anion channel), 227, 234–237
  aggregation at pH 5·5, 235
  akyrin binding, 235
  chemistry of, 234, 235
  binding of ferritin-lectin, 232
  in reconstituted systems, 235–237
Erythrocyte membrane proteins, 205–320
  actin, 279–282
  band 4·1, 282, 287–290
  electrophoretic analysis of, 211–215
  glyceraldehyde-3-phosphate dehydrogenase (band 6), 294
  HMM decoration of actin filaments, 280, 281
  IMP aggregation induced by lectin, 223
  IMP clustering during membrane fusion, 223
  IMP density, 221
  IMP linear arrays, 223
  IMP mobility, 222
  length of actin protofilaments, 282
  major integral proteins, 234–241
  membrane skeleton proteins, 260–294
  molecular properties of, 211–215
  nature of IMPs, 219–222
  particle aggregation studies, 222–226
  polymerization state *in vivo*, 281
  protease digestion of, 219, 222, 223, 227
  skeletal proteins polymerize muscle G-actin, 283
  torin and cylindrin, 295–303
Erythrocyte membrane skeleton, 241–260
  association of spectrin, actin and band 4·1, 282–290
  attachment to integral proteins, 291–294
  cell shape determination by, 243
  comparison with cytoskeleton, 242, 243
  electron microscopic observations on, 255–260
  negative staining of, 259
  protein components of, 243
  protein phosphorylation, 245
  reconstitution of, 287, 288
  scanning electron microscopy of, 257, 259
  thin sectioning of, 255–257
Eye lens
  gap junctions of, 354, 355

F

Fascia adherens
  morphology, 339–341
  protein composition, 339, 342
Ferritin
  cationized, 228
  conjugates with lectins, 232
Ferritin antibody
  against hapten-labelled erythrocytes, 227, 233

Ferritin antibody (*cont.*)
  against spectrin, 223, 233, 255
  against troponin, 41
  against erythrocyte membrane band 4·1 protein, 288
Fibroblast
  contractile proteins of, 64
Fourier-Bessel transformation, 7
  of tropomyosin tactoids, 26
Freeze fracture
  of developing gap junctions, 354
  distribution of intramembrane particles, 218, 219
  of erythrocyte membrane, 215–234
  etching of actin filaments, 71
  etching of erythrocyte membrane surfaces, 227–234
  etching of studies employing cationized ferritin, 228
  P- and E-fracture faces, 218
  of gap junctions, 348, 351, 355, 356
  pastic distortion of IMPS, 219, 221
  of septate junctions, 357
  of tight junctions, 356

G

Gadolinium ions
  effect on G-actin, 16–18
Gap junctions, 348–356
  antibody against 28 000 dalton peptide, 356
  in cardiac intercalated discs, 354
  detergent extraction, 354
  development of, 353, 354
  in eye lens epithelia, 354, 355
  freeze fracture of, 348, 351, 355, 356
  formation hormonally dependent, 354
  hexagonal lattice of, 348, 351–353
  isolation and purification, 354, 355
  metabolic turnover, 356
  models of, 353
  morphology, 348–354
  negative staining of, 348, 352
  protein composition, 354–356
  thin sections of, 348, 350
Glyceraldehyde-3-phosphate dehydrogenase
  band 6 of erythrocyte membrane, 294
  decoration of band 3 protein, 234

Glyconnectin
  erythrocyte band 4·1 protein binding, 294
  in erythrocyte membrane skeleton, 243
  sialoglycoprotein (PAS2), 294
Glycophorin
  cationized ferritin binding by, 228
Glycophorin A, 212, 221, 237–241
  binding of ferritin-lectin, 232
  chemistry of, 238, 239
  electron microscopy in reconstituted systems, 239–241
  En(a-) antigen deficiency in man, 308
  incorporation into liposomes, 239
Glycophorin B, 212, 237, 238
Glycophorin C, 212, 238
  presence in membrane skeleton, 238

H

Haemolytic anaemias, 303–316
  membrane skeleton defects, 304
  in *Mus musculus* mutants, 307, 308
Hemidesmosome, 337
Hereditary elliptocytosis, 303–305
  abnormal spectrin band 2 phosphorylation, 305
  absence of erythrocyte membrane band 4·1 protein, 305
Hereditary pyropoikilocytosis, 304–307
Hereditary spherocytosis, 303, 305
  defective spectrin, 304

I

Immunoelectron microscopy
  of muscle, 126
Immunofluorescence
  of myofibrils, 126
  of tropomyosin in non-muscle cells, 65
Influenza virus receptor, 237
Insect flight muscle
  Z-disc, 58, 59
Intramembrane particles (IMP)
  conformation of, 221
  identity with anion channel protein (band 3), 219
  mobility on removal of spectrin and actin, 222

Intramembrane particles (*cont.*)
  nature of in erythrocyte, 219–222
  proteolytic destruction of, 219
  restraint by membrane skeleton, 223
  revealed by freeze fracture, 218

L

Lanthanide ions
  actin tube formation by, 16–18
Lectin receptor, 237
Leucocyte
  actinomyosin, 64
Liposomes
  glycophorin A tryptic peptide incorporation, 239
Limulus polyphemus
  sperm actin filaments, 69
  striated muscle, 151–156

M

Macula adherens (spot desmosome), 333–337
  morphology of, 333, 334
  protein composition, 335–337
Metal shadowing
  of antibody-spectrin complex, 278, 279
  of crosslinked myosin dimer, 134
  invertebrate striated muscle thick filaments, 154
  of myosin filament, 106
  of myosin molecule, 107, 108, 113, 116
  non-muscle myosin, 151
  rotary shadowing of IMPs, 221
  of spectrin, 267–279
  of tropomyosin fibrils, 21
  of troponin T, 43
  of Z-disc, 58
Microtubules
  at synapses, 342, 343
Microvillus
  decoration of actin filaments, 67
  terminal web of, 68, 69
Monoclonal antibodies
  to synaptic junction proteins, 347
  to specific antigenic sites on spectrin, 278

Mus musculus
  spectrin deficient mutants of, 307, 308
Muscle
  cross-bridges between myosin and actin, 102
  I-segment polarity of decoration, 166
  invertebrate catch muscle, 157–161
  mechanism of contraction, 162, 164, 165
  rigor mortis and ATP depletion, 165, 183, 191
  sarcoplasmic Ca^{2+} level, 164
  sliding filament model, 99, 100
  tropomyosin movement, 45, 46
  X-ray diffraction of, 40, 100, 101
  Z-line, 20, 39, 156
  Z-region, 31
Muscle A-band, 31, 39, 100
  in crustacea, 156
Muscle I-band, 30, 39, 100
  insect flight muscle, 156
Muscle M-line, 129
  in crustacea, 156
  in insect flight muscle, 156
Muscle proteins *see* Actin, Actin and thin filaments, Thick filaments, Actin-myosin interaction, Myosin, Tropomyosin, Troponin
Muscle Z-disc, 57–63, 86, 87
  α-actinin, 63
  of barnacle scutal depressor muscle, 151, 152
  of cardiac muscle, 61
  in insects and vertebrates, 58–63
  metal shadowed, 58
  positively stained, 58
  of skeletal muscle, 61
  and tropomyosin, 59
  Z-filament connections, 59, 61
Muscular dystrophy, 309–316
  biochemical and biophysical studies, 311–313
  decrease in spectrin extractability, 312
  erythrocyte echinocytic morphology, 313
  freeze fracture morphology of erythrocyte membrane, 314, 315
  increased erythrocyte osmotic fragility, 312
  membrane fluidity in, 313

Muscular dystrophy (*cont.*)
  scanning electron microscopic studies of erythrocytes, 313, 314
Myotonic muscular dystrophy, 311
Myosin
  aggregation of molecules to form filament, 104, 105
  ATPase of, 64
  attachment to tropomyosin, 56, 57
  Ca^{2+} binding, 116
  depolymerization of filament, 103
  diversity of structure, 115–118
  filament bare zone, 103–106
  flexible attachment of head and tail, 107
  flexible hinge in tail, 109, 110
  fluorescent antibody labelling of light chains, 113
  heavy meromyosin (HMM), 109–111
  HMM and S-1 ATPase, 110
  heavy meromyosin subfragment-1 (S-1), 109–111
  heavy meromyosin subfragment-2 (S-2), 109–112
  lability of non-muscle myosin, 150
  light meromyosin (LMM), 109–112
  metal shadowing of molecule, 107, 108, 113, 116
  molecule and filament, 103–106
  monomer-polymer transformation, 134, 135
  negative staining of filament, 103, 104, 106
  negative staining of molecule, 107, 113
  non-muscle, 115
  phosphorylation of light chain, 116
  phosphorylation in smooth muscle, 150
  polypeptide composition of, 109, 112
  prolonged digestion of, 110
  protease digestion fragments, 109–112
  S-1 decorated actin, 113
  S-1 ATPase, actin activation, 112
  smooth muscle, 115
  structure and function of head, 112–115
  structure of molecule, 107–118
  X-ray diffraction of molecules, 107

N

Negative staining
  of actin, 3–5
  of actin-tropomyosin-troponin, 40
  of actin tubes, 16–18
  of α-actinin, 63
  of cryosections of muscle, 121–123
  of decorated actin, 175
  distortion of myosin during drying, 113
  of erythrocyte membrane skeleton, 259
  of gap junctions, 348, 352
  of *Limulus* sperm actin filaments, 69
  of muscle A-segments, 121–124
  of myosin filament, 103, 104, 106
  of myosin molecule, 107, 113
  of spectrin, 267
  selection of best specimens, 57
  of thick filaments, 136
  of torin and cylindrin, 295–303
  of tropomyosin crystals, 23–25
  of tropomyosin tactoids, 26–28
  of troponin-tropomyosin tactoids, 35–39

O

Optical diffraction
  of actin paracrystals, 31, 32
  of G-actin microcrystals, 16–18
  of muscle A-segments, 124
  of sectioned muscle, 40

P

Papain
  digestion of myosin, 109, 110
Phospholipid
  distribution in erythrocyte membrane, 264
  interaction with spectrin, 264
Plasma membrane intercellular junctions, 331–359
  cell-cell adhesion junctions, 333–342
  cell-cell communicating junctions, 342–356
  desmosome formation, 333
  fascia adherens, 339–342
  gap junctions, 348–356
  macula adherens, 333–337

Plasma membrane intercellular junctions (*cont.*)
  septate junctions, 357, 358
  synaptic junctions, 342–347
  tight junctions (*zonulae occludentes*), 356, 357
  Zonula adherens (belt desmosome), 337–339
Positive staining
  of Z-disc, 58
  with phosphotungstic acid, 39
Profilin
  actin complex, 19
Protease
  cleavage of erythrocyte band 3 protein, 253
  cleavage of spectrin, 263
  digestion of erythrocyte membrane proteins, 219

R

Rotational diffusion
  of erythrocyte membrane proteins, 253

S

Sarcoma cells
  contractile proteins of, 64
Scanning electron microscopy
  of erythrocytes from muscular dystrophy patients, 313, 314
  of erythrocyte membrane skeleton, 257, 259
  of erythrocyte shape changes, 245
Scanning transmission electron microscopy (STEM)
  of thick filament, 120, 121
Sodium dodecyl sulphate-polyacrylamide gel electrophoresis (SDS-PAGE)
  of actin paracrystal proteins, 47
  of desmosome polypeptides, 336
  of erythrocyte membrane proteins, 211–215
  impurities in myosin, 126
  muscle actin:myosin ratio, 120

  nomenclature of erythrocyte proteins, 212–214
  of skeletal muscle myosin, 112
  of spectrin deficient mutants, 307
  of spectrin tryptic fragments, 278
Septate junctions, 357, 358
  freeze fracture of, 357
Sialoglycoproteins *see also* Glycophorins
  staining of, 212
Sickle cell anaemia
  and erythrocyte membrane, 308, 309
Slime mould
  cytoplasmic actin, 64, 65
Smooth muscle
  thick filaments of, 145–151
Spectrin, 260–279
  ankyrin binding, 292
  association with actin and band 4·1, 282–290
  binding to integral membrane proteins, 291–294
  cleavage by trypsin, 263, 264
  deficiency in mice, 307, 308
  in erythrocyte membrane skeleton, 243
  ferritin antibody labelling of, 255, 257
  fibrous aggregate formation, 267
  gel filtration of, 263
  interaction with actin, 265
  interaction with anionic phospholipids, 264
  interaction with muscle G-actin, 288–290
  low angle shadowing of, 267–279
  molecular shape, 267–269
  molecular weight of, 263
  monoclonal antibodies to, 278
  negative staining of, 267
  non-erythroid, 317
  oligomeric forms, 263, 275
  phosphorylation of, 264, 265, 267
  protein chemistry of, 260–267
  separation of polypeptide chains, 264
Split membrane analysis, 219
Synaptic junctions, 342–347
  detergent extractions, 346
  enzymic activities of, 347
  ferritin-Con A labelled, 346
  glycoproteins of, 346
  morphology, 342–345
  protein composition, 343, 346, 347

T

Tactoids
 of light meromyosin, 137–141
Tannic acid
 fixation of erythrocyte membrane, 255–258
Thick filaments
 A-band antibody labelled, 121
 A-segments, 121–124
 arrangement in myofibril lattice, 131–133
 bare zone, 121
 C-protein, 126, 127, 129
 C-protein 43 nm repeat, 127, 129
 cardiac C-protein, 126
 diversity in invertebrates, 155, 156
 filament formation, 134
 groupings of myosin tails in, 124
 helical myosin cross-bridges, 118–120
 invertebrate catch muscle, 157–161
 invertebrate cross-bridges, 154
 invertebrate striated muscle, 151–157
 LMM assemblies, 136–141
 M-line, 129–131
 models for invertebrate muscle, 154–155
 models for vertebrate muscle, 142–145
 myosin:paramyosin ratio in catch muscles, 160
 myosin rod aggregates, 141
 myosin S-2 aggregates, 141
 myosin synthetic assemblies, 134–142
 native filaments, 118–134
 non-muscle, 145–151
 non-myosin components of, 124–131
 number of myosin molecules per filament, 120
 organization of myosin molecules, 118–124
 paramyosin in invertebrate muscle, 157
 paramyosin of molluscan catch muscle, 157–160
 paramyosin synthetic assemblies, 157–159
 role of non-myosin proteins, 129
 rotational symmetry of cross-bridges, 120
 STEM study, 120, 121
 structure of paramyosin filament, 159
 synthetic assemblies of intact myosin, 134–136
 synthetic assemblies of myosin fragments, 136–141
 vertebrate smooth muscle, 145–151
 vertebrate striated muscle, 118–145
 X-ray diffraction of, 124
Thin filaments *see* Actin and thin filaments
Tight junctions (*zonulae occludentes*), 356, 357
 freeze fracture of, 356
Torin, 295–298
 isolation and purification, 296
 negative staining of, 297
 SDS-PAGE of, 296, 298
 subunit composition, 296–298
Translational diffusion
 of membrane proteins, 249
Triton X-100, 241, 251
 extraction of erythrocyte membrane, 243
 extraction of sickle cell ghosts, 309
Tropomyosin, 81, 82
 action of divalent cations, 23
 association with actin, 30–34
 of blood platelets, 65–67
 the coiled-coil, 25–28
 crystals, fibres and tactoids, 21–25
 crystal structure, 28–30
 distribution of troponin T, 43
 fluorescent-antibody labelling, 30, 31
 interaction with troponin, 35–39
 location of cysteine residues, 27–30
 molecular overlap in tactoids, 28
 movement of, 45, 46
 myosin attachment site, 56, 57
 negatively-stained tactoids, 26–28, 35
 in non-muscle cells, 65–67
 non-muscle tactoids of, 66, 67
 role in regulation, 45–57, 83–86
 the 40 nm stripe, 39–44
Troponin, 33–44, 82, 83
 antibodies to components, 42
 Ca^{2+} sensitivity, 34, 37, 39
 fluorescent-antibody labelling, 36
 interaction with tropomyosin, 35–39
 isolation and components of, 34–35
 role in regulation, 45–57, 83–86
 troponin C, 34–39

Troponin (cont.)
  troponin I, 34
  troponin T, 34–39
Trypsin
  cleavage of myosin, 109, 110
  cleavage of spectrin, 263, 264
  treatment of erythrocyte ghosts, 280

V

Vertebrate muscle
  striated muscle myosin, 107–112
  striated muscle thick filaments, 118–145
  Z-disc, 59–63

X

X-ray diffraction
  of actin crystals, 79
  of actin + troponin T gels, 47
  of antibody-labelled muscle, 126
  of contracting frog striated muscle, 192–194
  of F-actin, 10–13
  of frog sartorius muscle, 11
  of gap junctions, 353
  of glycerinated muscle, 40, 41
  of insect flight muscle in rigor, 188, 189
  of invertebrate muscle, 124, 145
  invertebrate striated muscle, 152–155
  of invertebrate thick filament, 155
  of LMM tactoids, 139
  of molluscan catch muscle, 157, 160
  of muscle, 40, 100, 101, 118–120
  of myosin molecules, 107
  of *Mytilus edulis* retractor muscle, 11
  and regulation of contraction, 45, 46
  of resting muscle, 131
  of smooth muscle, 146
  of thick filaments, 124
  of tropomyosin, 25, 30, 33
  and troponin model, 53
X-ray scattering
  of myosin subfragment-1 (S-1), 113

Z

Zonula adherens (belt desmosome), 337–339
  morphology, 337–339
  protein composition, 339